Land-Use Planning: Sustainability and Environment

Land-Use Planning: Sustainability and Environment

Edited by **Alex Vedder**

Callisto Reference
New York

Published by Callisto Reference,
106 Park Avenue, Suite 200,
New York, NY 10016, USA
www.callistoreference.com

Land-Use Planning: Sustainability and Environment
Edited by Alex Vedder

International Standard Book Number: 978-1-63239-713-3 (Hardback)

Printed in the United States of America.

Contents

Preface

This book has been a concerted effort by a group of academicians, researchers and scientists, who have contributed their research works for the realization of the book. This book has materialized in the wake of emerging advancements and innovations in this field. Therefore, the need of the hour was to compile all the required researches and disseminate the knowledge to a broad spectrum of people comprising of students, researchers and specialists of the field.

Land is a limited resource and as the world population is rising consistently, judicious use of this resource is of prime importance. Land-use requires changing the already existing natural environmental setting into fields, pastures, buildings, road networks, etc. These activities impact the environment adversely and pose many problems such as soil erosion, salination, soil degradation and desertification. This book includes theories for sustainable management of land. It covers researches contributed by veterans from all parts of the world to provide a global and comprehensive overview of the subject. This book is compiled in such a manner, that it will provide in-depth knowledge about the theory and practice of environmental land-use, planning and management. It will serve as a valuable reference guide for readers.

At the end of the preface, I would like to thank the authors for their brilliant chapters and the publisher for guiding us all-through the making of the book till its final stage. Also, I would like to thank my family for providing the support and encouragement throughout my academic career and research projects.

Editor

1

Simulation *vs.* Definition: Differing Approaches to Setting Probabilities for Agent Behaviour

Fraser J. Morgan [1,*], Philip Brown [2] and Adam J. Daigneault [1]

[1] Landcare Research New Zealand, Private Bag 92170, Auckland Mail Centre, Auckland 1142, New Zealand; E-Mail: daigneaulta@landcareresearch.co.nz

[2] Landcare Research New Zealand, P.O. Box 69040, Lincoln 7640, New Zealand; E-Mail: brownp@landcareresearch.co.nz

* Author to whom correspondence should be addressed; E-Mail: morganf@landcareresearch.co.nz

Academic Editors: James Millington and John Wainwright

Abstract: While geographers and economists regularly work together on the development of land-use and land-cover change models, research on how differences in their modelling approaches affects the results is rare. Answering calls for more coordination between the two disciplines in order to build models that better represent the real world, we (two economists and a geographer) developed an economically grounded, spatially explicit, agent-based model to explore the effects of environmental policy on rural land use in New Zealand. This inter-disciplinary collaboration raised a number of differences in modelling approach. One key difference, and the focus of this paper, is the way in which processes that shape the behaviour of agents are integrated within the model. Using the model and a nationally representative survey, we compare the land-use effects of two disciplinary-aligned approaches to setting a farmer agent's likelihood of land-use conversion. While we anticipated that the approaches would significantly affect model outcomes, at a catchment scale they produced similar trends and results. However, further analysis at a sub-catchment scale suggests the approach to setting the likelihood of land-use conversion does matter. While the results outlined here will not fully resolve the disciplinary differences, they do outline the need to account for heterogeneity in the predicted agent behaviours for both disciplines.

Keywords: agent-based; behaviour; agricultural; farmer; survey; econometrics; social networks; farm conversion

1. Introduction

With an increase in demand for strong, evidence-based environmental policy and management, scientists have called for methods to accurately capture the complex nature of socio-ecological systems [1,2]. This call is driven by the need to understand the likely consequences and trade-offs of proposed policies on economic outcomes, land use, and social well-being [3]. A modelling approach is well suited to this task because the social, economic, and geographic factors that determine the choice and impact of land use are in themselves complex [4–7].

Land use and land cover change (LULCC) models represent a well-developed approach to modelling and understanding processes that shape the environment [8–10] and have developed alongside our understanding of wider economic and social systems. As with most modelling approaches, early implementation of LULCC models focused on mathematical programming and rational utility theory, *i.e.*, individuals are assumed to maximise profits [11–15]. These approaches are still common, and while these models capture trends in LULCC, they may fail to reflect accurately the underlying processes driving the change in LULCC [2].

More specifically, more economically focused LULCC models focus on management practices that maximise net returns for a given land use while omitting key spatial, bio-physical, and social details [2,16–21]. Such abstractions ignore the processes, people, and space within the model, thus making the "optimally derived" solution unrealistic [22]. Geographically defined LULCC models, on the other hand, typically account for heterogeneity across space and individuals, but often simplify the level of economic behaviour [23–27]. As such, geographic models are typically structured to include simplified economic approaches and to exclude explicit representations of land and commodity markets [23].

Geographers and economists have rarely collaborated in undertaking these analyses, leading to calls for modellers from these two disciplines to coordinate efforts in order to build models that better represent the real world [23,28].

LULCC is a complex, adaptive process that can also be explained through the use of computational tools such as agent-based models (ABMs) [2,29,30]. ABMs are well suited to analysing decentralised, autonomous decision making such as that underlying LULCC because they represent complex spatial interactions under heterogeneous conditions [30–32]. In addition, the ABM approach accounts for space, distance, and time in decision making.

However, capturing the social and economic behaviour of farmers via ABMs to analyse LULCC is not without its own complexities and limitations [27,33–36]. For example, Burton [37] outlines numerous social processes that should be evaluated when assessing farmer behaviour, including cultural embeddedness [38], social networks, and technology transfer [39,40], and the dichotomy between social and economic approaches to farming [41,42]. Therefore, capturing the heterogeneity of farmer behaviour is essential when modelling rural land-use change. While this notion is widely supported [38,43], there is significant variation in how heterogeneity in farmer decision making is accounted in ABMs. Examples of such heterogeneity include: variation in different production strategies [35,44,45]; dealing with external factors [46,47]; and simulating key parts of the farming process [48–50]. In all cases, this variation depends on the objective of the ABM [51].

To answer these calls, the authors (two economists and a geographer) developed an economically grounded, spatially explicit ABM to explore the effects of environmental policy on rural land use in New Zealand. The Agent-based Rural Land Use New Zealand (ARLUNZ) is capable of analysing the impact of a variety of policies on land use, net revenue, and environmental indicators such as greenhouse gas (GHG) emissions, nutrient loadings, and soil erosion [36].

This inter-disciplinary collaboration required that two differences in approach be resolved. The first is a disciplinary perspective on how individual agents enter into the model. Geographers have traditionally had a strong preference for defining types of agents within a population according to a typological framework [35,38,43,44,46] to limit complexity while still moving agents towards their predefined goals [52–55]. While economists recognise the need to limit computational complexity [34], they have also called for empirical calibration and validation of decision-making hypotheses through surveys, interviews, participatory modelling, and experimental economics [33,56,57]. Because we have access to a large-scale, nationally representative survey that accounts for demographics, social processes, and land use, we side with economists and rely on empirical distributions of farmer and forester characteristics to simulate a population of agents [58].

The second disciplinary disagreement—and a primary focus of this paper—is the way in which processes that shape the behaviour of agents are integrated within the model. Irwin [23] observes that the methods used for modelling land-use change vary significantly: economists tend to focus on econometric analyses, while geographers tend to base their analyses on simulations.

Farmers' information networks are framed around their social interactions and play a role in shaping their decision making processes [39,40,59,60]. Through the nationally representative survey, we could define the observable effects of each farmer's networks into the agent-based model by directly affecting the likelihood of a certain type of behaviour, in this case land-use conversion. Conversely, we could simulate the agent's interactions with their networks and observe how these interactions affect the agent's likelihood of land-use conversion.

Consequently, this paper analyses how each approach affects the resulting land use, net revenue, and environmental outputs at a catchment scale. We hypothesise that the two approaches will produce significant differences for each of these metrics.

We note that these disagreements relate to representation of people and the empirical characterisation of agents within ABMs [1,51,57,58,61,62]. Specifically, the disagreements relate to how empirical data is used to capture and define the bounds of decision making available to the agents. Greater variety of on-farm management options (e.g., reducing stocking rates, fencing streams, and planting riparian buffers) and more information being made available to farmers (e.g., climate, biophysical and soils data) increase the complexity associated with defining farmer agents. Because of the significant empirical data required to inform the use of on-farm management options and to account for additional information through climate and biophysical models, we constrain farmer decision making in this manuscript to complete farm conversion from enterprise to enterprise.

The remainder of this paper is organised as follows: Section 2 describes the methodology used in this research and the approaches used to define an agent's likelihood of land-use conversion; Section 3 summarises the experimental design for the research specifically the region the model has been applied on; Section 4 presents the results from the experiment; and Section 5 concludes.

2. Methods

The ARLUNZ model was designed to analyse complex environmental issues in the rural landscape, to provide information about how farmers will adapt to change, and to inform policy that seeks to address vulnerability to resource scarcity. Specifically, ARLUNZ focuses on variability in decision making among farmers, moving away from a representative decision-making agent to a spatial and behaviourally heterogeneous population of farmers whose decision making reflects the real world.

ARLUNZ is written in Version 5.0.5 of NetLogo [63] using the GIS, String, and Shell extensions. Python 2.7 is used to facilitate a loose coupling [64] between ARLUNZ and a modified version of the New Zealand Forest and Agriculture Regional Model (NZFARM) that provides economic information. NZFARM is a non-linear, partial equilibrium mathematical programming model of New Zealand land use operating at the sub-catchment scale [65]. The version used within ARLUNZ has been refined to produce an economically optimised result for each farm rather than an optimised landscape for a sub-catchment [66–69].

Morgan and Daigneault [36] provides detailed information on the design, structure, outputs, and parameterisation of ARLUNZ and its coupling with NZFARM as well as its use to estimate the impacts of climate change policy on land use in New Zealand. In addition, Table S1 contains an ODD+D description [70] for the ARLUNZ model.

2.1. Survey Research

Some parametrisation in the ARLUNZ model is based on the Survey of Rural Decision Makers (SRDM), a nationally representative survey of land owners and other decision makers [71]. The survey was conducted online between March and July 2013.

The survey gathered up to 192 data points on each respondent, land characteristics and use, current farm practice, demographics, succession plans, professional networks, sources of information regarding best practice, management objectives, income, risk tolerance, and values [71]. The questionnaire was developed in consultation with the Ministry for the Environment, the Ministry of Primary Industries, Dairy New Zealand, Beef + Lamb New Zealand, HortNZ, regional councils, AgResearch, the New Zealand Institute for Economic Research, farmer discussion groups, and other stakeholders.

The sample was drawn from the AssureQuality AgriBase database [72]. Developed in 1993 to track foot and mouth disease, AgriBase records detailed information on privately held rural land across New Zealand. Inclusion in AgriBase is voluntary and entries are updated irregularly. As such, the median address was entered into the database seven years before the survey and some of the individuals contacted for the survey had left farming, making the true response rate difficult to ascertain. However, a total of 1564 responses were collected, yielding a response rate of at least 21%. Participation was incentivised via a donation made to a charity of each respondent's choice and an invitation to view summary results online after the survey had closed. The primary decision-maker for each property was asked to complete the survey, which, on average, took approximately 20 min to complete.

Summary statistics for the variables of interest are shown in Table 1. The mean property comprises 486 hectares, although this high average is driven by a handful of very large Sheep and Beef stations. The average age of respondents is 56.5, consistent with ages reported in New Zealand's Agricultural Census [73]. The average farmer has 25 years of experience. One-third of farmers hold university degrees, while 27.8% have completed diplomas or post-secondary technical training in farming and/or farm management.

Table 1. Summary Statistics from the Survey of Rural Decision Makers.

Variables	Mean	Std. Dev.	Min.	Max.
intend to intensify over the following 5 years (1–10)	2.678	3.046	0	10
intend to de-intensify over the following 5 years (1–10)	3.569	3.452	0	10
effective land quantity (hectares)	486.440	1932.137	2	34,000
age (years)	56.471	10.098	24	87
experience (years)	25.100	15.812	1	66
high school education (dummy)	0.393	0.488	0	1
diploma/tech training (dummy)	0.278	0.448	0	1
university or higher (dummy)	0.329	0.470	0	1
importance of being highly productive (1–10)	6.535	2.787	0	10
profitable business (dummy)	0.785	0.411	0	1
respondent exceeds median # of farm/farmer visits (dummy)	0.487	0.500	0	1
risk tolerance (1–10)	5.437	2.403	0	10
enterprise = sheep and beef (share)	0.444	0.497	0	1
enterprise = dairy (share)	0.209	0.407	0	1
enterprise = deer and other livestock (share)	0.035	0.183	0	1
enterprise = horticulture and viticulture (share)	0.107	0.309	0	1
enterprise = arable (share)	0.030	0.171	0	1
enterprise = dairy support (share)	0.045	0.207	0	1
enterprise = forestry (share)	0.079	0.270	0	1
enterprise = other enterprise (share)	0.052	0.222	0	1
number of land uses on this operation (#)	1.684	0.884	1	5
region = Auckland (share)	0.031	0.173	0	1
region = Bay of Plenty (share)	0.054	0.226	0	1
region = Canterbury (share)	0.178	0.382	0	1
region = Gisborne (share)	0.024	0.154	0	1
region = Hawke's Bay (share)	0.084	0.277	0	1
region = Marlborough (share)	0.057	0.232	0	1
region = Manuwatu-Whanganui (share)	0.066	0.249	0	1
region = Northland (share)	0.053	0.224	0	1
region = Otago (share)	0.128	0.334	0	1
region = Southland (share)	0.086	0.280	0	1
region = Tasman and Nelson (share)	0.067	0.250	0	1
region = Taranaki (share)	0.043	0.203	0	1
region = Waikato (share)	0.074	0.262	0	1
region = Wellington (share)	0.036	0.186	0	1
region = West Coast (share)	0.020	0.139	0	1

The importance of being highly productive was self-evaluated on a scale of 0 to 10, where 0 indicates that being highly productive is "not at all important' and 10 indicates that being highly productive is "extremely important". The mean score for the importance of being highly productive is 6.53. Risk tolerance is measured by the question "Are you a person who is generally prepared to take risks?", where 0 indicates "don't like to take risks" and 10 indicates "fully prepared to take risks". The mean score for risk tolerance was 5.44, indicating a moderate level of risk-taking.

Some 78.5% of respondents report that their business is either profitable or that it breaks even. Having a "large" professional network is defined by visiting more than the median number of farms (*i.e.*, five) or meeting more than the median of farmers to discuss productivity (*i.e.*, four); by definition, half of the survey respondents meet these criteria. The main reported farm enterprises by area include sheep and beef (44%), dairy (21%), horticulture and viticulture (11%), forestry (8%), dairy support (4%), deer and other livestock (3%), and arable (3%). The average number of different enterprises on the farm is 1.68, although a small number of farms have as many as five different enterprises. Respondents in Canterbury (the region on which this paper focuses) account for 17.2% of the total sample.

2.2. Defining the Likelihood of Land-Use Conversion

Decision making within the model rests entirely with the farmer agent. The economic component of the model returns the net revenue-maximising land-use for each farm along with the expected net revenue for each enterprise that could be undertaken on each farm. If the economically optimal land use differs from the enterprise currently undertaken, then the farmer agent chooses whether or not to convert through an evaluation against each farmer's "likelihood of land-use conversion".

The likelihood of land-use conversion is thus fundamental to decision making within the model. We adopt several methods for evaluating this value. The first (the homogeneous approach) ignores individual attributes of farmers and assigns an identical likelihood to each farmer. The second (the network approach) allows farmers to interact with peers and to make decisions that are informed by peer performance, either via networks or imitation. The third (the survey approach) uses empirical data from the Survey of Rural Decision Makers to predict this likelihood based on individual characteristics.

2.2.1. Homogeneous Approach

For this approach, which we class as the baseline, we ignore all farmer attributes and define the likelihood of land-use conversion at 0.2 (or 20%) for all farmer agents. We use this approach to represent a common type of economic LULCC model that uses a single rational, profit-maximising agent to make decisions.

2.2.2. Network Approach

This approach uses social and geographic networks to shape the farmer agent's likelihood of land-use conversion. Farmer's information networks are framed around their social [39,40,59] and geographical [74,75] interactions and play a role in decision-making processes [39,50,75]. Two theoretical frameworks inform how networks influence farmer decision making—endorsements and imitation. While endorsements and imitation in social networks are understood, the scale and impact that these processes have on decisions are difficult to quantify in the farming context.

Studies have found that the proximity to the people in one's network is not as important as the stature of those people [59,76]. Therefore we assume that two-way interactions such as endorsements are preferred by farmers and provide a higher level of information acceptance compared with one-way interactions such as imitation. Consequently, we specify that endorsements obtained through social networks have higher weightings than those obtained through imitation.

Specifying these weightings for these network types required additional experimentation as there was no empirical data on the level of acceptance of information obtained through them. Sociological opinion in New Zealand suggests that the relative weighting of the information provided via farmer networks should range between 0.05 and 0.15 [77]. Based on this, we defined the weightings for the likelihood of information uptake as 0.10 for Endorsement and 0.05 for Imitation. To explore the influence that these weightings have on the outputs of the model, we undertook a local sensitivity analysis (Supplementary Material 2). We found that the model outputs are relatively insensitive to small variations in the weightings used for both endorsements and imitation.

Endorsements work on the concept that information about a product, process, or person (*i.e.*, the endorsed) is transferred from one individual (*i.e.*, the endorser) to another individual (*i.e.*, the receiver) through a social process. The information that is transferred by the endorser is subjective and is validated by the receiver based on his or her understanding of the endorser and the product, process, or person. Thus, endorsements capture a "*subjective but socially embedded agent's reasoning process about cognitive trajectories aimed at achieving information and preferential clarity over another, endorsed agent*" ([78]; p. 1).

With endorsement in ARLUNZ, each farmer agent incorporates information on the success of the farming operation for ten farmer agents who are located closest to the decision maker and who undertake the same enterprise as the farmer agent. Each farmer agent learns the profitability/ha of each of the farmer agents within his or her social network; using these values, a mean profitability/ha value is derived for the farmer agent's network and is then compared to farmer agent's profitability/ha value. If the farmer agent's profitability/ha is higher than the mean profitability/ha of the farmer agent's social network, then his or her likelihood of land-use conversion is decreased by 0.10 percentage points to 10%. If the farmer agent's profitability/ha is lower than the mean profitability/ha of the farmer agent's social network, then his or her likelihood of land-use conversion increases by 0.10 percentage points to 30%. The ARLUNZ model assumes that each farmer in the social network has identical stature.

The theory of Social Learning [79,80] describes imitation as a process in which a person observes another person being rewarded for understandable and reproducible behaviour. The original person might then imitate that behaviour to try to achieve the same reward [81]. Imitation transfers knowledge through a one-way network in which information is "absorbed" from the person's surroundings and then used to inform the decisions they make. Farming practices are visible to all, particularly so to farmers in close proximity because of the regular exposure [75,82].

With imitation in ARLUNZ, the farmer agent incorporates information from the farms that are geographically adjacent to his or her own farm, regardless of the enterprise undertaken. If the economic component of the model proposes a change in land use, then each farmer agent in the geographic network that undertakes the proposed land use provides his or her profitability/ha value. Using these values, a mean profitability/ha value is derived for the farmer agent's network, and this figure is compared with the farmer agent's profitability/ha value. If the farmer agent's profitability/ha is higher than that of his or her geographic network, then his or her likelihood of land-use conversion is decreased by 0.05 percentage points to 15%. If the farmer agent's profitability/ha is higher than that of his or her geographic network, then the agent's likelihood of land-use conversion is increased by 0.05 percentage points to 25%.

2.2.3. Survey Approach

This approach is based on empirical data from the Survey of Rural Decision Makers, which accounts for the decision to de-intensify land use as well as the decision to intensify land use based on the predicted net revenue of each enterprise.

The perceived likelihood of changing your current land use to more intensive or less intensive uses over the following five years was evaluated using an 11-point scale, with 0 representing "extremely unlikely" and 10 representing "extremely likely". The average reported likelihood of intensification was 2.68, which we interpret to mean that there is a 26.8% probability of intensifying in the next five years, on average. Similarly, the average reported likelihood of de-intensification was 3.57, which we interpret to mean there is a 37.5% probability of de-intensifying in the next five years, on average.

Importantly, survey participants were asked about intensification and de-intensification, which could mean a change in management on the farm (such as an increase in the number of livestock per hectare) rather than wholesale conversion of a farmer's land use. As these are the best empirical indicators of intentions, however, we ignore this possibility in the analysis that follows.

For the purposes of this research, we ranked the three most common land uses based on the intensiveness of their land use (Figure 1). Dairy farming represents the most intensive land use, followed by sheep and beef farming. Forestry is the least intensive land use.

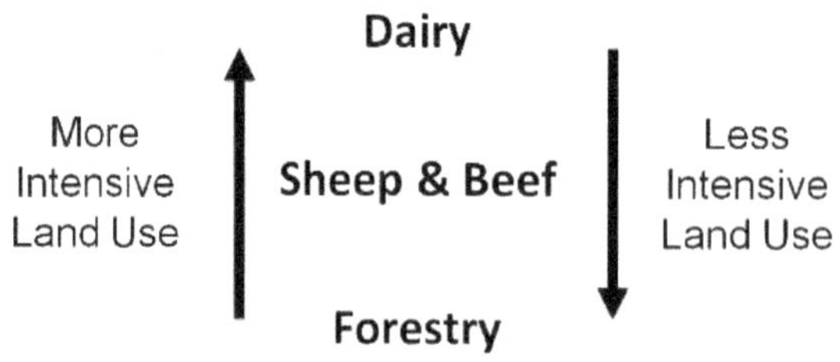

Figure 1. Land uses within the ARLUNZ model ranked by the intensiveness of their land use.

Using attributes defined by geospatial information (specifically, predominant land use and farm size) and empirical data from the Survey of Rural Decision Makers (specifically, age, experience, education level, importance of productivity, profitability, and network size), we define the likelihood of a farmer intensifying or de-intensifying his or her land use econometrically. Specifically, the probabilities of moving from a low-intensity activity to a high-intensity activity over the subsequent five years and vice versa are estimated using Tobit models in which the dependent variables are censored at 0 and 10. Specifically,

$$y_i^* = \mathbf{X}_i\boldsymbol{\beta} + u_i, u \sim N(0, \sigma^2) \tag{1}$$

where y_i^* is a latent variable equal to the observed variable, y_i, only when the latent variable falls between the values of 0 and 10, $\mathbf{X}$ is a vector of explanatory variables, and the error term, u, is normally distributed. Thus, we have:

$$y_i = \begin{cases} 10 \text{ if } y_i^* \geq 10 \\ y_i^* \text{ if } 0 < y_i^* < 10 \\ 0 \text{ if } y_i^* \leq 0 \end{cases} \tag{2}$$

In contrast to the ordinary least squares with censored data, the tobit estimator is consistent [83].

Table 2 presents the tobit estimates based on the national level data. The $\boldsymbol{\beta}$ are interpreted as the expected change in the uncensored latent variable, *i.e.*, the uncensored likelihood of intensification or de-intensification associated with a marginal change in an explanatory variable. For example, increasing age of the decision maker by 1% reduces the predicted (uncensored) perceived likelihood of intensification by 0.034 points on the 11-point scale, an effect that is statistically significant at the 0.01 level. Similarly, having a diploma or technical training increases the predicted (uncensored) perceived likelihood of intensification by 0.570 points, also significant at the 0.01 level. Neither of these explanatory variables has a statistically distinguishable effect on the perceived likelihood of de-intensification.

Table 2. Predicted intensification and de-intensification for the survey approach (tobit model).

Variables	Intensify	De-Intensify
log of effective land quantity	0.206 *	0.150
	(0.105)	(0.118)
log of age	−3.374 ***	−0.0697
	(0.767)	(0.923)
log of experience	0.380 ***	0.352 **
	(0.145)	(0.164)
diploma/tech training	0.570 *	−0.337
	(0.329)	(0.370)
university or higher	0.295	−0.159
	(0.323)	(0.373)
importance of being highly productive	0.140 **	−0.0216
	(0.0605)	(0.0657)
profitable business	−0.842 **	−0.342
	(0.373)	(0.410)
respondent exceeds median # of farm/farmer visits	1.272 ***	0.787 **
	(0.286)	(0.328)
risk tolerance	0.179 ***	0.0248
	(0.0619)	(0.0705)
Constant	6.874 **	−3.812
	(3.330)	(3.988)
Enterprise dummies	YES	YES
Region dummies	YES	YES
Observations	1,507	1,507
McFadden's adjusted R-squared	0.0449	0.0182

Note: * $p < 0.10$; ** $p < 0.05$; *** $p < 0.01$.

To use the likelihood of land-use conversion values as defined above, we take a random draw from a uniform distribution between 0 and 1. If the value of the random draw is less than the farmer's likelihood of land-use conversion, then the proposed land-use change is accepted by the farmer agent and the farm is immediately converted to the proposed enterprise. In addition to converting between enterprises, the farmer agent also realises the predicted net revenue for that land use as defined by the economic component. If the random draw exceeds the likelihood of land-use conversion, then the incumbent enterprise is retained until the next time step of the model.

3. Experimental Section

To illustrate the variation in the predicted land use, economic outcomes, and environmental impacts caused by various approaches to assigning farmers' likelihood of land-use conversion, we explore the effects on landowners in the Hurunui-Waiau catchment in the Canterbury region of New Zealand's South Island (Figure 2). These catchments have a large and diverse set of land uses that are expected to see significant changes in the future.

Figure 2. Location of the Hurunui-Waiau catchment. The catchment is located within the Canterbury region of New Zealand's South Island. Planners in the catchment anticipate significant changes in land use over the next several decades.

To provide a sample that encompasses the range of possible outcomes, the model was run using 50 simulations for each of the three approaches being investigated (*i.e.*, the homogeneous approach, the network approach, and the survey approach). The results in the following section are based on the averaged values over all 50 simulations for each of the three approaches.

The model covers a time horizon of 50 years with ten incremental time steps, each of which represent five years. The scenario assumes a real annual increase in farm commodity prices (*i.e.*, milk, meat, and timber) of 2%, which is commensurate with the last 50 years of commodity prices [84]. Climate and available technology are held constant over the entire model simulation. We note that the economic picture is consistent across scenarios and that the strength of milk prices relative to meat and timber prices produces a trend toward dairy conversion.

The land-use map used in the model was captured in June 2010 [85] (Figure 3a), and although the map includes seven different land uses, the model focuses on the three key enterprises that represent 94% of the productive land available within the catchment: dairy, sheep and beef, and plantation forestry. The cadastral land parcel boundaries used are derived from Land Information New Zealand and represent the cadastral structure of the catchments as at August 2012 (Figure 3b), which was the closest database to the 2010 land use map. Farmer agents are only created for farms exceeding 100 ha in order to focus on commercially operated enterprises. Productivity zones are delineated by land use classification [86] and slope and are classified into productivity zones—flats, foothills, or hills (Figure 3c). Any land owned by the Crown (e.g., native forest) is assumed to be non-productive in use and hence no farmer agent is created.

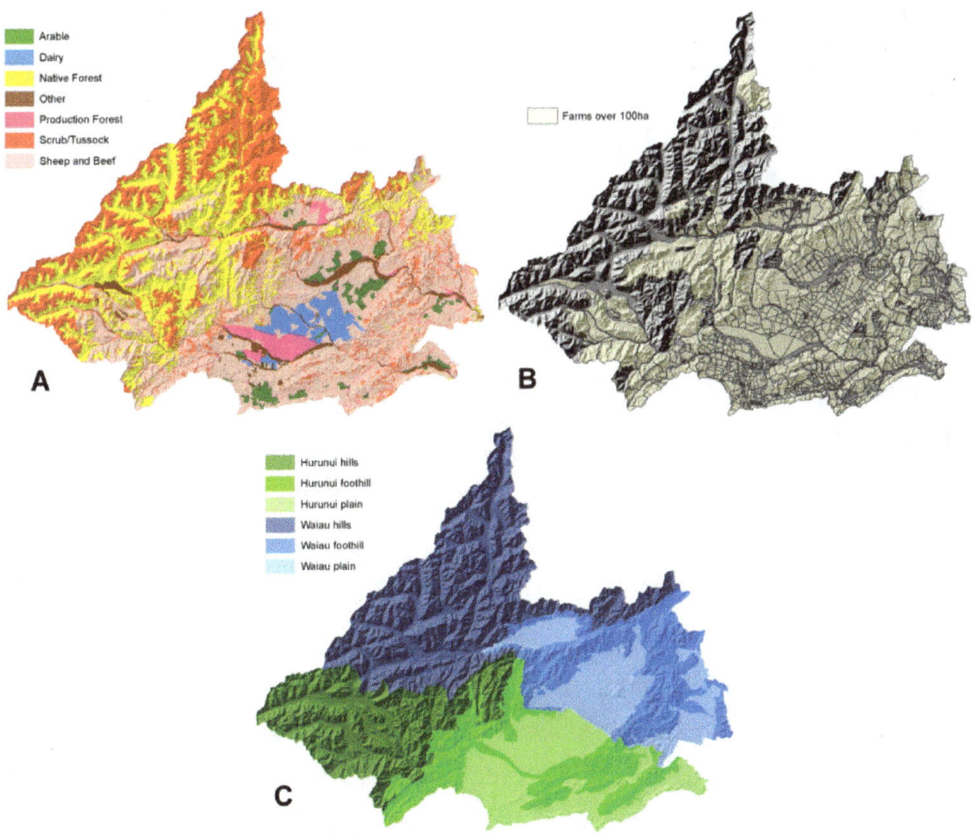

Figure 3. Detailed map of the Hurunui-Waiau catchment by (**a**) 2010 land use; (**b**) 2012 cadastral land parcels; and (**c**) productivity zone. The data layers are used within the model to define the initial land use, farm locations and extents, and the expected productivity for each farm.

The landscape provides a range of geospatial information about the catchment being modelled, such as cadastral boundaries, initial land use, and productivity zones. Using the cadastral boundaries, a farmer agent is generated at the centre of each cadastral parcel and makes decisions for the entire cadastral parcel (*i.e.*, farm). This agent holds a range of social and economic attributes such as age, education, and the size of social networks. These attributes are defined empirically for all farmer agents through the Survey of Rural Decision Makers. For the homogeneous and network approaches, each agent's initial likelihood of land-use conversion is set to 20%; in the case of the survey approach, each agent's likelihood of land-use conversion is defined econometrically as described above.

For the survey approach, there is significant variation in the farmer agents' likelihood of land-use conversion between each of the three enterprises. Table 3 summarises this variation over the 50 randomly generated populations used in the model. Farmers who undertake forestry have the lowest likelihoods of land-use conversion. These values reinforce two characteristics forestry, namely, its low intensity (an average likelihood of de-intensification at 0% highlights a lack of less-intensive options) and that farmers undertaking forestry are less likely to move to more intensive land uses (13% compared with 26% and 31% for sheep and beef and dairy farmers, respectively).

Table 3. Mean values across the 50 simulations for the survey approach using ARLUNZ. Mean values, standard deviation, and confidence intervals for this figure are available in Supplementary Material 2.

	Forestry	Sheep and Beef	Dairy
Intensify	12.99%	26.12%	31.06%
De-Intensify	0.00%	29.10%	26.99%

The simulated likelihoods of land-use conversion to more and less intensive enterprises for sheep and beef farmers are more balanced (26.12% *vs*. 29.10%, respectively), but sheep and beef farmers are more likely to de-intensify their land use over the next five years. The simulated likelihoods of land-use conversion for dairy farmers highlight the production-focused approach commonly associated with the enterprise: alongside a 27% probability that they will de-intensify their land use, there is a 31% chance that they would further intensify their land use over the next five years. However, the model does not include a more intensive land use, so while this intention is accounted for within the model, it is not currently utilised.

4. Results

In this section, we compare results obtained after defining the likelihood of land-use conversion in each of the ways described above. At the catchment level, we project that the area of both dairy and forestry will increase over time (Figure 4). At the production zone level, dairy is estimated to increase in both the plains and foothills, while forestry is estimated to expand in the foothills and hills. The area of sheep and beef farms is estimated to decline in all three productivity zones.

For the homogenous approach, the area allocated to dairy increases from the initial 16,900 ha to 100,450 ha over the 50-year period (with a 95% Confidence Interval, hereafter 95% CI, of 1670 ha). This expansion is split between the highly productive plains region of the catchment and the less productive foothills.

Although such an expansion in dairy is large, it is not unrealistic. For example, the area of land allocated to dairy in Canterbury increased by 172% between 1996 and 2008, and it is projected to expand by an additional 51% by 2020 [87]. Moreover, the Hurunui-Waiau catchment have already witnessed conversion to dairy as forests in the highly-productive flat areas of the catchment reached harvest age. Third, there are ongoing discussions of implementing the Hurunui Water Project, which would expand the area of irrigated land by an additional 41,500 ha, bringing the total irrigated area of the Hurunui-Waiau catchment to over 72,000 ha [88].

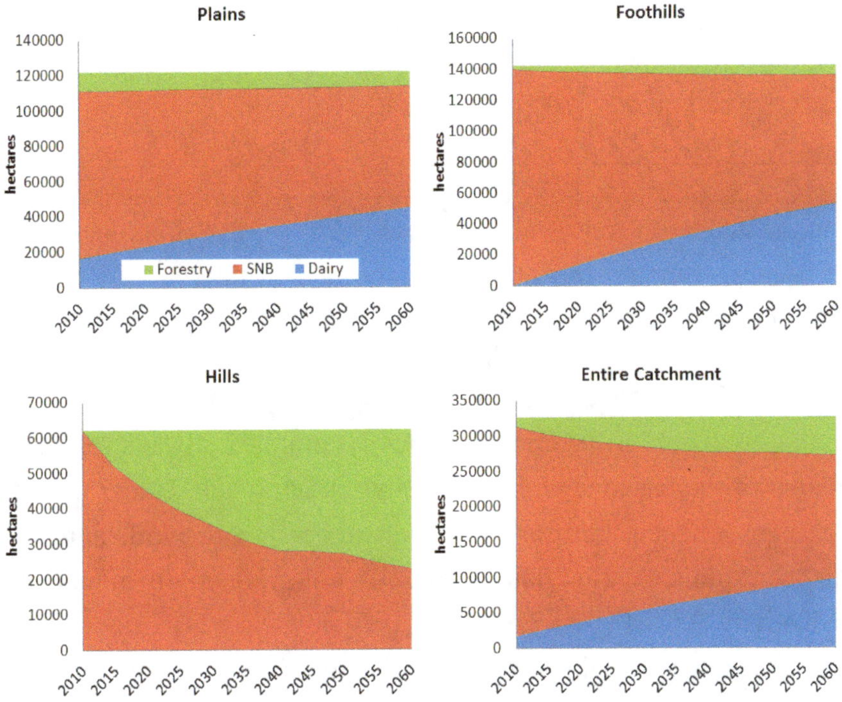

Figure 4. Regional land use area for "Homogenous" baseline projection. Mean values, standard deviation, and confidence intervals for this figure are available in Supplementary Material 3.

The modelled change in land use relative to the homogeneous approach is shown in Figure 5. For both dairy and sheep and beef operations, both approaches to defining the likelihood of land-use conversion trend positively and begin to converge by 2060 (Figure 5, C4). Divergence between the network and survey approaches by 2060 can only be found in forestry and then only for the network approach which results in a level of land use similar to the homogeneous baseline.

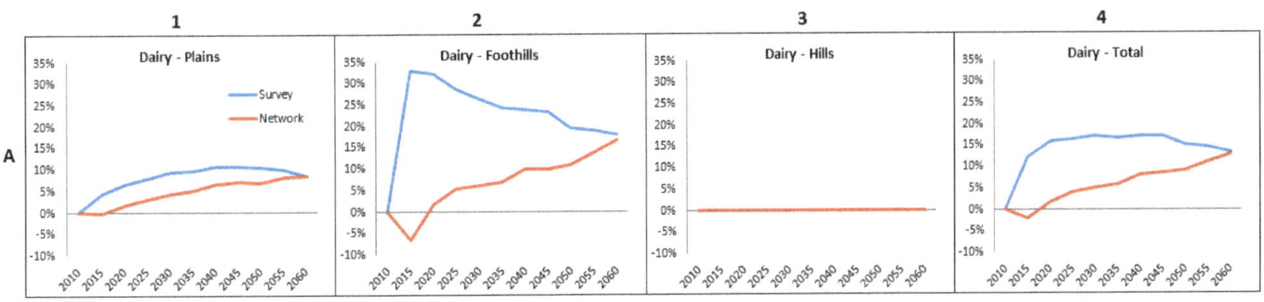

Figure 5. *Cont.*

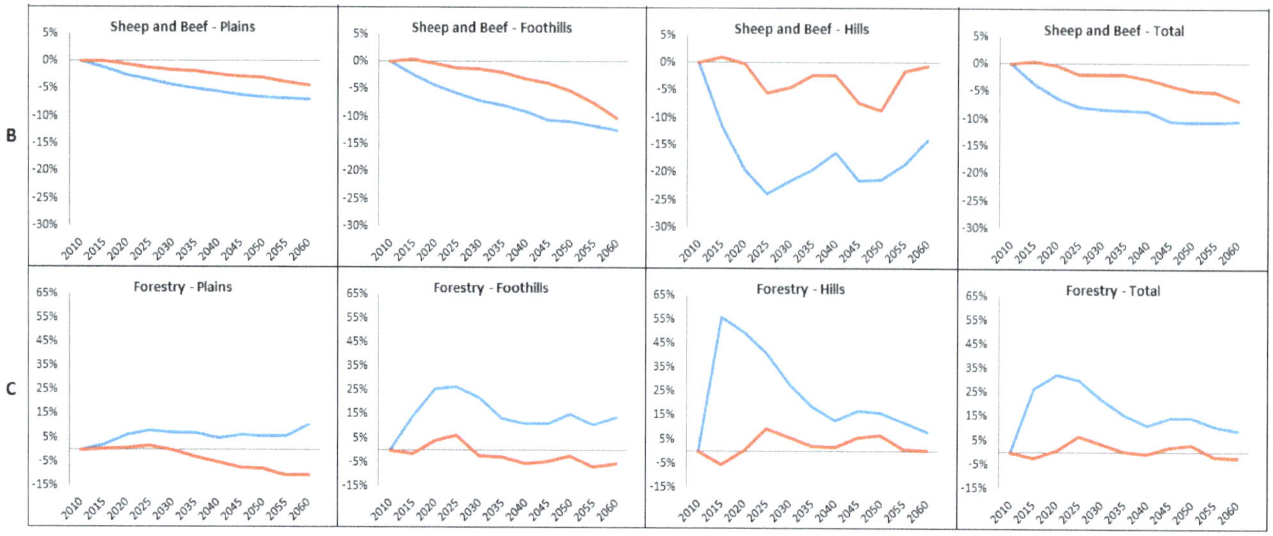

Figure 5. Relative change in regional land use area from "Homogenous" baseline projection. Mean values, standard deviation, and confidence intervals for this figure are available in Supplementary Material 3.

Using a sub-regional scale by productivity zone, the network approach results in impacts that are similar to the homogenous approach. In addition, the survey approach always results in a more rapid change in land use relative to the baseline and to the network approach. Under the survey approach, farmer agents make changes to their farms sooner than under the network approach, which may explain, which explains the rapid growth in dairy in the foothill productivity zone (Figure 5, A2) and forestry in the hills productivity zone (Figure 5, C3).

In the model, forestry at a sub-regional scale is significantly affected by the change in approach to define the farmer's likelihood of land-use conversion. For example, the area of forestry within the plains productivity zone significantly reduces for the network approach when compared with both the survey approach and the homogenous baseline (Figure 5, C1). This outcome is because the more intensive land uses on the plains (such as dairy or sheep and beef) are more profitable, which is translated into the farmer's networks, particularly their geographic network. The increased profitability within their geographic network alters the behaviour of the forestry agents to increase the likelihood of their conversion to more intensive land uses.

4.1. Farm Net Revenue

For reference, all monetary figures from the model are reported in 2012 New Zealand dollars. Farm net revenue is estimated to increase over time under all approaches (Table 4), increasing from \$153 million/yr in 2010 to between \$745 and \$807 million/yr by 2060 (95% CIs of \$8 million and \$6.8 million respectively). The increase over time stems from the assumption that all commodity prices and yield combinations (*i.e.*, farm revenue per hectare) increase by 2%/yr and the expansion of dairy. Interestingly, while the survey and network approaches vary, the difference in net revenue for the two approaches at a catchment scale over time is negligible (\$807 *vs.* \$804 million/yr by 2060 respectively, with 95% CI's of \$6.8 million/yr and \$8.2 million/yr respectively).

Table 4. Total annual farm profits for Hurunui-Waiau catchment (million NZD/yr). Mean values, standard deviation, and confidence intervals for this figure are available in Supplementary Material 3.

Year	Total			Dairy			Sheep and Beef			Forestry		
	Homo	Networks	Survey	Homo	Networks	Survey	Homo	Networks	Survey	Homo	Networks	Survey
2010	$153	0.0%	0.0%	$30	0.0%	0.0%	$115	0.0%	0.0%	$7	0.0%	0.0%
2020	$214	0.9%	6.6%	$83	1.2%	15.3%	$110	1.8%	−6.8%	$21	0.0%	25.0%
2030	$308	3.1%	7.8%	$156	4.3%	15.7%	$118	1.7%	−10.3%	$34	2.9%	19.0%
2040	$421	4.5%	8.7%	$246	7.2%	16.0%	$126	1.6%	−10.5%	$49	−2.1%	9.3%
2050	$568	4.9%	8.4%	$369	7.1%	14.0%	$139	0.0%	−14.9%	$60	1.6%	13.0%
2060	$745	7.3%	7.7%	$514	11.4%	12.7%	$153	−3.4%	−15.9%	$78	−1.3%	9.3%

While net revenue at a catchment scale is similar, the distribution in net revenue between the three enterprises differs. Total dairy profits are estimated to increase from $30 million in 2010 to between $514 and $589 million (95% CIs of $9.1 million and $8.3 million respectively) in 2060 (Table 4). The largest increases are estimated to occur under the survey approach, where annual profits are, on average, 15% larger than under the homogenous approach.

The expansion of forestry in the catchment also yields increased net revenue from an initial $7 million in 2010 to between $77 and $86 million (95% CI of $3.4 million and $3.4 million respectively) in 2060. Again, the survey approach is estimated to yield increased profits (10% or more/yr) relative to the homogenous approach. Profits in sheep and beef farming are estimated to increase from $115 million in 2010 to between $132 and $153 million (95% CI of $2.9 million and $3.3 million respectively) in 2060 even though the area devoted to this land use is expected to contract because of yield and commodity price changes that increases the per hectare revenue of sheep and beef farming.

Interestingly, the distribution in net revenue for sheep and beef under the network approach is estimated to be higher than under the homogenous approach up until 2060 (Table 4). The opposite is found for the survey approach, which sees a reduction in the level of profit for sheep and beef compared with the baseline.

4.2. Environmental Outputs

As land-use change impacts key environmental indicators, the ARLUNZ model captures a range of environmental outputs such as GHG emissions, forest carbon sequestration, nitrogen (N) and phosphorus (P) loss [36]. As seen in Table 5, the expansion of dairy—which often has a higher stocking rate per hectare than sheep and beef—causes an 18% increase in livestock GHG emissions compared with the homogeneous approach. The network and survey approaches show additional livestock GHG emissions of 6% and 3%, respectively.

This growth in livestock GHGs are offset by increases in carbon sequestration through forestry under two of the three approaches. Specifically, the networks approach shows a reduction in forest carbon sequestration of 2% relative to the baseline while the survey approach shows a 9% increase in the amount of carbon sequestered. This result highlights the detrimental effects of the network approach on forestry in more marginal production zones. Even so, the overall expansion of forestry in all three approaches results in a net GHG reduction of between 50% and 68% between 2010 and 2060.

Table 5. GHG outputs for Hurunui-Waiau catchment (tons/yr). Mean values, standard deviation, and confidence intervals for this figure are available in Supplementary Material 3.

Year	Livestock GHGs			Forest Carbon Sequestration			Net GHGs		
	Homogenous	Network	Survey	Homogenous	Network	Survey	Homogenous	Network	Survey
2010	988,619	988,619	988,619	−200,686	−200,686	−200,686	787,933	787,933	787,933
2020	1,009,062	1,027,099	1,007,832	−517,043	−502,681	−677,464	492,020	524,418	330,368
2030	1,052,232	1,080,009	1,060,729	−682,562	−697,085	−833,184	369,670	382,924	227,545
2040	1,080,496	1,125,738	1,113,336	−807,773	−792,856	−896,421	272,723	332,881	216,915
2050	1,134,640	1,179,598	1,164,413	−810,228	−821,304	−930,562	324,412	358,294	233,851
2060	1,166,265	1,238,713	1,201,127	−867,407	−847,212	−947,868	298,859	391,501	253,259

In contrast, the expansion of dairy farming in the plains and foothills, results in a large increase in nutrient loadings that could impact the environmental quality of the local waterways (Table 6). We estimate that N and P will increase by 86% and 43%, respectively, over the next 50 years under the homogenous baseline. The networks approach predicts increases in both N and P by an additional 8% relative to the baseline, while the survey approach predicts additional increases of 8% and 5%, respectively. Consequently, while the expansion of dairy farming in the catchment produces economic benefits, the negative impacts on water quality are non-negligible.

Table 6. Nitrogen and Phosphorus outputs for Hurunui-Waiau catchment (tons/yr). Mean values, standard deviation, and confidence intervals for this figure are available in Supplementary Material 3.

Year	Nitrogen			Phosphorus		
	Homogenous	Network	Survey	Homogenous	Network	Survey
2010	4039	4039	4039	37	37	37
2020	4899	4970	5171	40	41	41
2030	5701	5882	6115	44	45	45
2040	6339	6652	6899	46	49	49
2050	7009	7370	7599	50	53	53
2060	7517	8136	8111	53	57	56

5. Conclusions

We anticipated that the two different approaches to defining the likelihood of land-use change would significantly affect model outcomes. However, the results from this experiment show that the approach has a limited effect at a catchment scale for both the distribution of land use and the resulting total net revenue. Nevertheless, the differing distributions of land use across productivity zones suggest that the approach to estimating the likelihood of land-use change ultimately does matter.

Simulating the social process of information transfer between agents through the network approach resulted in less economically optimal land use for all three enterprises. For dairy, the network approach predicts less conversion vis-à-vis the survey approach. Still, results for both approaches converge by 2060.

The network approach to defining the likelihood of land-use change profoundly impacts model outcomes for sheep and beef farmers in the hill productivity zone. Within this zone, it appears that the level of productivity is very similar at a per hectare level for each sheep and beef farmer. This similarity of profitability for these farmer agents reduces the likelihood of conversion to the more profitable enterprise of forestry within the network approach. For the small amount of forestry on the highly productive plains in the initial state of the model, the network approach enabled the farmer agents to easily recognise significant increases in average net revenue if they converted to different enterprises, primarily dairy. This resulted in a steady decrease in the area of forestry on the plains. In the foothill zone, many sheep and beef farms convert to dairy.

Defining the agents' likelihood of land-use conversion through the survey approach provides greater deviation from the baseline for all sub-regional results. This deviation is particularly evident in the early steps of the model simulation, where variations in area of up to 55% compared with the baseline are observed. This result highlights the design of the survey approach, where land that suits the two ends of the spectrum of land-use intensity—dairy and forestry—quickly sees the influx of new farms, in contrast to the slow changes that occur under the network approach.

The results for the survey approach highlight two interesting observations. First, even with the significant increase in net revenue when converting forestry to other enterprises in the plains productivity zone, the amount of forestry in this zone increases under the survey approach. This result is embedded in the initial land-use conversion values defined for a move to a more intensive enterprise. For example, regardless of the significant increase in net revenue, the survey found that foresters are unlikely to convert to a more intensive enterprise, highlighting the static nature of the survey approach, which—unlike the dynamic network approach—is unable to account for changes in preference over time.

A second interesting result is the fact that substantively different approaches to defining the likelihood of land-use conversion produce similar trends. This observation raises questions that cannot be fully answered in this paper: If the attributes of a population can be reliability and accurately captured through surveys, can the agent-based model that uses these data be simplified by removing the need to simulate social processes? And would this facilitate greater uptake among end users because the process used to define behaviour within an agent-based model is easier to understand?

Malanson and Walsh [29] recently noted the problems of calibration and validation stemming from complex interactions in agent-based models. While agreeing that the challenge for applied agent-based models is in correctly parametrising the agents, we found that the survey approach provided enough detail to generate reliable populations of agents and did not over parametrise the model.

We believe that model design should be informed by model purpose. Where comprehensive surveys are available, we advocate using empirical data to define an agent's likelihood of a type of behaviour. However, if there is limited information about the size and importance of a farmer's social and geographical networks or if a key purpose of the model is to account for the effects of these networks, then developing the structure within the model to simulate, test, and document the effects of changes in the networks is preferred.

For LULCC models—especially those in applied settings—exploring the impacts of different approaches to modelling the likelihood of land-use conversion is critical. Understanding the effects of key model parameters on economic, social, and environmental factors will facilitate the continued acceptance of ABM of LULCC among end users and will improve the results generated by both the economic and geographic ABMs of LULCC. While the results outlined here will not fully resolve the disciplinary differences, they do outline the need to account for heterogeneity in the predicted agent behaviours for both disciplines.

Acknowledgements

The development of ARLUNZ was supported by Landcare Research (CF1112-83-02 and CF1011-83-11) and by the New Zealand Ministry of Business, Innovation, and Employment (C09X1307). The Survey of Rural Decision Makers was supported by the Ministry for the Environment (20443) and the Ministry of Business, Innovation, and Employment (C09X1003). The founding sponsors had no role in the design of the study; in the collection, analyses, or interpretation of data; in the writing of the manuscript, and in the decision to publish the results.

The authors acknowledge the contributions of Oshadhi Samarasinghe and Suzie Greenhalgh, who supported development of the ARLUNZ model, and Florian Eppink for assistance in R. Anne Austin, Suzie Greenhalgh, and William Wright commented on drafts of this paper. Errors and omissions are the responsibility of the authors alone.

Author Contributions

Fraser Morgan had the original idea for the paper, developed and ran the experiments, and led the preparation of the manuscript. Philip Brown led the development and analysis of the survey of rural decision makers and the writing in relation to the survey and statistics derived from the survey. Adam Daigneault analysed the results from the experiments and provided assistance with the economic components of the model. All authors participated in the writing and revision of the manuscript. In addition, all authors approved the final manuscript.

Conflicts of Interest

The authors declare no conflict of interest.

References

1. Smajgl, A.; Brown, D.G.; Valbuena, D.; Huigen, M.G.A. Empirical characterisation of agent behaviours in socio-ecological systems. *Environ. Model. Softw.* **2011**, *26*, 837–844.
2. Parker, D.C.; Manson, S.M.; Janssen, M.A.; Hoffmann, M.J.; Deadman, P. Multi-agent systems for the simulation of land-use and land-cover change: A review. *Ann. Assoc. Am. Geogr.* **2003**, *93*, 314–337.

3. Villamor, G.B.; van Noordwijk, M.; Troitzsch, K.G.; Vlek, P.L. Human decision making for empirical agent-based models: Construction and validation. In Proceedings of the International Environmental Modelling and Software Society (iEMSs)—2012 International Congress on Environmental Modelling and Software, Leipzig, Germany, 1–5 July 2012.
4. Hersperger, A.M.; Gennaio, M.-P.; Verburg, P.H.; Bürgi, M. Linking land change with driving forces and actors: Four conceptual models. *Ecol. Soc.* **2010**, *15*, 1.
5. Lambin, E.F.; Turner, B.L.; Geist, H.J.; Agbola, S.B.; Angelsen, A.; Bruce, J.W.; Coomes, O.T.; Dirzo, R.; Fischer, G.; Folke, C.; *et al.* The causes of land-use and land-cover change: Moving beyond the myths. *Glob. Environ. Chang.* **2001**, *11*, 261–269.
6. Rindfuss, R.R.; Walsh, S.J.; Turner, B.L.; Fox, J.; Mishra, V. Developing a science of land change: Challenges and methodological issues. *Proc. Natl. Acad. Sci. USA* **2004**, *101*, 13976–13981.
7. Brown, D.G.; Verburg, P.H.; Pontius, R.G., Jr.; Lange, M.D. Opportunities to improve impact, integration, and evaluation of land change models. *Curr. Opin. Environ. Sustain.* **2013**, *5*, 452–457.
8. Agarwal, C.; Green, G.M.; Grove, J.M.; Evans, T.P.; Schweik, C.M. *A Review and Assessment of Land-Use Change Models: Dynamics of Space, Time, and Human Choice*; U.S. Department of Agriculture Forest Service, Northeastern Forest Research Station: Burlington, VT, USA, 2002.
9. Baker, W.L. A review of models in landscape change. *Landsc. Ecol.* **1989**, *2*, 111–133.
10. Veldkamp, A.; Lambin, E.F. Predicting land-use change. *Agric. Ecosyst. Environ.* **2001**, *85*, 1–6.
11. Chuvieco, E. Integration of linear programming and GIS for land-use modeling. *Int. J. Geogr. Inf. Syst.* **1993**, *7*, 71–83.
12. Longley, P.; Higgs, G.; Martin, D. The predictive use of GIS to model property valuations. *Int. J. Geogr. Inf. Syst.* **1994**, *8*, 217–235.
13. Sklar, F.H.; Costanza, R. The development of dynamic spatial models for landscape ecology: A review and prognosis. In *Quantitative Methods in Landscape Ecology*; Tuner, M.G., Gardner, R.H., Eds.; Springer-Verlag: New York, NY, USA, 1991; pp. 239–288.
14. Weinberg, M.; Kling, C.L.; Wilen, J.E. Water markets and water quality. *Am. J. Agric. Econ.* **1993**, *75*, 278–291.
15. Leggett, C.G.; Bockstael, N.E. Evidence of the effects of water quality on residential land prices. *J. Environ. Econ. Manag.* **2000**, *39*, 121–144.
16. Parker, D.C.; Filatova, T. A conceptual design for a bilateral agent-based land market with heterogeneous economic agents. *Comput. Environ. Urban Syst.* **2008**, *32*, 454–463.
17. O'Sullivan, D.; Haklay, M. Agent-based models and individualism: Is the world agent-based? *Environ. Plan. A* **2000**, *32*, 1409–1425.
18. Nolan, J.; Parker, D.; van Kooten, G.C.; Berger, T. An overview of computational modeling in agricultural and resource economics. *Can. J. Agric. Econ.* **2009**, *57*, 417–429.
19. Caldas, M.; Walker, R.; Arima, E.; Perz, S.; Aldrich, S.; Simmons, C. Theorizing land cover and land use change: The peasant economy of Amazonian deforestation. *Ann. Assoc. Am. Geogr.* **2007**, *97*, 86–110.
20. Irwin, E.G.; Geoghegan, J. Theory, data, methods: Developing spatially explicit economic models of land use change. *Agric. Ecosyst. Environ.* **2001**, *85*, 7–24.

21. Heckelei, T.; Britz, W.; Zhang, Y. Positive mathematical programming approaches—Recent developments in literature and applied modelling. *Bio-Based Appl. Econ.* **2012**, *1*, 109–124.
22. Evans, T.P.; Sun, W.; Kelley, H. Spatially explicit experiments for the exploration of land-use decision-making dynamics. *Int. J. Geogr. Inf. Sci.* **2006**, *20*, 1013–1037.
23. Irwin, E.G. New directions for urban economic models of land use change: Incorporating spatial dynamics and heterogeneity. *J. Reg. Sci.* **2010**, *50*, 65–91.
24. Janssen, M.; Jager, W. The human actor in ecological-economic models. *Ecol. Econ.* **2000**, *35*, 307–310.
25. Magliocca, N.R.; Brown, D.G.; Ellis, E.C. Cross-Site comparison of land-use decision-making and its consequences across land systems with a generalized agent-based model. *PLoS One* **2014**, *9*, e86179.
26. Valbuena, D.; Verburg, P.H.; Bregt, A.K.; Ligtenberg, A. An agent-based approach to model land-use change at a regional scale. *Landsc. Ecol.* **2010**, *25*, 185–199.
27. Malanson, G.P.; Verdery, A.M.; Walsh, S.J.; Sawangdee, Y.; Heumann, B.W.; McDaniel, P.M.; Frizzelle, B.G.; Williams, N.E.; Yao, X.Z.; Entwisle, B.; *et al.* Changing crops in response to climate: Virtual Nang Rong, Thailand in an agent based simulation. *Appl. Geogr.* **2014**, *53*, 202–212.
28. Farmer, J.D.; Foley, D. The economy needs agent-based modelling. *Nature* **2009**, *460*, 685–686.
29. Malanson, G.P.; Walsh, S.J. Agent-based models: Individuals interacting in space. *Appl. Geogr.* **2015**, *56*, 95–98.
30. Matthews, R.B.; Gilbert, N.G.; Roach, A.; Polhill, J.G.; Gotts, N.M. Agent-based land-use models: A review of applications. *Landsc. Ecol.* **2007**, *22*, 1447–1459.
31. Heard, D.; Dent, G.; Schifeling, T.; Banks, D. Agent-Based models and microsimulation. *Ann. Rev. Stat. Appl.* **2015**, *2*, 259–272.
32. An, L. Modeling human decisions in coupled human and natural systems: Review of agent-based models. *Ecol. Model.* **2012**, *229*, 25–36.
33. Heckbert, S.; Baynes, T.; Reeson, A. Agent-based modeling in ecological economics. *Ann. N. Y. Acad. Sci.* **2010**, *1185*, 39–53.
34. Kaye-Blake, B.; Schilling, C.; Post, E. Validation of an agricultural MAS for southland, New Zealand. *J. Artif. Soc. Soc. Simul.* **2014**, *17*, 5.
35. Valbuena, D.; Verburg, P.H.; Veldkamp, A.; Bregt, A.K.; Ligtenberg, A. Effects of farmers' decisions on the landscape structure of a Dutch rural region: An agent-based approach. *Landsc. Urban Plan.* **2010**, *97*, 98–110.
36. Morgan, F.J.; Daigneault, A.J. Estimating impacts of climate change policy on land use: An agent-based modelling approach. *PLoS One* **2015**, *10*, e0127317.
37. Burton, R.J.F. *Strategic Decision-Making in Agriculture: An International Perspective of Key Social and Structural Influences*; AgResearch: Lincoln, New Zealand, 2009; p. 162.
38. Bakker, M.M.; van Doorn, A.M. Farmer-specific relationships between land use change and landscape factors: Introducing agents in empirical land use modelling. *Land Use Policy* **2009**, *26*, 809–817.
39. Maertens, A.; Barrett, C.B. Measuring social networks' effects on agricultural technology adoption. *Am. J. Agric. Econ.* **2013**, *95*, 353–359.

40. Ramirez, A. The influence of social networks on agricultural technology adoption. *Procedia Soc. Behav. Sci.* **2013**, *79*, 101–116.
41. Smithers, J.; Furman, M. Environmental farm planning in Ontario: Exploring participation and the endurance of change. *Land Use Policy* **2003**, *20*, 343–356.
42. Smithers, J.; Johnson, P. The dynamics of family farming in North Huron County, Ontario. Part I. Development trajectories. *Can. Geogr.* **2004**, *48*, 191–208.
43. Burton, R.J.F. The influence of farmer demographic characteristics on environmental behaviour: A review. *J. Environ. Manag.* **2014**, *135*, 19–26.
44. Millington, J.; Romero-Calcerrada, R.; Wainwright, J.; Perry, G. An agent-based model of mediterranean agricultural land-use/cover change for examining wildfire risk. *J. Artif. Soc. Soc. Simul.* **2008**, *11*, 4.
45. Valbuena, D.; Verburg, P.H.; Bregt, A.K. A method to define a typology for agent-based analysis in regional land-use research. *Agric. Ecosyst. Environ.* **2008**, *128*, 27–36.
46. Acosta-Michlik, L.; Espaldon, V. Assessing vulnerability of selected farming communities in the Philippines based on a behavioural model of agent's adaptation to global environmental change. *Glob. Environ. Chang.* **2008**, *18*, 554–563.
47. Ziervogel, G.; Bithell, M.; Washington, R.; Downing, T. Agent-based social simulation: A method for assessing the impact of seasonal climate forecast applications among smallholder farmers. *Agric. Syst.* **2005**, *83*, 1–26.
48. Berger, T. Agent-based spatial models applied to agriculture: A simulation tool for technology diffusion, resource use changes and policy analysis. *Agric. Econ.* **2001**, *25*, 245–260.
49. Deffuant, G.; Huet, S.; Bousset, J.P.; Henriot, J.; Amon, G.; Weisbuch, G. Agent-based simulation of organic farming conversion in Allier département. In *Complexity and Ecosystem Management: The Theory and Practice of Multi-agent Systems*; Janssen, M., Ed.; Edward Elgar Publishing: Cheltenham, UK, 2002; pp. 158–187.
50. Deffuant, G.; Skerratt, S.; Amblard, F.; Ferrand, N.; Chattoe, E.; Gilbert, N.; Weisbush, G. Agent-based simulation of decision process mixing rational reasoning and influences from socio-informational networks: Case studies of agri-environmental measures adoption by farmers. In Proceedings of the Fifth International Conference on Social Science Methodology, Cologne, Germany, 1 October 2000. [CD-ROM]
51. Robinson, D.T.; Brown, D.G.; Parker, D.C.; Schreinemachers, P.; Janssen, M.A.; Huigen, M.; Wittmer, H.; Gotts, N.; Promburom, P.; Irwin, E.; *et al.* Comparison of empirical methods for building agent-based models in land use science. *J. Land Use Sci.* **2007**, *2*, 31–55.
52. Gigerenzer, G.; Todd, P. *Simple Heuristics That Make Us Smart*; Oxford University Press: Oxford, UK, 1999.
53. Manson, S.M. Bounded rationality in agent-based models: Experiments with evolutionary programs. *Int. J. Geogr. Inf. Sci.* **2006**, *20*, 991–1012.
54. Simon, H.A. Behavioral economics and bounded rationality. In *Models of Bounded Rationality*; Simon, H.A., Ed.; MIT Press: Cambridge, MA, USA, 1997; pp. 267–433.
55. Tversky, A.; Kahneman, D. Rational choice and the framing of decisions. In *The Limits of Rationality*; Cook, K.S., Levi, M., Eds.; University of Chicago Press: Chicago, IL, USA, 1990; pp. 60–89.

56. Dawid, H.; Fagiolo, G. Agent-based models for economic policy design: Introduction to the special issue. *J. Econ. Behav. Organ.* **2008**, *67*, 351–354.
57. Janssen, M.A.; Ostrom, E. Empirically based, agent-based models. *Ecol. Soc.* **2006**, *11*, 37.
58. Berger, T.; Schreinemachers, P. Creating agents and landscapes for multiagent systems from random samples. *Ecol. Soc.* **2006**, *11*, 19.
59. Nelson, K.; Brummel, R.; Jordan, N.; Manson, S. Social networks in complex human and natural systems: The case of rotational grazing, weak ties, and eastern US dairy landscapes. *Agric. Hum. Values* **2014**, *31*, 245–259.
60. Manson, S.M.; Jordan, N.R.; Nelson, K.C.; Brummel, R.F. Modeling the effect of social networks on adoption of multifunctional agriculture. *Environ. Model. Softw.* **2014**, doi:10.1016/j.envsoft.2014.09.015.
61. Heckbert, S.; Bishop, I. Empirical calibration of spatially explicit agent-based models. In *Advanced Geosimulation*; Marceau, D., Benenson, I., Eds.; Bentham Books: Oak Park, IL, USA, 2011; pp. 92–110.
62. Kennedy, W.G. Modelling human behaviour in agent-based models. In *Agent-Based Models of Geographical Systems*; Heppenstall, A.J., Crooks, A.T., See, L.M., Batty, M., Eds.; Springer: Dordrecht, The Netherlands, 2012; pp. 167–179.
63. Wilensky, U. *NetLogo*; Center for Connected Learning and Computer-Based Modeling, Northwestern University: Evanston, IL, USA, 1999. Available online: http://ccl.northwestern.edu/netlogo (accessed on 24 September 2015).
64. Brandmeyer, J.E.; Karimi, H.A. Coupling methodologies for environmental models. *Environ. Model. Softw.* **2000**, *15*, 479–488.
65. Daigneault, A.; Greenhalgh, S.; Samarasinghe, O. A response to Doole and Marsh (2013) article: Methodological limitations in the evaluation of policies to reduce nitrate leaching from New Zealand agriculture. *Austr. J. Agric. Resour. Econ.* **2014**, *58*, 281–290.
66. Daigneault, A.; Greenhalgh, S.; Samarasinghe, O. Economic impacts of GHG and nutrient reduction policies in New Zealand: A tale of two catchments. In Proceedings of the Australian Agricultural and Resource Economics Society 2012 Conference (56th), Freemantle, WA, Australia, 7 February 2012.
67. Daigneault, A.; Greenhalgh, S.; Samarasinghe, O.; Jhunjhnuwala, K.; Walcroft, J.; de Oca Munguia, O. *Sustainable Land Management and Climate Change—Catchment Analysis of Climate Change: Final Report*; Ministry of Primary Industries: Wellington, New Zealand, 2012.
68. Daigneault, A.; McDonald, H.; Elliott, S.; Howard-Williams, C.; Greenhalgh, S.; Guysev, M.; Kerr, S.; Lennox, J.; Lilburne, L.; Morgenstern, U.; *et al. Evaluation of the Impact of Different Policy Options for Managing to Water Quality Limits*; Ministry of Primary Industries: Wellington, New Zealand, 2012.
69. Daigneault, A.; Samarasinghe, O.; Lilburne, L. *Modelling Economic Impacts of Nutrient Allocation Policies in Canterbury—Hinds Catchment: Final Report*; Landcare Research: Lincoln, New Zealand, 2013.
70. Müller, B.; Bohn, F.; Dreßler, G.; Groeneveld, J.; Klassert, C.; Martin, R.; Schlüter, M.; Schulze, J.; Weise, H.; Schwarz, N. Describing human decisions in agent-based models—ODD+D, an extension of the ODD protocol. *Environ. Model. Softw.* **2013**, *48*, 37–48.

71. Brown, P. Survey of Rural Decision Makers. Available online: www.landcareresearch.co.nz/science/portfolios/enhancing-policy-effectiveness/srdm (accessed on 20 March 2015).
72. AsureQuality New Zealand. AgriBase Database. Available online: http://www.asurequality.com/capturing-information-technology-across-the-food-supply-chain/agribase-database-of-new-zealand-rural-properties.cfm (accessed on 24 September 2015).
73. Statistics New Zealand. New Zealand Agriculture Production Surveys and Censuses. Available online: http://datainfoplus.stats.govt.nz/item/nz.govt.stats/6362a469-f374-412e-ac25-d76fd0962003/106/ (accessed on 24 September 2015).
74. Isaac, M.E.; Erickson, B.H.; Quashie-Sam, S.J.; Timmer, V.R. Transfer of knowledge on agroforestry management practices: The structure of farmer advice networks. *Ecol. Soc.* **2007**, *12*, 32.
75. Schmit, C.; Rounsevell, M.D.A. Are agricultural land use patterns influenced by farmer imitation? *Agric. Ecosyst. Environ.* **2006**, *115*, 113–127.
76. Banerjee, A.; Chandrasekhar, A.G.; Duflo, E.; Jackson, M.O. *Gossip: Identifying Central Individuals in a Social Network*; NBER Working Paper No. 20422. Available online: http://arxiv.org/pdf/1406.2293v2.pdf (accessed on 24 September 2015).
77. Small, B. Personal communication. AgResearch: Hamilton, New Zealand, 2013.
78. Alam, S.J.; Geller, A.; Meyer, R.; Werth, B. Modelling contextualized reasoning in complex societies with endorsements. *J. Artif. Soc. Soc. Simul.* **2010**, *13*, 6.
79. Bandura, A. *Social Learning Theory*; Prentice Hall: Englewood Cliffs, NJ, USA, 1977.
80. Bandura, A. *Social Foundations of Thought and Action: A Social Cognitive Theory*; Prentice Hall: Englewood Cliffs, NJ, USA, 1986.
81. Jager, W.; Janssen, M.A.; de Vries, H.J.M.; de Greef, J.; Vlek, C.A.J. Behaviour in commons dilemmas: Homo economicus and Homo psychologicus in an ecological-economic model. *Ecol. Econ.* **2000**, *35*, 357–379.
82. Gotts, N.M.; Polhill, J.G. When and how to imitate your neighbours: Lessons from and for fearlus. *J. Artif. Soc. Soc. Simul.* **2009**, *12*, 2.
83. Amemiya, T. Regression analysis when the dependent variable is truncated normal. *Econometrica* **1973**, *41*, 997–1016.
84. Ministry for Primary Industries. *Situation and Outlook for Primary Industries*; MPI Policy Publication: Wellington, New Zealand, 2013.
85. Hill, Z.; Lilburne, L.; Guest, P.; Elley, R.; Cuff, J. *Preparation of a GIS Based Land Use Map for the Canterbury Region*; Environment Canterbury Report: R10; Environment Canterbury: Christchurch, New Zealand, 2010. Available online: http://ecan.govt.nz/publications/Reports/gis-based-land-map-canterbury.pdf (accessed on 24 September 2015).
86. Lynn, I.H.; Manderson, A.; Page, M.; Harmsworth, G.; Eyles, G.; Douglas, G.; Mackay A.D.; Newsome, P.J.F. *Land use Capability Survey Handbook: A New Zealand Handbook for the Classification of Land*, 3rd ed.; AgResearch: Hamilton, New Zealand; Landcare Research: Lincoln, New Zealand; Institute of Geological and Nuclear Sciences: Lower Hutt, New Zealand, 2009.
87. Parliamentary Commission on the Environment. Water Quality in New Zealand: Land-Use and Nutrient Pollution. Available online: http://www.pce.parliament.nz/publications/all-publications/water-quality-in-new-zealand-land-use-and-nutrient-pollution/ (accessed on 29 September 2014).

88. Environment Canterbury. Hurunui Water Project—Waitohi Proposal. Available online: http://ecan.govt.nz/get-involved/consent-projects/past-notifications/hwp/Pages/waitohi.aspx (accessed on 29 September 2014).

2

Using an Agroecosystem Services Approach to Assess Tillage Methods: A Case Study in the Shikma Region

Hila Sagie * and Uri Ramon

Open Landscape Institute (Machon Deshe), Negev 2 St., Tel Aviv 66186, Israel;
E-Mail: uriramon@bezeqint.net

* Author to whom correspondence should be addressed; E-Mail: hilasagie23@gmail.com

Academic Editors: Benjamin Burkhard, Stefan Hotes and Hubert Wiggering

---

Abstract: The use of ecosystem services (ES) in agricultural management is expanding; however, its integration in decision making processes is still challenging. This project was formulated to examine the ES approach and its usefulness with regard to management dilemmas. The Shikma region, north of the Negev Desert, was chosen as a case study. The management issue identified was the effect of various alternatives (minimum-tillage, no-tillage, straw-mulch and stubble-grazing) on the supply of ES. The expert-based ES assessments' findings reveal that no-tillage has the potential to increase many agroecosystem services and be more profitable for the farmer and the public. However, trade-offs between different ES and among stakeholder groups make it difficult to reach an unequivocal conclusion. As we have found, the process of the study is as important as the results. Throughout the project, an effort was made to engage stakeholders and policy-makers and to define decision-making processes. The study suggests that the ES approach can be useful in expanding the scope of agricultural management beyond provisioning services and create collaborations among farmers, communities, national institutions and environmental organizations to advance conservation agriculture. The study provides guidelines for conducting a productive ES assessment process that will lead to enhanced awareness and implementation.

Keywords: no-tillage; soil erosion; ecosystem services; conservation tillage; multi-stakeholder analysis; straw mulch; stubble grazing; Negev Desert

---

1. Introduction

1.1. Agroecosystems

Agricultural ecosystems provide humans with food and forage, bio-energy and pharmaceuticals and are essential to human wellbeing. These systems also produce and influence a variety of Ecosystem Services (ES), such as regulation of soil and water quality, carbon sequestration, support for biodiversity and cultural services. Depending on management practices, agriculture may have a negative impact on these services and cause soil erosion, loss of wildlife habitat, nutrient runoff, sedimentation of waterways, greenhouse gas emissions, and an increase in pest and weed contamination [1]. These may ultimately undermine the natural base upon which agricultural livelihoods depend [2]. Research on agroecosystem services examines the trade-offs that may occur between provisioning services and other ecosystem services. As more effective methods for assessing and valuing ES become available, the potential for "win-win" scenarios increases. Under all scenarios, appropriate agricultural management practices are critical to realizing the benefits of ES and reducing the negative effects resulting from agricultural activities [1]. Agroecosystem services are defined as the contributions of ecosystem structure and function combined with other inputs, to human well-being "Salzau message" 2010 in [3]. "Other/additional inputs" refer to non-ecosystem-based anthropogenic contributions to ES, for example fertilizers, energy, pesticides, techniques, labor or knowledge use in human-influenced land use systems. A balanced assessment of ES provided by agriculture requires a systems-level socio-ecological understanding of related management practices at local to landscape scales [4].

1.2. Development of Conservation Tillage

The transition from the plow to various forms of conservation tillage began after World War II with the attempt to find ways to produce more crops while using less energy and with the development of non-selective herbicides which made the transition possible [5–7]. Only in the 1990s the concept of no-tillage as a method for soil and water conservation evolved as it became related to the network of ecological endeavor. In addition, at that time, farmer associations for no-tillage and soil conservation were forming, and governments began to support it. Research shows that the process of no-tillage development and uptake may be regarded as a dynamic process of co-creation of innovation including human actors (pioneering farmers, extension workers, scientists, government officials) and non-human actors (herbicides, no-till machinery, policy incentives, soil conditions, crops and yield) [6–8]. By 2010, no-till was practiced on about 111 million hectares globally, an annual growth rate of 6 million ha in many kinds of climates, soils and cropping conditions [9]. A review of conservation agriculture and ES shows that no-till technologies are known to be very effective [5,10] and the awareness of their advantages is growing. Nevertheless, many farmers still struggle with no-tillage adoption [8] and the body of research on constraints regarding its adoption is continuously growing [11–14].

In Israel, the development and spread of no-tillage began in the 1990s. Long-term experiments began in the Negev Desert to reduce energy expenditure and expenses. In the beginning of the millennium, no-tillage was implemented on a number of farms in the Jezreel Valley in northern Israel. In 2007, the Ministry of Agriculture took an important step and began providing annual monetary support for managing fields with soil and water conservation tillage regimes, and provided a 40% subsidy for no-tillage machinery [15]. The long-term experiments showed that no-tillage, including no-tillage with straw mulch, has the potential to increase many agroecosystem services while being economically viable for the farmer. Therefore, it was recommended that no-till farming should be implemented in the dryland fields of Israel [16,17]). Nevertheless, implementation of no-till is still relatively low among Israeli farmers [18].

1.3. Using the Ecosystem Services Approach

Understanding the impacts, of actual and potential changes at a site on ES is important for promoting better planning and management decisions to support ES delivery. The approach emphasizes the importance of comparing estimates for alternative states of a site (for example, before and after conservation agriculture) so that decision-makers can assess the net consequences of such a change, and hence the benefits for human well-being that may be lost through the change or gained by conservation [19]. The ES approach is becoming a dominant part of environmental discourse as well as of the academic literature [20]. The supporters of the approach emphasize its advantages in creating a common language between various stakeholders and decision-makers based on a common basis of data, assessments and preferences. However, in fact, a large part of the work in this field is academic, and it is often difficult to show an explicit impact on decision-making [21]. Only in a relatively small number of ES assessments is a profound process carried out that includes defining the need of the stakeholders on the site, involving the public, conducting ES assessments, and creating a participative discussion on the findings that will influence decision-makers in managing natural resources and landscapes more sustainably.

In this study, we present an example of how to conduct a cooperative process to incorporate the ES approach in agroecosystem management, and how the process and its outcomes can facilitate the enhancement of the overall ES supply for sustainable agriculture. The framework for the project was established jointly by OLI (the Open Landscape Institute) with the Shikma Park executive committee. The sustainable and participatory approach of the Shikma Park management program, where the study takes place, and the remarkable collaboration of the farmers and stakeholders in the area, provided an excellent environment for this study. The goals and focal points are: (1) to assess the effect of different tillage methods used in the dryland fields of Shikma Park, on agroecosystem services; (2) to involve farmers, stakeholders, government officials, experts and the public during the study process and following it, while creating dialogue, raising awareness, obtaining feedback and promoting the implementation of findings; (3) to influence policy and suggest appropriate institutions that can provide effective incentives to promote cropland management that enhances agroecosystem services.

2. The Study Region

Shikma Park is a geographical-organizational unit, composed of various types of open landscapes: agricultural fields, protected areas (at varying levels), and rural communities. It is similar to a "biosphere reserve" in its conservation and sustainable management approach without officially being one (see biosphere reserves in [22]. It is located in the semi-arid northwestern part of the Negev Desert in Israel (Figure 1). A significant part of the area is agricultural: 25,000 hectares of agricultural land, of which 40% is dryland agriculture, including fields that are alternately rain-fed and irrigated. The average annual precipitation varies between 200–400 mm. Wheat is the crop grown on most of the dryland areas in Israel [23] as well as in the dryland fields of Shikma Park, with rotations of barley and various legume types. Several tillage methods are used, including conservation tillage of various types, levels and distribution throughout the farmers' fields. Over the past few years, awareness of conservation agriculture in the northern Negev has increased, leading some of the farmers to modify their management practices and use reduced tillage (minimal, non-inversion tillage), and others to use no-tillage (zero-tillage) [24]. In Shikma Park most of the dryland fields are under reduced-till regimes, while others, about 25%, are under no-till regimes (some of the fields under no-till regimes combine occasional tillage every 4–5 years). Due to well acknowledged benefits of no-till systems [5,10], we attempted to understand, by using the ES approach, why there are still farmers who do not adopt no-till regimes.

Figure 1. Shikma Park.

3. Materials and Methods

The methods were developed during the project process according to the goals described above. It was aided by ES guidance manuals prepared by various international ES working groups such as: [19,25–27], combined with specific features compatible to the socio-political framework in Israel and Shikma Park. The preliminary steps were selecting the research site and creating the organizational structure of the project. Shikma Park was chosen due to the sustainable and collaborative approach of its management program and past collaboration between OLI and the Regional Executive Committee. A partnership was created with the Shikma Park management members, farmers and other stakeholders and experts from local and national levels. In addition, two committees were formed to support and escort the project: a steering committee, to provide feedback and approve the project after every stage, and a multi-disciplinary professional committee of experts—to help with professional ES assessments. Economists were added to the working group for conducting the economic assessment. The early engagement of stakeholders and decision-makers is a key component of an assessment, as it can help provide an accurate understanding on the economic, ecological, social and cultural dynamics of the site [28]. The means used to strengthen stakeholders' engagement included meetings in different forums, interviews, discussion groups, mutual fact-finding, steering committees, agricultural and environmental conferences and day-excursions to the fields. The project includes the following stages. The stages described can be used as guidelines for conducting a productive ES assessment process.

(1) **Defining agroecosystem management issues** currently on the agenda at the research site and **identifying management alternatives**. We began, as do most ecosystem assessments, with extensive consultations with stakeholders and specific user groups in Shikma Park to identify priorities for the assessment questions. These consultations are fundamental to both the relevance and the legitimacy of the process and play a key role in providing its structure and focus [29]. Following the consultations above, the management question and alternatives were identified.

(2) **Mapping stakeholders, policy makers and decision making processes** related to the alternatives identified, based on interviews with stakeholders, reports and literature.

(3) **The perspective of farmers and experts regarding the constraints to no-till adoption.** As part of the study, we conducted open-ended interviews with farmers and experts in order to get their views on what prevents wider adoption of no-tillage in light of findings that demonstrate the success of the method. Insights into farmers' decision-making together with barriers/drivers for implementation of sustainable soil management practices, can contribute to a better understanding of what is needed to foster better compatibility between the method and specific farms, and thus encourage adoption of these practices[14]. All farmers that cultivate land in Shikma Park (six farmers), four agricultural extension service instructors that work or worked in the past in the area, the head of the agricultural organization of the area, as well as several experts from the R&D, the Faculty of Agriculture and the Ministry of Agriculture were interviewed. All interviewees were asked about the constraints as well as about specific issues that relate to each expert that were necessary for the different stages of the project.

(4) **Preparing for ecosystem services assessment of each management alternative identified**. This stage included screening and prioritizing relevant ES and recruiting experts/professional working groups for the assessment. The first part was done by identifying ES relevant to the alternatives being assessed, from the broad list of ES that was formulated by the Israel National Ecosystem Assessment according to international ES work [27,30,31]. Then these ES were prioritized according to various parameters and indices including: relevance and importance to the site, existing tools or data for evaluation, level of complexity and resources needed, and the feasibility of assessment. The second part, of recruiting experts for the assessment, was done by using the "snowball effect" often used in qualitative social research: we began by identifying a preliminary list of experts to consult with on each of the services, these experts provided us with names of additional experts recommended by them, and those provided us with additional experts and so on. In this manner, we were able to reach a wide network of experts in each subject. In some cases, such as the assessment of soil erosion, a larger group of experts was formed to help and guide the assessment process. ES that were ranked high later underwent an in-depth professional assessment by the experts.

(5) **Conducting the ecosystem services assessment.** The assessment process was based on collection of existing data (from the literature and reports), interviews with stakeholders and experts, extrapolation from data collected in other locations, and when needed, a new interpretation of the data gathered on-site, or a small-scale scientific study. For example, a survey was held in order to identify the public's perspectives regarding their environment and their perceptions regarding the

transition to no-tillage (the survey will be elaborated in the following stage). The services were assessed quantitatively, qualitatively, socially, and economically where possible. The economic assessment was based on data gathered from the expert-based ES assessment reports, the farmers, agricultural reports and previous economic evaluations done for the Shikma-Besor Drainage Authority. It was done by using the accepted environmental economic methods including: evaluation of externalities, avoided damage cost, damage prevention cost and cost-benefit analysis. Each expert provided a detailed report of the service being assessed, describing the service and how it is affected by the different alternatives identified, and indicating the trend (increase or decrease of the service) and the level of confidence according to his/her professional opinion. The report was written according to an outline provided to them. Subsequently, the assessments were summarized and presented in a table (Table 2). In the scope of this paper, only the summarized assessments are presented. The final part of the assessment process was presenting the assessments to the relevant group of stakeholders and experts, receiving feedback and suggestions, and revising and implementing them where possible.

(6) **Conducting a public survey of the perceptions of the Shikma Park residents** regarding their environment, agriculture and conservation tillage. The goals of the survey were: to assess the public's attitudes towards agriculture and the environment, to raise local public awareness to ES and environmental issues, to expand and intensify the existing educational program, create community participation, and to assess cultural ES that would be difficult to assess otherwise. The survey was held as part of an educational program led by the regional Nature and Parks Authority and in cooperation with it. The survey included 246 respondents from the Shikma Park local communities and nearby towns. One of the questions asked in the survey was formulated to understand where the stakeholders stand regarding the transition to conservation tillage and who should pay for it.

(7) **Integrating & communicating the results**. In this stage the main findings (from all the stages above) were integrated in order to communicate the results to decision makers in a simple and coherent manner and provide them with an effective tool, while being transparent about the limitations and gaps of the findings. This was done by using tables and charts to simplify and focus the findings, while emphasizing the multi-stakeholder influence. The tools chosen for presenting the findings were *rose plots* and the *table of distribution of benefits among stakeholders* that were obtained from previous ES work done by [19,25]. Another important part was advising decision makers on mechanisms and research topics that need further investigation that can be used to facilitate the implementation of conservation tillage. This part will be described in the discussion Section.

4. Results

This section presents the outcomes of each stage described in the materials and methods section above.

4.1. Defining the Agroecosystem Management Question and Identifying Alternatives

The management question that was chosen was: "the effect of various tillage methods used in the dryland fields of Shikma Park on agroecosystem services". The following tillage alternatives were identified:

(1) Minimum-tillage—In this area, it refers to shallow plowing once a year; deep plowing is seldom used.
(2) Conservation-tillage—No-tillage including long-term no-tillage (12 years) and occasional tillage (shallow plowing or deep plowing every 4–5 years).

Additional alternatives include crop residues/straw mulch alternatives:

(1) Baling and selling of wheat straw (leaving 5–15 cm stubble)—Currently all fields fall under this category.
(2) Leaving straw mulch as ground cover (3,000 kg per ha, at least 30–40 cm high) as is customary in most no-till fields around the world and as was recommended in a long-term no-tillage study done nearby [16,17].

Crop residue management is another important topic in conservation agriculture. It involves maintaining a permanent organic soil cover by leaving the previous year's residue on the field. In general, crop residue on the soil helps maintain humidity and promote soil microbial activity, while increasing the organic carbon influx on the ground and improving the structure and quality of the soil. Straw cover protects the soil from raindrops and minimizes evaporation and the risk of runoff and soil erosion [16,17,24,32,33].

An additional alternative is stubble grazing by sheep in post-harvest wheat fields:

(1) Permit sheep stubble grazing in post-harvest fields
(2) Prevent sheep stubble grazing in post-harvest fields

Stubble grazing by livestock in post-harvest wheat fields is a common practice throughout the Middle Eastern dryland fields, including the Negev Desert where this study was conducted. Between June and September, livestock enters the post-harvest fields after grazing in other open landscapes and JNF (Jewish National Fund) forests from February to May. Wheat fields are another source of grazing land that compensates for the decreasing amount of pasture over the years. The Bedouin shepherds in the area have been grazing their sheep on the wheat fields for dozens of years. Some farmers continue with stubble grazing on their fields while others have stopped. The growing awareness of conservation agriculture that is leading to modifications of management practices by using no-till and crop residue methods, is threatening the practice of post-harvest stubble grazing [24].

The alternatives above are dealt with in the text separately and combined where needed. For example, leaving straw mulch in the field and stubble grazing cannot be combined, but no tillage can be combined with straw mulch or without.

4.2. Mapping Stakeholders and Decision-Making Processes

Table 1 shows the main stakeholders that were identified, whose decisions may have a direct impact on the choice between the management alternatives described above. The mapping of these stakeholders was conducted to identify the decision-makers whose engagement in the process is crucial.

Table 1. Mapping decision makers and their point of decision.

Decision Maker	Decision Issue	Interest/Perspective
Ministry of Finance	Budget allocated for conservation tillage	Internalization of externalities
Israel Land Authority	Land leasing regime and terms	Securing the value of the land
Ministry of Agriculture	Extent and type of incentive/subsidy programs for soil conservation	Agricultural sustainability and profitability
Drainage and River Authority	Drainage basin master-plan Organizational support for soil conservation	Regulation of soil erosion, water runoff, river sediments
Shikma Park Executive Committee	Representation of farmers vis-à-vis the institutions	Sustainability, economic profit, conservation
JNF Forest Managers	Part of the committee that regulates grazing in the park	Grazing in the forests and their vicinity as a tool to prevent wildfires and conserve soil
Farmers	Selecting the tillage method to implement, request for government support, permit grazing	Economic profit, long term sustainability
Shepherd	Remain in permanent dwellings or move with the herd to graze in agricultural and forest areas	Feed for their sheep, economic profit, preserving their pastoral lifestyle

4.3. Farmers' and Experts' Perspectives on the Constraints to No-Till Adoption

Below are the main constraints that were identified in the interviews. It is interesting to note that most of the constraints that were mentioned by the farmers and the agricultural extension instructors were also mentioned by the Ministry of Agriculture experts, Research and Development stations and faculty of Agriculture; therefore, we did not separate the constraints according to the specific stakeholders that mentioned them.

- High cost of equipment—government subsidies are insufficient and the machinery is difficult to maintain; once broken it is hard to fix.
- There is a lack of scientific experiments and research on the method and what does exist is often dubious due to small sample size and the different soils and topographic conditions of the plots.
- A research and development network and agricultural advice are lacking; a private network is partially replacing the previous government network.
- Some of the farmers are conservative and skeptical regarding the no-till approach.
- There is insufficient knowledge regarding weed and pest control resulting from no-tillage.
- Leaving straw mulch is problematic in Israel due to the high price received for selling it.
- Lack of strong evidence for economic profit from the transition.
- Lastly, an important driver of change was identified: as the water supply for agriculture from treated wastewater increases, potential land for no-till farming is decreasing, as more fields will be irrigated.

Table 2. The effect of the transition from minimum tillage to no-tillage on ecosystem services delivery based on experts' assessments.

Ecosystem Services	Benefit of the Service	Trend: Direction and Strength (Increase or Decrease)	Level of Confidence	Description/Details	Main Sources, Experts and Professionals	On-Site Economic Assessment (Economic Team: Gadi Rozenthal and Hadar Fuchs-Rubel)
Agricultural crops	Food and animal feed	↑	Medium	According to experts and farmers in the area: better crop yields in no-till farming combined with straw mulch. According to others: there is no certain increase in crop yields but economic profit is not negatively affected.	[16–18,34]; Farmers and economic team.	The farmer is likely to enjoy an increment of 20 NIS*/dunam**/year for the amount and quality of the crops and 6 NIS/dunam/year for the straw.
Soil carbon sequestration greenhouse gas regulation	Climate regulation	↑	Low	Preventing soil erosion increases gas exchange in the soil. In addition, no-till saves energy from machinery and consequently reduces emissions. However, on-site calculations show that since the no-till fields use additional chemical fertilizers and herbicides it is hard to specify an increase in climate regulation in the short term.	Team leaders: Yoni Weitz and Dr. Jose Gruenzweig. [33,35,36].	Fossil fuel consumption decrease of 2.5 liter/dunam/year equivalent to 5.9 kg/dunam/year. Value of decrease in CO_2 emission equivalent to 0.65 NIS/dunam/year in case of no chemical fertilizer addition.
Water cycle Regulation	Soil moisture retention	↑	High	Decrease of runoff, water percolation in soil, additional water available for plant uptake.	[16,17]	Higher yield from crops due to additional water available for the plant. The public benefit is 4 NIS/dunam/year.
Disease and pest regulation	Plant pest and disease outbreak regulation	↓	High	Increase in rodent activity and a minor and unstable effect on insects, pests and diseases. Increase in uncertainty regarding crops due to lack of knowledge and guidance on the subject.	Team leader: Ofer Mendelson. [37,38].	Increment of 6.7 NIS/dunam/year for pesticides.
Alien or invasive species regulation	Weed regulation	↓	High	Increase in weed range, subsequently regulated by additional herbicides, which lead to higher costs and the risk of developing herbicide resistance. Increase in uncertainty regarding crops due to lack of knowledge and guidance on the subject.	Team leader: Ofer Mendelson in consultation with Prof. Baruch Rubin, Local farmers and extension instructor Uzi Naftaliahu.	Increment of 4 NIS/dunam/year for herbicides (not including future risk of herbicide resistance, which will create additional costs).
Soil erosion regulation	Maintaining soil quality and quantity, and regulation of wind erosion	↑	High	Global research and experiments in the Negev Desert show an explicit trend of soil erosion prevention in no-tillage systems. On-site assessment of soil erosion risk under "business as usual" scenario shows: between 0.15-0.61 ml/year. However, according to experts' feedback: it could reach between 1-2 ml/year or even 4.	Team leader: Alon Yaron. [9,10,16,17].	1,341 NIS/dunam/year can be saved by preventing erosion [39,40].
Symbolic interactions with the ecosystem	Cultural identity	↓	Low	A threat to the farmer's identity as a ploughman and to the Bedouins' identity as pastoralists grazing their sheep for many generations.	Interviews with local farmers, shepherds and experts. [41,42].	-
Symbolic interactions with the ecosystem	Heritage and sustainability values	↑	High	Strong cultural links between local residents and agriculture. Conservation tillage allows protection of soil and agriculture for future generations.	Survey: "Local residents' perceptions of the environment in Shikma Park".	-
Biodiversity and habitat	Improvement of biodiversity status	↑	High	Improvement in: species richness and diversity, amount of organic matter in soil, intensity of biological activity in soil and landscape complexity. May also have a negative effect because of additional use of pesticides and herbicides.	Dr. Amir Perelberg; [9,10].	-

*NIS—New Israeli Shekel; **1 dunam equals 0.1 hectares.

4.4. Preparing for the Ecosystem Services Assessment

In this stage we screened and prioritized relevant ES according to the methodology described in the section above. Nine services were selected for assessment (see Table 2). In order to conduct the assessment, a team of experts in the various fields, were recruited. These included economists, geomorphologists, Ministry of Agriculture experts, agricultural extension service instructors, drainage authority officials, ecologist, grazing experts, social study experts, *etc.* Each expert led the assessment of the ES related to his/her field.

4.5. Conducting ES Assessments

ES assessments are used to compare alternative states of the site under different management systems in order to eventually help decision-makers identify what could be the most plausible state of the site [28]. Table 2 shows the increase or decrease in the various ES with the transition from minimum tillage to no-tillage as well as the confidence level (low, medium, high), main sources and experts, description of the effect and the on-site economic assessment. The assessments were done by the recruited experts mentioned in the previous stage, based on existing data (academic literature and local reports), stakeholder and expert interviews, field excursions, public survey, expert forums, and economic evaluation where possible.

It should be noted that ES assessments were also conducted for the two other alternatives described above: leaving straw mulch cover *versus* removing it and permitting stubble grazing in post-harvested fields *versus* preventing it. In the scope of this paper, these will be presented only in short in rose plots (Figures 3 and 4) and in Table 3.

4.6. Educational Program and Public Survey: Residents' Views on the Environment in Shikma Park

Another part of the assessment process was a survey of public perceptions of the environment and agriculture. Numerous scholars have emphasized that understanding how humans perceive and value ES is as fundamental to ecosystem management as understanding how ecological functions generate these services, and is a crucial step toward successfully integrating the ES concept in policy [43–45]. The results of over 70% of the respondents show that the local residents value the open landscape and its aesthetic, environmental and cultural significance and use it for recreation and tourism. They appreciate the agricultural fields and their significance for food production, water infiltration, protection of open landscapes from development and as part of the residents' identity and connection to the landscape. Regarding the question who should pay for the transition to conservation tillage: about 80% were found to be willing to adopt conservation methods and 11% to do so even at their own expense. It is important to mention that most of the residents in the area live in kibbutzim, collective communities that combine socialist and Zionist values; therefore, the whole community is affected by the profit and loss from agricultural activity.

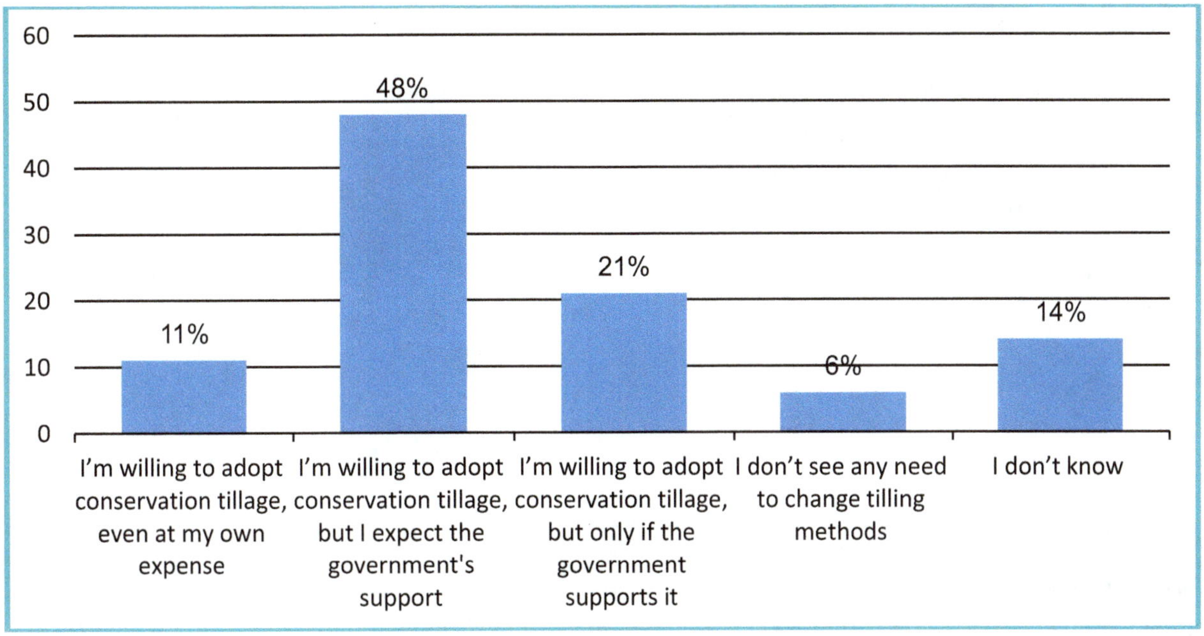

Figure 2. The perspectives of the public regarding the adoption of conservation tillage and who should pay for it.

4.7. Integrating & Communicating the Results

After completing individual assessments of the effect of each alternative state on overall ES, an integrated assessment was prepared in rose plots (Figures 3 and 4) and Table 3 based on the assessment of the experts shown in Table 2. Figure 3 presents the impact of "no-tillage without straw" and "no-tillage with straw" on ES delivery in reference to the existing most common management—minimum-tillage. Figure 4 shows the impact of the transition between grazing alternatives on ES delivery. The scale of the measurement includes 5 levels of impact on ES, while in the reference management all ES received level 3. The impact of each management alternative on each ES is measured according to the reference alternative: an ES that increased received the value 4 or 5 (depending on the degree of increase); an ES that decreased received 1 or 2; and ES that remained unchanged received 3.

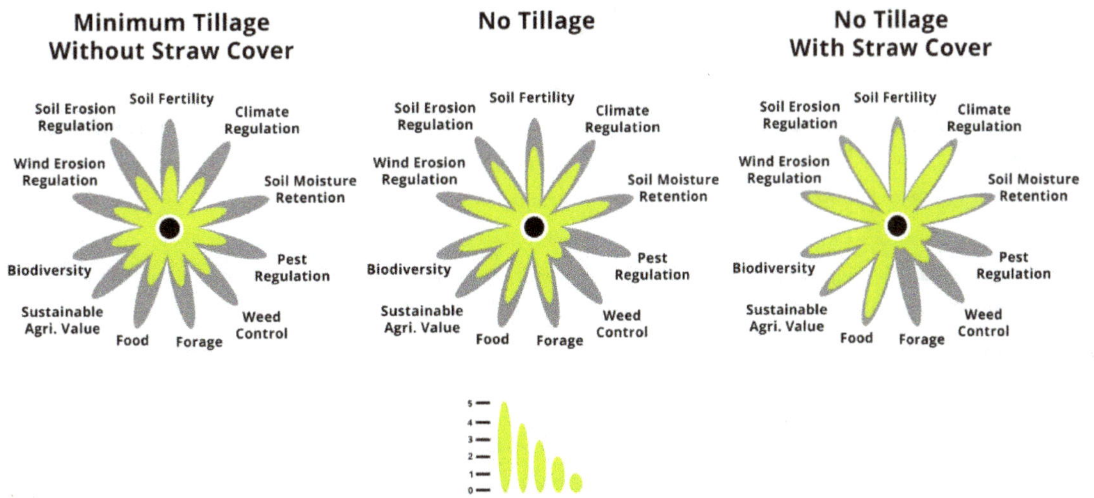

Figure 3. Rose plots showing a comparison of the impact on ecosystem service delivery of no-tillage alternatives in reference to minimum tillage, based on experts' ES assessments.

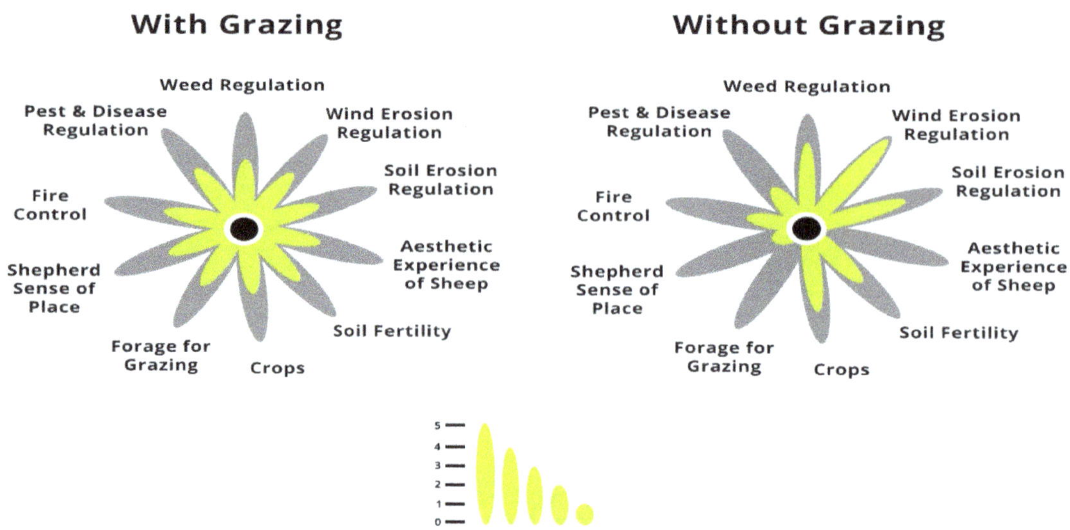

Figure 4. Rose plots showing a comparison of the impact on ecosystem service delivery of "without grazing alternative" in reference to "with grazing alternative", based on experts' ES assessments.

Table 3. Distribution of benefits among stakeholders.

Decision	Goal	Winners	ES decreased	Losers
Increasing one service at the expense of other services				
Straw mulch	Maintain humidity and promote soil microbial activity, while increasing the organic carbon influx on the ground and improving the structure and quality of the soil. Increasing crop quantity and quality.	**The public and the state:** Soil conservation for future generations. **The farmers:** Extended soil moisture retention in the growing season, particularly useful in drought years. Soil conservation for future generations.	Provision of animal feed	**Farmers:** Economic loss due to not selling the straw, which is very profitable in Israel. **Consumers**: May need to import straw or buy from another area in Israel. **Milk consumers:** May be negatively affected by the price increase because of high straw import prices.
Competition among different users for limited services The decision to prevent stubble grazing creates additional winners and losers.				
Preventing stubble grazing	To enable straw mulch cover that improves soil and water conservation. To prevent the potential damages caused by unregulated grazing such as: soil quality degradation, water runoff and introduction of noxious weeds via the sheep's wool and droppings. To prevent potential disputes between shepherds and farmers.	**Farmers:** Improving weed control, preventing soil erosion. Preventing the possibility of negative relationships with shepherds (some farmers complained of this). **The state and the public**: soil conservation for future generations.	The service of feed for grazing will no longer be available, cultural services such as shepherd identity, heritage and sense of place are harmed, pest and disease regulation, fire control, and sheep as an aesthetic function in the landscape.	**Shepherds:** Lack of grazing lands, additional cost of buying feed instead of grazing. The effect on their Bedouin-pastoral lifestyle, identity and tradition of many years. **Farmers**: Loss in income from rent paid by shepherds. **Foresters:** might cause a negative effect on fire control in forests due to the combined system of grazing in forests prior to grazing in the wheat fields. Without stubble grazing the shepherds may stop forest-grazing as well. **Public:** Lack of sheep in the agricultural landscape may affect the scenery and the anemone bloom.

An important component, which is not expressed in the rose plots, is the multiple stakeholders that are affected by the management dilemma, which makes the question of who is harmed and who benefits from the increase and decrease of each ES very important. The multi-stakeholder approach aims to identify all the groups of people affected by the current state of a piece of land and its future use [26]. In order to provide decision-makers with the information on how various stakeholders will be affected by the transition, and the time span of the impact (short *versus* long-term), the following table was prepared (based on [46], p. 40 and [25], p.60):

5. Discussion

The expert-based ES assessments indicate an improvement in most ES with the transition to no-till, and a further improvement with straw mulch as is seen in Figure 3. The main trade-off to be considered with the transition to no-till is the advantage of soil and water retention (a service that is expressed in the long-term protection of land for future generations) *versus* the main disadvantages: a reduction in weed and pest control and loss of profit from selling the straw. The reduction of weed and pest control may cause a series of events: additional use of herbicides and pesticides, which imply a possible development of herbicide-resistant weeds and pesticide-resistant pests that may cause additional damage to crops, economic loss for the farmer, and further harm to the environment and ES in surrounding ecosystems.

The decision of the farmer to prevent stubble grazing is one of the outcomes of leaving straw mulch, but the decision can be made based on other considerations, regardless of whether the farmer decides to leave mulch or not. Figure 4 and Table 3 show that four of the services increase without grazing (regulation of soil erosion by water and wind, weed control and crops) and five of them decrease (forage for grazing, fire control, the shepherd's sense of place, aesthetic view of sheep as part of the landscape and pest control). In addition, one ES remained unchanged—quality of soil. This is explained by the fact that the soil is under conflicting impact—on the one hand, the sheep enrich the soil with their droppings but on the other hand, their trampling may increase soil erosion. Table 3 shows that when shifting to no-tillage with straw mulch the farmers and the state & public are the main "winners". This agrees with the results of the cost-benefit analysis that shows that over the long-term, it is advantageous for the farmer and society as a whole to make the transition, and with the public survey results that shows public support for the transition. The main "losers" are the shepherds that would not be able to graze and the farmers who, in the short-term, will not profit from leasing the land to the shepherds, nor from selling straw. It is important to note that in interviews with shepherds and experts, continuation of grazing was questioned unrelated to the issue of straw mulch, but because the younger generation might not want to maintain the shepherd lifestyle. In addition, Bedouin grazing on Jewish farms is also affected by political forces beyond the scope of the farmer's decision. There are also additional "losers" such as straw consumers, who may need to import straw, as Shikma Park produces a large part of the straw in Israel. This in turn, may affect milk consumers by raising the price of milk. Therefore, there are obviously no clear-cut answers. It can only be stated, that according to current prices, the economic assessment shows that in the short term it may be more advantageous to keep on selling the straw.

In order to deal with the issues raised above, a few possible mechanisms have been identified, that may help to enhance overall ES supply and be more suitable to the region and the stakeholders, with further investigation. Examples include occasional tillage (as recommended by the Ministry of Agriculture) which would contribute to pest and weed control and allow application of organic fertilizers in the soil. In addition, practicing IPM (integrated pest management) could further assist pest and weed regulation in a manner less harmful to the environment. Partial straw cover—leaving a cover of straw only about 10–15 cm high (as opposed to 30 cm or more), as is done on several fields today—will allow farmers to continue selling straw and still enjoy some of the benefits of straw mulch. In terms of moderate grazing, in cases when grazing continues, farmers should be aware of the mode of grazing (amount of sheep per hectare and post-fleece introduction of the goats) to minimize its negative effects [24,42,47]. In an attempt to find ways to ease the economic barriers for the farmers, or mitigate the fear of unstable income, the following mechanism was proposed: instead of financial support (grants for purchasing necessary equipment) as the case is today, an insurance plan for agricultural hazards and risks that may result from the transition to no-tillage would be formulated. These mechanisms were presented to various key policy makers in the Ministry of Agriculture (including the Division of Research, Finance and Strategy), who gave a positive feedback, and agreed to examine the issue further. During the study, a list of topics that need further examination was identified and communicated to governmental and academic research institutions. Several conservation tillage studies on the topics identified such as soil degradation risk assessment, farmer constraints in adopting no-tillage, solutions for weed and pest control and the effect on biodiversity have already begun in different parts of the country. The studies include either farmers, extension workers or ministry of agriculture representatives in the teams in order to improve the future implementation of the results [6–8].

One of the main challenges in the course of the study was the agroecosystem services assessment process. This was due to the following reasons: lack of detailed and organized documentation by the farmers regarding inputs and outputs of different tillage systems, variation in physical conditions among the farms being studied, as well as economic and social conditions, biased information provided by different stakeholders and experts to promote their agenda, limited scientific research done in the region, difficulties in the implementation of research conclusions from other places in the world because of different physical conditions and scale, difficulty in quantifying the findings without a large scale-long-term scientific research and lack of economic tools for providing value for un-marketed goods. These challenges were partly dealt with by providing the best assessments possible under the existing limitations, *i.e.*, in some cases only the main trends and insights were reported, and also by making complementary on-site research efforts, conducting economic assessments and convening expert forums. For example, in the assessment of soil erosion regulation, which is a crucial issue in this context, assessing the damage caused by erosion under the "business as usual" scenario and the benefit in regulating it was difficult; therefore, specific expert forums were convened to advise on the issue.

6. Conclusions

This study proposes a few guidelines for conducting a productive ecosystem services assessment process that will lead to enhanced awareness and implementation of policies that will enhance ES supply. During the project, the complex system of conservation tillage decision making and its effect on ES was studied and documented. As we have seen, the process is as important as the results. We believe that providing an unequivocal answer regarding which of the alternatives is preferable should not be the ultimate goal of the ES process. Rather, it should be to provide decision-makers with findings that could help them make better decisions, which would consider benefits to the environment and the stakeholders involved and to raise public awareness of ES and the risks involved in their degradation. The research shows that by conducting a participatory assessment process with a wide range of stakeholders and experts, conducting a multi-stakeholder analysis, involving the public through interviews, a public survey, meetings and an educational program, *etc.*, and communicating the results to policy makers, these goals can be met. Further outcomes of the study showed that the process and its deliberations enriched and deepened the agroecosystem management discourse and contributed to expanding the scope beyond provisioning services (crops and feed) to soil regulation (and the risks of soil erosion) as well as to other services such as carbon sequestration, biodiversity, water cycle regulation, pest and weed control, cultural services and sustainability values. The study emphasizes the potential of the ES approach to create collaborations between farmers, regional councils, environmental organizations and policy makers (such as the Ministry of Agriculture and Drainage Authority) to promote the initiation of effective research and policies which will maximize the overall ES supply and benefit all. We believe that the network of people working for the implementation of conservation tillage and soil conservation has been expanded and unified through this study. We hope that the seed has been planted for further promotion of conservation tillage implementation and further use of the ES approach in agroecosystems.

Acknowledgments

This work was supported through a framework made available by the Open Landscape Institute, and the funding of the Open Landscape Conservation Fund of the Israel Land Authority, Ministry of Environmental Protection, and the "Bracha" Fund. We would like to thank the team that worked on the project: Yoav Sagi, head of the Open Landscape Institute, Kutzi Weil and Haya Erez, Shikma Park management board members, especially Haim Hermoni, who escorted the process, and Alon Shuster, the head of the board. The professional teams who conducted the ES assessments: Gadi Rozenthal and Hadar Fuchs-Rubel—economic assessment, Alon Yaron in consultation with Prof. Yonatan Laron—soil and water regulation, Ofer Mendelson in consultation with Prof. Baruch Rubin—pest and weed regulation, Dr. Amir Perelberg—biodiversity, Yoni Weitz and Dr. Jose Gruenzweig—climate regulation, Dr. Iris Schoenbaum in consultation with Dr. Yan Landau—ES related to grazing, Rachel Evron and Nature Park Authority—public survey and educational program, and Yonatan Yaacobi—Statistical analysis. We thank the farmers and shepherds of Shikma Park, and the agricultural extension instructors, Uzi Naftaliahu and Idan Richker, whom without this project could not have taken place. We would like to thank the numerous scientists and experts that we consulted with from the faculty of agriculture, Volcani Institute, Soil Erosion Research Station, and all the universities in Israel. Special thanks to the Ministry of Agriculture for the collaboration, particularly the Soil Conservation and Drainage Department. In addition, we thank the Ministry of Environmental Protection, the Drainage and River Authority, Society for the Protection of Nature in Israel, Israel Nature and Parks Authority, Jewish National Fund (Forestry department), Esther Lachman for English editing, and the many others who took part in the research. For more detailed information about the project, see the Open Landscape Institute website: http://www.deshe.org.il/?CategoryID=334.

Author Contributions

Uri Ramon, together with the head of the Open Landscape Institute had the original idea for the initiation of the project and led the development of the project and the methodology. Hila sagie coordinated the project, led the public survey, and the preparation of the manuscript. All authors participated in the writing and revision of the manuscript. In addition, all authors approved the final manuscript.

Conflicts of Interest

The authors declare no conflict of interest.

References

1. Power, A.G. Ecosystem services and agriculture: Tradeoffs and synergies. *Phil. Trans. R. Soc. B* **2010**, 365, 2959–2971.
2. TEEBAF. *The Economics of Ecosystems and Biodiversity (TEEB) for Agriculture and Food—Concept Note*; TEEB: Geneva, Switzerland, 2014.
3. Burkhard, B.; De Groot, R.; Costanza, R.; Seppelt, R.; Jørgensen, S.E.; Potschin M. Solutions for sustaining natural capital and ecosystem services. *Ecol. Indic.* **2012**, *21*, 1–6.
4. Robertson, G.P.; Gross, K.L.; Hamilton, S.K.; Landis, D.A.; Schmidt, T.M.; Snapp, S.S.; Swinton, S.M. Farming for ecosystem services: An ecological approach to production agriculture. *BioScience* **2014**, *64*, 404–415.
5. Lal, R.; Reicosky, D.C.; Hanson J.D. Evolution of the plow over 10,000 years and the rationale for no-till farming. *Soil Tillage Res.* **2007**, *93*, 1–12.
6. Coughenour, C.M. Innovating conservation agriculture: The case of no-till cropping. *Rural Sociol.* **2003**, *68*, 278–304.
7. Schneider, F.; Steiger, D.; Ledermann, T.; Fry, P.; Rist, S. No-tillage farming: Co-creation of innovation through network building. *Land Degrad. Dev.* **2012**, *23*, 242–255.
8. Coughenour, C.M.; Chamala, S. *Conservation Tillage and Cropping Innovation: Constructing the New Culture of Agriculture*; John Wiley & Sons: Ames, IA, USA, 2007.
9. Derpsch, R.; Friedrich, T.; Kassam, A.; Hongwen L. Current status of adoption of no-till farming in the world and some of its main benefits. *Int. J. Agric. Biol. Eng.* **2010**, *3*, 1–26.
10. Palm, C.; Blanco-Canqui, H.; DeClerck, F.; Gatere, L.; Grace, P. Conservation agriculture and ecosystem services: An overview. *Agric. Ecosyst. Environ.* **2014**, *187*, 87–105.
11. Lichtenberg, E. *Adoption of Soil Conservation Practices: A Revealed Preference Approach*; WP 01-12; Department of Agricultural and Resource Economics, The University of Maryland, College Park, MD, USA, 2001.
12. Knowler, D.; Bradshaw, B. Farmers' adoption of conservation agriculture: A review and synthesis of recent research. *Food Policy* **2007**, *32*, 25–48.
13. Wauters, E.; Mathijs, E. The adoption of farm level soil conservation practices in developed countries: A meta-analytic review. *Agric. Resour. Gov. Ecol.* **2014**, *10*, 1–12.

14. Werner, M.; Steinmann, H.H.; Schlatter, N.; Spiegel, H.; Wauters, E.; Ruysschaert, G. *Farmers' Rationality in Soil Management: Which Factors Influence Implementation of Sustainable Management Practices in Soil Conservation?—A Case Study in Germany and Austria*; Schobert, H., Riecher, M.C., Fischer, H., Aenis, T., Knierim A., Eds.; IFSA Book of Abstracts: Cottbus–Senftenberg, Germany, 2014.
15. Ministry of Agriculture Website. Available online: http://www.moag.gov.il (accessed on 31 March 2015).
16. Bonfil, D.J.; Mufradi, I.; Klitman, S.; Asido, S. Wheat grain yield and soil profile water distribution in a no-till arid environment. *Agron. J.* **1999**, *91*, 368–373.
17. Bonfil, D.J.; Rubin, B.; Shtienberg, D.; Mufradi, I.; Asido, S.; Kitain S.; Naftalyahu, U.; Vaza, A. No-till and straw mulching: Perennial summary report. *Gan Sadeh Vameshek* **2004**, *8*, 13–23. (In Hebrew).
18. Yaakobi, B.; Eizenkot, A. *The Valley District—Conservation Tillage in Dryland Farming*; Summary Report of the Accumulative Experience of Conservation Tillage up to 2013; Ministry of Agriculture and Rural Development: Rishon Le Zion, Israel, 2013. (In Hebrew).
19. TESSA—Toolkit for Ecosystem Service Site-Based Assessment. 2013. Available online: http://tessa.tools (accessed on 30 October 2014).
20. Fisher, B.; Turner, R.K.; Morling P. Defining and classifying ecosystem services for decision making. *Ecol. Econ.* **2009**, *68*, 643–653.
21. De Groot, R.S.; Alkemade, R.; Braat, L.; Hein, L.; Willemen, L. Challenges in integrating the concept of ecosystem services and values in landscape planning, management and decision making. *Ecol. Complex.* **2010**, *7*, 260–272.
22. Gordon, U.; Safriel, U. *Program for Ecosystem Assessment of Dryland Biosphere Reserves*; Blaustein Institutes for Desert Research, Ben Gurion University of the Negev: Beer Sheva, Israel, 2010.
23. Klein, J.D.; Mufradi, I.; Cohen, S.; Hebbe, Y.; Asido, S.; Dolgin, B.; Bonfil, D.J. Establishment of wheat seedlings after early sowing and germination in an arid Mediterranean environment. *Agron. J.* **2002**, *94*, 585–593.
24. Stavi, I.; Barkai, D.; Kandikar, R.I.; Zaady, E. No adverse effect of moderate stubble grazing on soil quality and organic carbon pool in dryland wheat agro-ecosystems. *Agron. Sustain. Dev.* **2015**, in press.
25. The Economics of Ecosystems and Biodiversity (TEEB). *Guidance Manual for TEEB Country Studies*, Version 1.0; United Nations Environment Programme; UNEP: Berlin, Germany, 2013.
26. ELD Initiative—Economics of Land Degradation. *Practitioner's Guide Based on the Massive Open Online Course*; 2014. Available online: http://www.eld-initiaitve.org (accessed on 18 November 2014).
27. Mace, G.M.; Bateman, I.; Albon, S.; Balmford, A.; Brown, C.; Church, A.; Haines-Young, R.; Pretty, J.N.; Turner, K.; Vira, B.; *et al.* Conceptual Framework and Methodology. In *UK National Ecosystem Assessment*; Technical Report; UK National Ecosystem Assessment: Cambridge, UK, 2011; Chapter 2, pp. 10–26.

28. Peh, K.S.H.; Balmford, A.; Bradbury, R.B.; Brown, C.; Butchart, S.H.M.; Hughes, F.M.R.; Stattersfield, A.J.; Thomas, D.H.L.; Walpole, M.; Bayliss, J.; *et al.* TESSA: A toolkit for rapid assessment of ecosystem services at sites of biodiversity conservation importance. *Ecosyst. Serv.* **2013**, *5*, 51–57.
29. Tomich, T.P.; Garbach, K.; Cantor A. A framework for assessment of California agroecosystems: Pitfalls and possibilities. In Proceedings of the Workshop and Policy Round Table on: California Agroecosystem Services: Assessment, Valuation and Policy Perspectives, University of California at Davis, CA, USA, 29 September 2007; pp. 1–17. Available online: http://aic.ucdavis.edu/research/sustainability/agroecosystems/AGECO2_papers.pdf (accessed on 10 March 2015).
30. MEA—Millennium Ecosystem Assessment. *Ecosystems and Human Well-Being, Our Human Planet—Summary for Decision Makers*; Reid, W.V., Ed.; Island Press: Washington, DC, USA, 2005.
31. UNEP (United Nations Environment Programme). *Developing and Mainstreaming Ecosystem Service Indicators for Human Wellbeing: Gaps, Opportunities and Next Steps*; Report from the Workshop on Ecosystem Service Indicators; UNEP World Conservation Monitoring Centre: Cambridge, UK, 2009.
32. ICARDA—International Center for Agriculture Research in the Dry Areas. *Conservation Agriculture: Opportunities for Intensified Farming and Environmental Conservation in Dry Areas*; A Synthesis of Research and Trials with Smallholder Farmers in Drylands Systems; Benefits and Constraints to Adoption; ICARDA: Alepo, Syria, 2012. Available online: ftp://ftp.fao.org/ag/agp/ca/CA_CoP_May12/Conservation_agriculture_ICARDA.pdf (accessed on 13 February 2015).
33. Eshel. G.; Lifschitz, D.; Bonfil, D.J.; Sternberg, M. Carbon exchange in rainfed wheat fields: Effects of long-term tillage and fertilization under arid conditions. *Agric. Ecosyst. Environ.* **2014**, *195*, 112–119.
34. Verhulst, N.; Govaerts, B.; Verachtert, E.; Castellanos-Navarrete, A.; Mezzalama, M.; Wall, P.; Deckers, J.; Sayre, K.D. Conservation agriculture, improving soil quality for sustainable production systems? In *Advances in Soil Science: Food Security and Soil Quality*; Lal, R., Stewart, B.A., Eds.; CRC Press: Boca Raton, FL, USA, 2010; pp. 137–208.
35. Stavi, I.; Argaman, E. No-till systems: Gains and drawbacks for carbon sequestration, ecosystem services and environmental health. *Carbon Manag.* **2014**, *5*, 123–125.
36. Sombrero, A.; de Benito, A. Carbon accumulation in soil. Ten-year study of conservation tillage and crop rotation in a semi-arid area of Castile-Leon, Spain. *Soil Tillage Res.* **2010**, *107*, 64–70.
37. Johnson, D. Effect of alternative tillage systems on rodent density in the Palouse region. *Northwest Science* **1987**, 61, 37–40.
38. Kassam, A.H.; Friedrich, T.; Derpsch, R. Conservation agriculture in the 21st century: A paradigm of sustainable agriculture. In Proceedings of the European Congress on Conservation Agriculture, Madrid, Spain, 6–8 October 2010; pp. 1–46.
39. Shahaf, N.; Halamish, N; Alexandrov, Y. *Drainage and River Authority Shikma-Bsor*; Bsor-Shikma Master Plan Report; 2011. Available online: http://www.besor.org.il (accessed on 20 February 2014).

40. Hadas, E.; Turzion, Y.; Eizenkot A.; Zaidenberg, R. Soil erosion prevention—Cost benefit analysis. *Nir va Telem.* **2009**, 13, 17–27. (In Hebrew).
41. Stavi, I.;Kressel G.; Gutterman, Y.; Degen, A. Flock use among Bedouin in 'Spontaneous' settlements in the Negev Desert, southern Israel. *Nomadic Peoples* **2006**, *10*, 53–69.
42. Landau, S.; Bonfil, D.J.; Barkai, D.; Yonatan, R.; Mufradi, I.; Dvash, L.; Brosh, A. A system analysis of the interaction between wheat crop management and sheep grazing in aftermath stubble. *Options Med.* **2005**, *61*, 153–160.
43. Menzel, S.; Teng, J. Ecosystem services as a stakeholder-driven concept for conservation science. *Conserv. Biol.* **2010**, *24*, 907–909.
44. Swinton, S.M.; Lupi, F.; Robertson, G.P.; Hamilton, S.K. Ecosystem services and agriculture: cultivating agricultural ecosystems for diverse benefits. *Ecol. Econ.* **2007**, *64*, 245–252.
45. Sagie, H.; Morris, A.; Rofè, Y.; Orenstein, D.E.; Groner, E. Cross-cultural perceptions of ecosystem services: A social inquiry on both sides of the Israeli–Jordanian border of the Southern Arava Valley Desert. *J. Arid Environ.* **2013**, *97*, 38–48.
46. World Resources Institute (WRI). *Ecosystem Services: A Guide for Decision Makers*; WRI" Washington, DC, USA, 2008. Available online: http://pdf.wri.org/ecosystem_services_guide_for_decisionmakers.pdf (accessed on 14 April 2015).
47. Schoenbaum, I.; Kigel, J.; Barkai, D.; Landau, S. Weed infestation of wheat fields by sheep grazing stubble in the Mediterranean semi-arid region. *Crop Pasture Sci.* **2009**, *60*, 675–683.

3

Growing City and Rapid Land Use Transition: Assessing Multiple Hazards and Risks in the Pokhara Valley, Nepal

Bhagawat Rimal [1,*], Himlal Baral [2], Nigel E. Stork [3], Kiran Paudyal [4] and Sushila Rijal [1]

1 Everest Geoscience Information Service Center, Kathmandu, Nepal; E-Mail: sushilarijal@ymail.com

2 Centre for International Forestry Research (CIFOR), P.O. Box 0113 BOCBD, Bogor 16000, Indonesia; E-Mail: h.baral@cgiar.org

3 Environmental Futures Research Institute, Griffith School of Environment, Nathan Campus, Griffith University, 170 Kessels Road, Nathan, QLD 4111, Australia; E-Mail: nigel.stork@griffith.edu.au

4 School of Ecosystem and Forest Sciences, Faculty of Science, the University of Melbourne, 221 Bouverie St., Carlton, VIC 3010, Australia, E-Mail: kpaudyal@student.unimelb.edu.au

* Author to whom correspondence should be addressed; E-Mail: bhagawatrimal@gmail.com

Academic Editor: Pavlos S. Kanaroglou

Abstract: Pokhara is one of the most naturally beautiful cities in the world with a unique geological setting. This important tourist city is under intense pressure from rapid urbanization and population growth. Multiple hazards and risks are rapidly increasing in Pokhara due to unsustainable land use practices, particularly the increase in built-up areas. This study examines the relationship among urbanization, land use/land cover dynamics and multiple hazard and risk analysis of the Pokhara valley from 1990 to 2013. We investigate some of the active hazards, such as floods, landslides, fire, sinkholes, land subsidence and earthquakes, and prepare an integrated multiple hazard risk map indicating the highly vulnerable zones. Land use and land cover maps from 1990 and 2013 from Landsat images (30 m resolution) have been prepared and analyzed for the spatial dynamics of urbanization and the transition of land use and land cover. In the 23-year period, the built-up area more than doubled from 24.03 km^2 to 54.20 km^2. Although the landscape in the urban, peri-urban and rural areas appears to be fragmented, different drivers play pivotal roles in landscape change in these areas. The results provide substantial information for establishing innovative action plans for disaster risk management in the valley. Recommendations are made for the most suitable places for future urban expansion in the valley. This study is important for raising awareness among policy makers and other public officials to include multiple hazard risk mitigation in land use policies and plans. Establishing connections between urban expansions, escalating population growth and multiple hazards and risk assessment will also improve in modelling the latent impact of future catastrophes and emergency preparedness.

Keywords: urbanization; land use/land cover change; multiple hazard; Pokhara valley; assessments; Nepal

1. Introduction

Disasters are split into two groups; natural hazards and human-induced disasters [1]. Natural hazards and human-induced disasters are frequent occurrences and include floods, cyclones, droughts, tidal surges, tornadoes, earthquakes, river erosion, fire, infrastructure collapse, high arsenic contents of ground water, water-logging, water and soil salinity, epidemics and various forms of pollution [2]. Over recent decades, the frequency of natural hazards has grown worldwide, with the number of disasters quadrupling in the last 30 years, resulting in escalating human cost and economic losses [3,4]. Developing countries bear the highest burden in terms of the number of human lives lost and the proportion of gross domestic product lost as a result of disasters [4]. The largely unplanned expansion of cities to accommodate rapid population growth, combined with inappropriate land use planning and the failure of urban authorities to regulate building standards contribute to the vulnerability of urban populations [5].

Urbanization is a complex spatial process that converts land from rural uses to urban uses and results in impacts on the structure, function and dynamics of ecosystems, as well as the livelihoods of humans [6]. The rapid conversion of agricultural land has become a serious issue for sustainable urban development in developing countries [7]. Urbanization often increases the socioeconomic activities and revolutionizes the land use practice according to time [8,9]. Land use/land cover change is directly proportional to increases in natural and socio-economic activities. The land use/land cover pattern of a region is the outcome of natural and socio-economic factors and their utilization by man in time and space [9]. Land use is the human use of land involving the management and modification of the natural environment or wilderness into a built environment, such as fields, pastures and settlements [10]. It is a continuous and inevitable process due to economic development, exploitation of natural resources and rapid population growth. Multiple hazard risk assessment refers to the process of identifying the occurrence, probability and estimation of the potential damage of various types of disasters within a specified spatio-temporal framework.

Nepal is ranked as the 20th most disaster-prone country in the world and ranked 11th for earthquake vulnerability, fourth for climate change vulnerability and 30th for flood vulnerability [11]. The recent devastating earthquake at 11.56 a.m. on 25 April 2015, with a magnitude of 7.8 on the Richter scale, was followed by subsequent aftershocks of 7.3 on 12 May 2015. The epicenter was the Barpak Village

Development Committees (VDCs) of the Gorkha district and Sunkhani VDC of the Dolakha district and was the most powerful earthquake in Nepal for 80 years. The earthquake caused more destruction to the east of the epicenter in Barpak VDC, Gorkha. Pokhara, being located to its west, witnessed less loss and damage from the earthquake. Arwa, Mijure and Purunchaur were the more affected VDCs of Kaski district. The earthquake created an unprecedented situation, affecting the country's most populous areas and the economic and administrative centers, along with numerous villages in the districts.

The Pokhara valley is an intermountain fluvial basin occupying the midsection of the Seti River in the Lesser Himalayas of Nepal. It is filled by a large volume of layered clastic deposits of gravel, silt and clay of Quaternary age, deposited from the Annapurna mountain range probably by a series of catastrophic debris flows [12,13].

The Seti watershed, and the Pokhara sub-metropolitan city section, in particular, has been exposed to frequent multiple hazard risks in the last few decades, ranging from avalanches and glacial lake outburst floods (GLOF) in the Himalayas to landslides and edge-falls, floods, sinkholes, land subsidence, fire and earthquakes. The watershed is characterized by deposits from past catastrophic floods. The fragile geology, dense settlement with weak structures in the municipal areas, unstable steep slopes and deep gorges and heavy rainfall are responsible for the debris flows. These are drivers of such hazards that may result in the remarkable loss of lives and property, threatening development activities, the environment and human civilization as a whole. Additionally, excessive and unbalanced land use and land cover change have contributed to increased susceptibility to the environmental risks of earthquakes, landslides, floods, fire and other environmental and industrial hazards, which are likely to become more inevitable, intense and frequent in the near future.

This study is based on 21 VDCs and two municipalities of Seti watershed within Pokhara valley of Kaski district. This paper aims to qualitatively characterize the historical patterns of land use and land cover, to examine the relationship between urbanization; land use and land cover dynamics and hazard analysis between the time period of 1990–2013 using GIS and remote sensing techniques.

2. Methodology

2.1. Study Area

This study area comprises the entire 976.37 km^2 Seti watershed region within the Kaski district. The Pokhara Valley lies in the Mahabharat Range, Midlands and the Great Himalayan Range of Nepal between longitudes 83°48'E and 84°13'11"E and latitudes 28°4'39"N and 28°36'18"N. The total area covered by the administrative division is 1152.19 km^2. High mountains are in the northern part, whereas the middle range mountains occupy the southern portion of the Seti watershed. The maximum altitude is 8091 from mean sea level, mount Annapurna, the tenth highest mountain in the world (Figure 1). The streams of the Seti watershed exhibit dendritic, parallel and sub-parallel drainage with three to five orders of tributaries. Most of the streams and gullies are very active, reflecting the tectonic activity of the rising Himalayas. The rainy season is June to September. The mean annual rainfall at Lumle station is 5700 mm and at Pokhara station is 3967 mm [14]. The intense rainfall causes severe landslides and floods in the region. The high topographic variation from south to north in the watershed basin and from valley bottom to hill and mountain tops in the sub-basins creates a unique mixture of climatic conditions within the Seti watershed; the climatic conditions of the central part of the basin range between sub-tropical to cool temperate, whereas climatic variations in the northern part range from temperate to alpine.

The rocks of the Seti watershed are predominately metamorphic with some sedimentary and igneous. A regional-scale anticlinorium passes through Pokhara and extends from northwest to southwest. The rocks of the area are dominated by northeast to southwest trending mineral and stretching lineation and northwest to southeast trending minor fold axes [15].

The study area has been divided into three regions that represent urban areas defined by urban administrative boundaries, the peri-urban area outside of the urban area and rural areas elsewhere in the watershed.

Pokhara city is the headquarters of the western development region, with many governmental offices situated in the area. It is the focal point for the different districts of the Gandaki zone for education, health, business, electricity, transportation, communication and tourism. The city is well-situated with colleges, universities and schools, as well as hospitals.

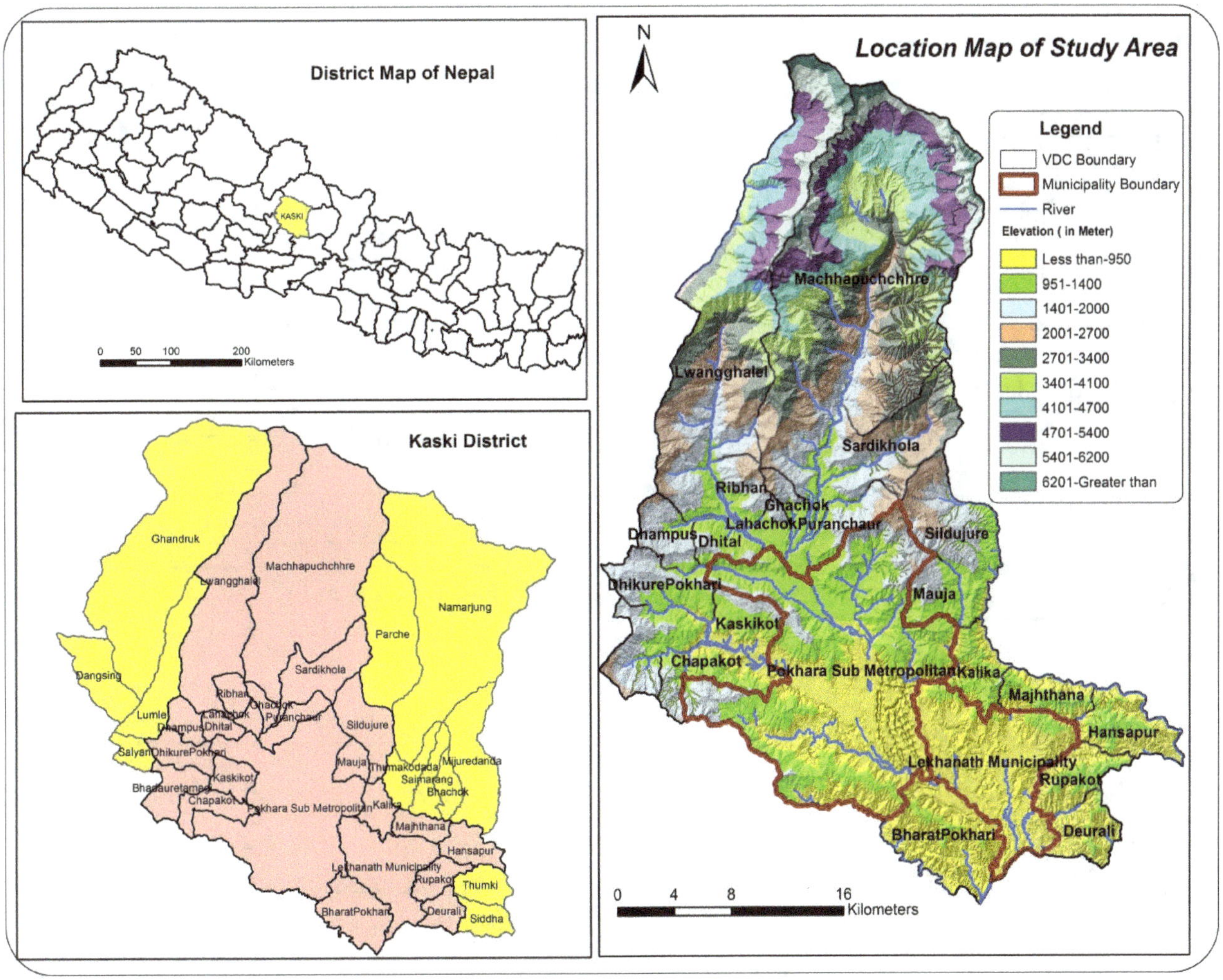

Figure 1. Location map of the study area.

2.2. Methods

The study is based on both primary and secondary information (Figure 2). Secondary data were collected from the Central Bureau of Statistics (CBS), land policy documents, topographic maps, land use plans, flood risk maps, images and other relevant maps available in published and unpublished research papers, articles, books, government documents and official records obtainable in the district office, sub-metropolitan office and municipality, as well as VDC profiles and other resources. A topographical map published by Survey Department, Government of Nepal, 1998, for Pokhara with a scale of 1:25,000 was used for the reference map for the analysis of images. The map was based on an aerial photograph at a scale of 1:50,000 taken in 1996.

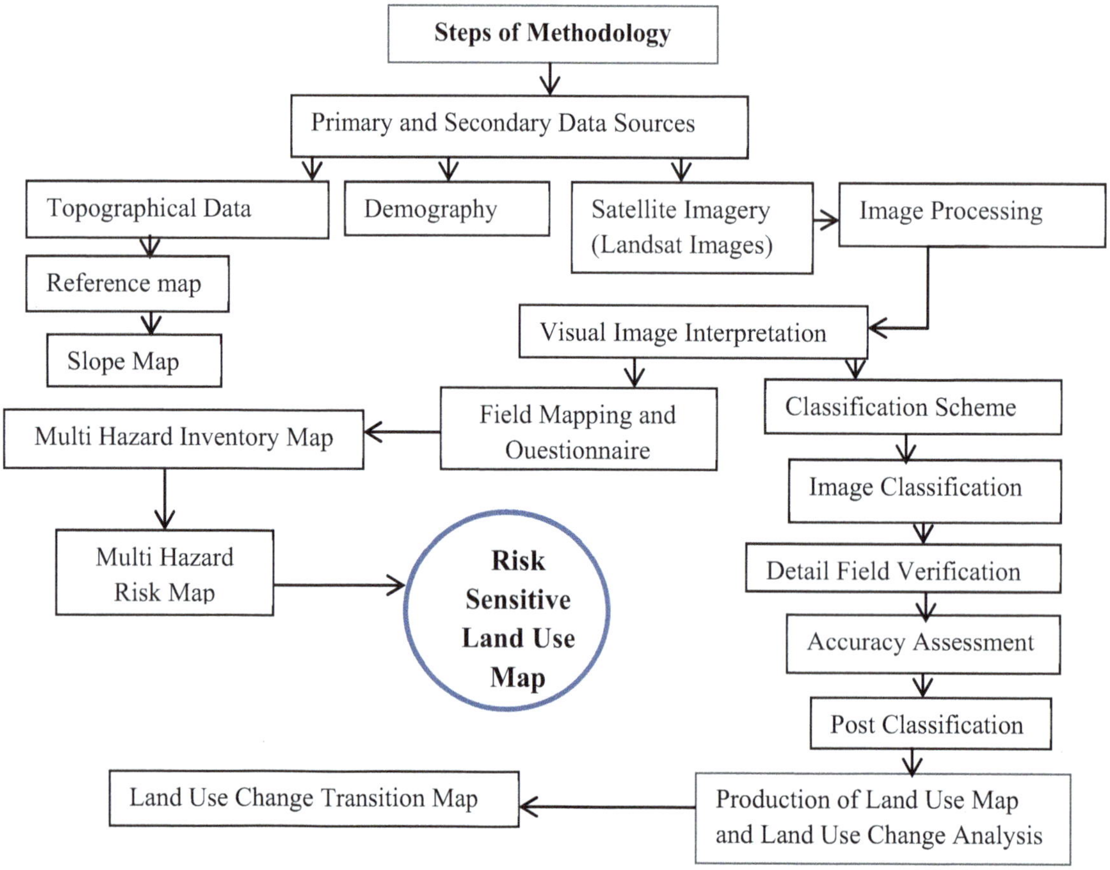

Figure 2. Flow chart showing the methodological approach.

Intensive fieldwork was performed to acquire ground-truth first-hand information and to verify secondary data. No field visits were conducted in grassland, snow-covered areas and glacier areas in the northern part of the study area. Landscape observations, an expert consultation and informant survey (CIS), such as from businesspersons, school teachers, university lecturers and people involved in traditional occupations, were made to obtain the information on the occurrence of historical land use change events and data on damage resulting from fire, landslide, flood and earthquake hazards. A total of 50 key informants and experts were consulted. The snowball technique was used to identify them.

The highly vulnerable zones in terms of fire, flood, landslide, earthquake and sinkholes were explored through primary and secondary data acquisition and analysis with the help of multiple hazard inventory mapping. Sinkhole hazards were identified through primary data collection. Locations of individual sinkholes were captured using the Global Positioning System (GPS). For fire hazard assessment, the locations of petrol stations in the dense settlements that pose a considerable threat were also located using GPS. Landslide and flood hazard-prone areas were identified through past research, field visits using GPS, along with consultations with local people and relevant experts. The earthquake hazard was assessed through secondary information. The collected information has been plotted to prepare a multiple hazard risk map. After finalization of the risk area, the risk sensitivity land use map has been prepared to show areas of high risk and areas more suitable for future urban expansion.

Two cloud-free Landsat images were used to classify the study area: Landsat Image 5 (with Path/Row 142/40) of November 1990 and Landsat 8 (with Path/Row 142/40) of October 2013. All data were projected to the Universal Transverse Mercator (UTM) projection system, which is the World Geodetic System 1984. The IDRISI Selva version image processing software was used for the image analysis. According to the land use classification scheme, an unsupervised and a supervised approach with the maximum likelihood parameter (MLP) system were applied to improve the accuracy of the land use classification for the 1990 and 2013 images. The classification system used is a slightly modified classification system for remotely-sensed data as recommended by Anderson *et al.* 1976 [16]. The Land Change Modeler (LCM) system was applied to analyze the changes, transition and exchanges of land use and land cover (LULC). The transition probability may be accurate on a per category basis.

3. Results and Discussion

3.1. Population Growth and Urbanization

Though the urban history of Nepal is relatively short, the rate of urban development is quite high. In the early period, Pokhara was essentially a market center situated in the broad valley of Seti and its tributaries. Because of its strategic location between the mountains and *Tarai* (plain), it became an important east-west staging point on the Trans-Himalayan trade route [8,17,18]. The history of Pokhara city begins with the periodical religious gatherings during the medieval period and permanent *bazaar* (market places) after the unification of Nepal in 1768. Enchanting natural beauty, tourism activities and its unique social and cultural heritage are responsible for the present development of the town population of the Pokhara sub-metropolis. Discussants also note that access to such facilities is the main cause of migration to and the population growth of Pokhara.

The Pokhara valley is going through an unprecedented rate of population growth. The Central Bureau of Statistics [14] recorded 446,764 as the total population of the valley among which, 255,465 (prior to annexation with 10 VDCs) people settle in the Pokhara sub-metropolitan city and 58,816 in Lekhnath Municipality alone. In the Pokhara sub-metropolitan city, the urban area had covered 6.33% in 1977, but has increased to 51.42% in 2010, whereas cultivated land has decreased from 60.73% down to 20.27% during the same period [19].

There has been a gradual increase in population in the Seti watershed area from 252,038 in 1991 to 335,695 in 2001 and 446,764 in 2011. The Pokhara sub-metropolitan city and Lekhnath municipality contain the highest and the second highest population counts, respectively. The total populations of the newly-annexed Pokhara sub-metropolitan area for the years 1991, 2001 and 2011 were 139,988,

208,692 and 313,841, respectively, and for Lekhnath in 1991, 2001 and 2011 were 30,107, 41,369 and 58,816. Population pressure is very high in the city area. The population census in 2011 showed that 70.25% of the population settled in the Pokhara sub-metropolitan city alone (including the newly-adjoined 10 VDCs) (Figure 3).

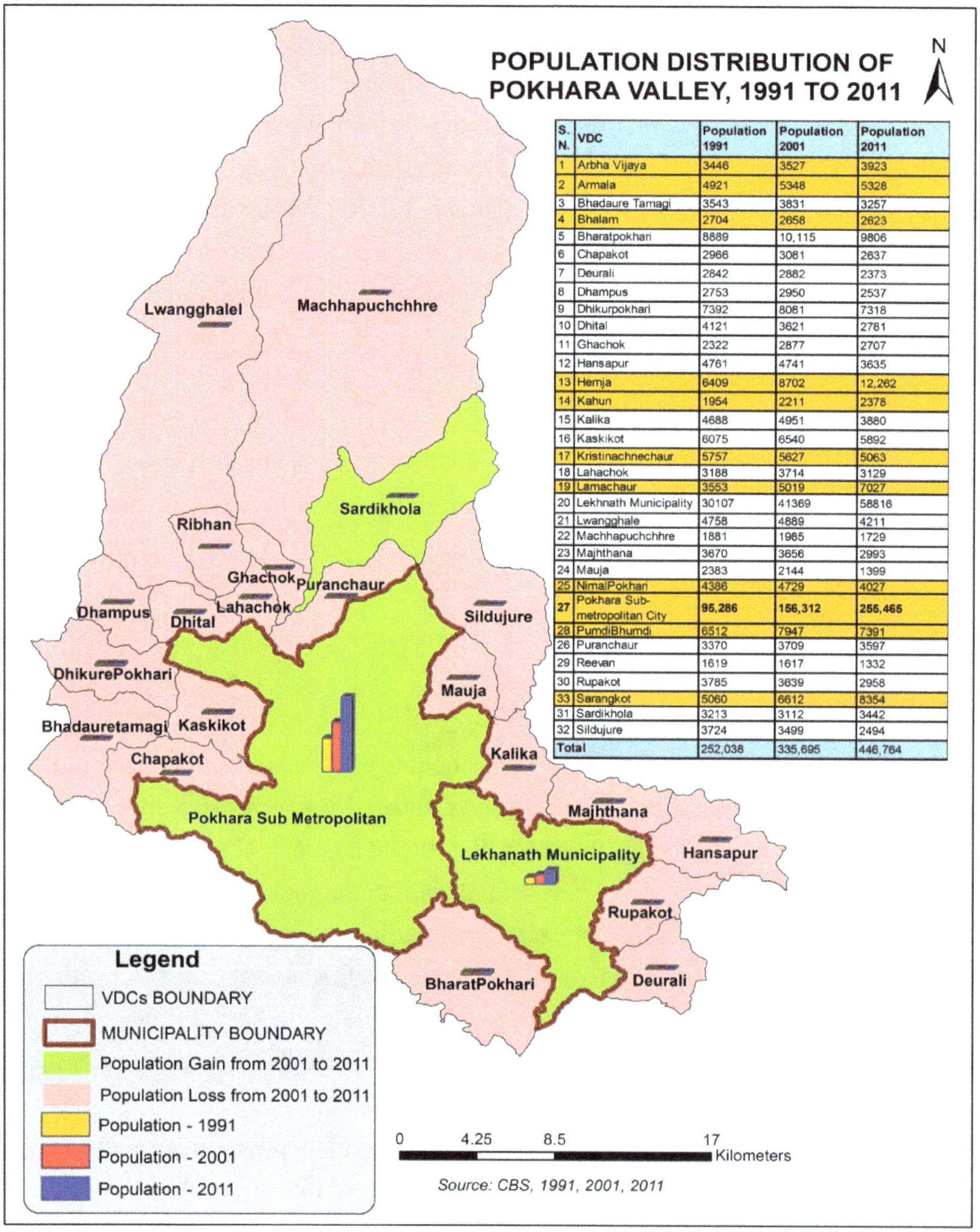

S. N.	VDC	Population 1991	Population 2001	Population 2011
1	Arbha Vijaya	3446	3527	3923
2	Armala	4921	5348	5328
3	Bhadaure Tamagi	3543	3831	3257
4	Bhalam	2704	2658	2623
5	Bharatpokhari	8889	10,115	9806
6	Chapakot	2966	3081	2637
7	Deurali	2842	2882	2373
8	Dhampus	2753	2950	2537
9	Dhikurpokhari	7392	8081	7318
10	Dhital	4121	3621	2781
11	Ghachok	2322	2877	2707
12	Hansapur	4761	4741	3635
13	Hemja	6409	8702	12,262
14	Kahun	1954	2211	2378
15	Kalika	4688	4951	3880
16	Kaskikot	6075	6540	5892
17	Kristinachnechaur	5757	5627	5063
18	Lahachok	3188	3714	3129
19	Lamachaur	3553	5019	7027
20	Lekhnath Municipality	30107	41369	58816
21	Lwangghale	4758	4889	4211
22	Machhapuchchhre	1881	1985	1729
23	Majhthana	3670	3656	2993
24	Mauja	2383	2144	1399
25	NimalPokhari	4386	4729	4027
27	Pokhara Sub-metropolitan City	**95,286**	**156,312**	**255,465**
28	PumdiBhumdi	6512	7947	7391
26	Puranchaur	3370	3709	3597
29	Reevan	1619	1617	1332
30	Rupakot	3785	3639	2958
33	Sarangkot	5060	6612	8354
31	Sardikhola	3213	3112	3442
32	Sildujure	3724	3499	2494
Total		252,038	335,695	446,764

Figure 3. Population distribution map. Note: The total area of the sub-metropolitan city was 55.31 km^2, which expanded to 226 km^2 by 2014 after the annexation of the other 10 adjoining VDCs, namely Arbha Vijaya, Armala, Bhalam, Hemja, Kahun, Kristinachnechaur, Lamachaur, Nimal Pokhari, PumdiBhumdi and Sarangkot. The population sum of the annexed 10 VDCs and Pokhara sub-metropolitan city is mentioned as the total population of Pokhara sub-metropolitan city in this study.

According to Rimal [8,9], multiple forces have contributed to driving people from rural villages and districts into the city area of Pokhara. The major causes of population migration in the city are greater economic opportunities (28%), accessibility to public service (22%), the influence of population growth and migration (13%), globalization (10%), physical conditions (7%), as well as tourism activities, government plans and policies, the land market and the political situation (all 5%). Rapid population growth has significantly influenced the utilization of land and other natural resources in this region, resulting in land use/cover changes.

3.2. Land Use/Land Cover Change (1990–2013)

Land use/land cover (LULC) change is quantified for the last 23 years in the vicinity of the fast urban growing city of Pokhara. This research is pivotal, since it shows the relationship between the urbanization process and the transition of various land use classes at different spatial and temporal scales.

There have been extensive changes in different land use classes between 1990 and 2013 (Table 1) due to diverse proximate and other underlying causes. The high rate of migration and population growth has directly contributed to rapid urban growth. Urban coverage of the Seti watershed has increased by 60% from 24.03 km² in 1990 (Figure 4a) to 54.20 km² in 2013 (Figure 4b) which is a remarkable rate of increase (Table 1). According to the analysis, 29.19 km² of cultivated land have been converted into built-up area (Figure 5). Easy road access to local villages has also fostered urban growth. Not surprisingly, the area of cultivable land is declining. One result of the rapid movement of people to city areas is that forest coverage for private forest and community forestry has increased remarkably in the rural area. The random flows of the rivers convert the cultivated land into sand and *vice versa*. Cultivated land occupied 349.63 km² (30.27%) in 1990, decreasing to 320.81 km² (27.77%) in 2013.

Table 1. Land use/land cover change, 1990–2013.

Land Cover Type	Land Cover 1990		Land Cover 2013		Magnitude (2013–1990) Change	
	Area in km²	Area in %	Area in km²	Area in %	Change Area in km²	Change Area in %
Urban/Built-up	24.03	2.08	54.20	4.69	30.17	2.61
Water Body	44.44	3.85	45.79	3.96	1.35	0.12
Open Field	3.97	0.34	2.63	0.23	−1.34	−0.12
Forest Cover	470.80	40.75	479.41	41.50	8.61	0.75
Cultivated Land	349.63	30.27	320.81	27.77	−28.82	−2.50
Sandy Area	6.23	0.54	7.36	0.64	1.13	0.10
Glacier	42.36	3.67	46.12	3.99	3.77	0.33
Snow Cover	1.34	0.12	0.50	0.04	−0.84	−0.07
Barren Land	105.46	9.13	99.28	8.59	−6.18	−0.54
Grass Land	102.33	8.86	95.49	8.27	−6.84	−0.59
Swamp Area	4.62	0.40	3.61	0.31	−1.00	−0.09
Total	**1155.19**	**100.00**	**1155.19**	**100.00**		

A total of 8.19 km² and 6.63 km² of forest has been converted into grassland and cultivated land, respectively. Despite this transformation, forest coverage overall has increased slightly. Table 1 shows that in 1990, forest covered 470.80 km^2 (40.75%), increasing to 479.41 km^2 (41.50%) in 2013. Grassland and barren land both decreased in the area. Similarly, snow-covered area decreased, but the area covered by glaciers increased by 3.99% over the time period. This signifies less snow in the Himalayas and rapid melting and formation of larger glaciers in the region. Furthermore, transformation of the open field into cultivated land is due to encroachment for food security and the rise in the number of urban slum dwellers. The open field land class, although small in area, encompasses an area in the western part of the Pokahra sub-metropolitan city that is the proposed site for the Pokhara International Airport. The open field and swamp areas in the region have also decreased, while water body and sandy areas have slightly increased. The total 44.44 km^2 of water body and 6.23 km^2 of sandy area increased to 45.79 km^2 and 7.63 km^2, respectively, in 2013. The total 3.97 km^2 of open area and 4.62 km^2 of swamp area has transformed to 2.63 km^2 and 3.61 km^2 in 2013, respectively (Table 1).

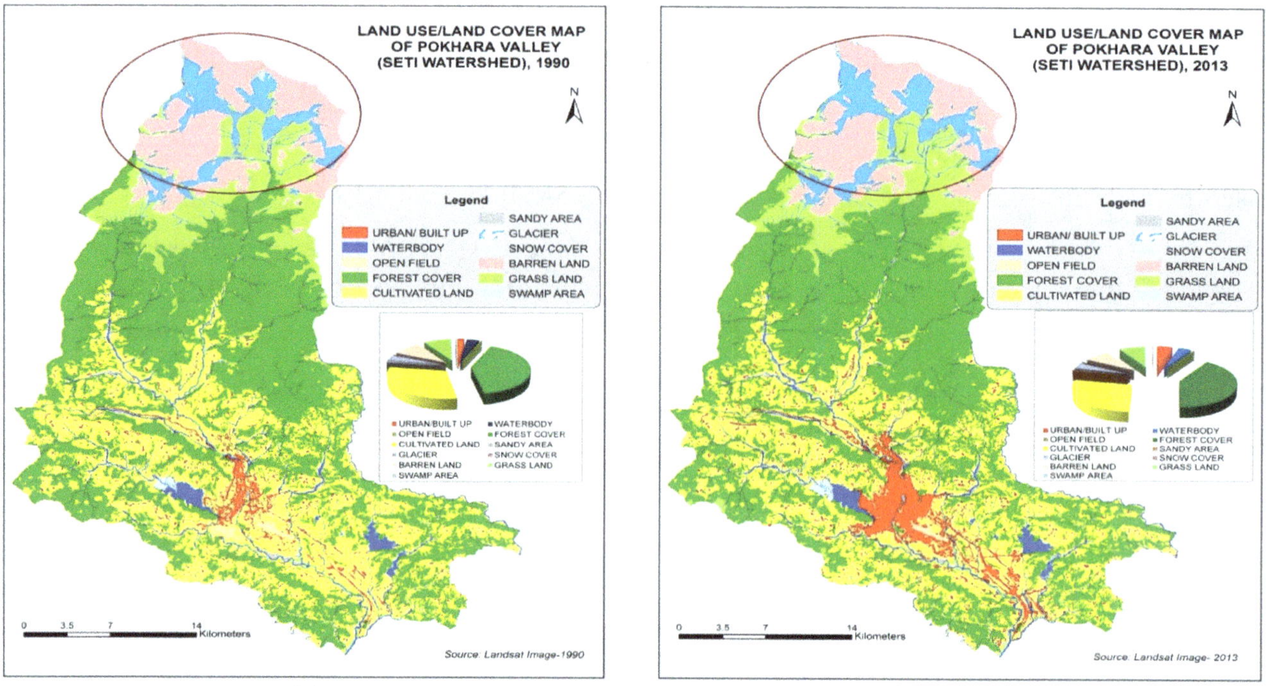

(a) (b)

Figure 4. (**a**) LULC map of Pokhara valley, 1990; (**b**) LULC map of Pokhara valley, 2013.

Figure 5 below shows the transition of different non-urban land use classes to urban/built-up. A large area (29.19 km^2) of cultivated land has been converted into urban/built-up. Similarly, the area transformed to urban/built-up from water body, open field, forest cover, sand, grassland and swamp is 0.43km^2, 1.09 km^2, 0.12 km^2, 0.51 km^2, 0.18 km^2 and 0.14 km^2, respectively.

Almost the entire urban expansion has taken place in the Pokhara sub-metropolitan area and Lekhnath municipality and their buffer areas. The area has seen drastic land use/land cover changes in the last two decades. Land cover changes are demonstrated in terms of decreasing cultivated areas and forest with respect to increasing urban/built-up. It is predictable that the cultivated land has been squeezed and the urban/built-up area has been intensified. Among the remaining agricultural areas, many are at high risk of edge fall, sinkholes, landslides and floods. Scattered settlements remain amid the agricultural areas. The loss of agricultural areas and the risk of frequent serious disasters pose a severe challenge to food security. Much of the city's rapid growth in population has been accommodated in informal settlements with a high risk of environmental impairments and hazards. The rapid increase in urban development is itself a risk factor. Besides, urban expansion in the high risk and vulnerable areas is yet another factor jeopardizing the safety of human life.

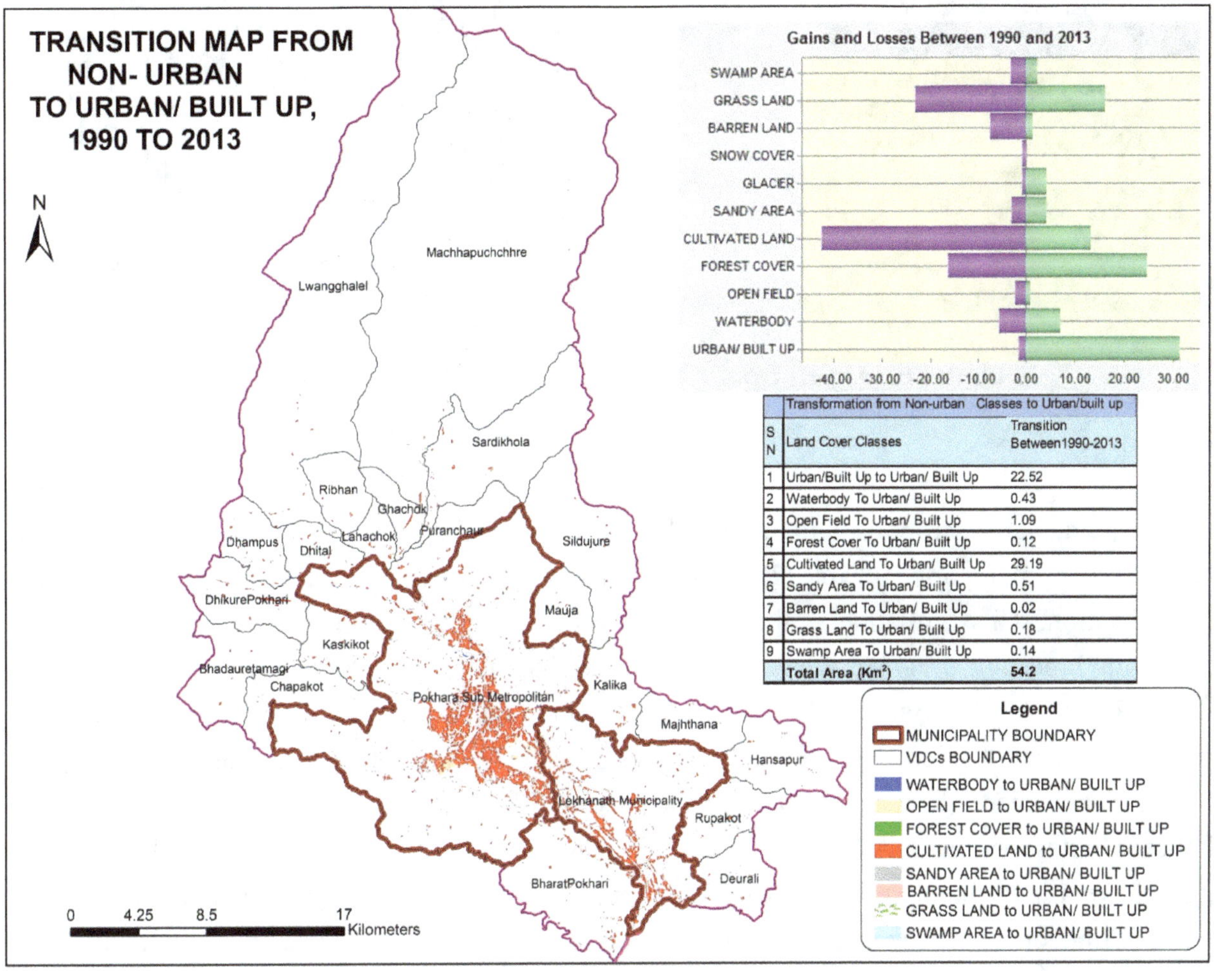

Transformation from Non-urban Classes to Urban/built up		
SN	Land Cover Classes	Transition Between1990-2013
1	Urban/Built Up to Urban/ Built Up	22.52
2	Waterbody To Urban/ Built Up	0.43
3	Open Field To Urban/ Built Up	1.09
4	Forest Cover To Urban/ Built Up	0.12
5	Cultivated Land To Urban/ Built Up	29.19
6	Sandy Area To Urban/ Built Up	0.51
7	Barren Land To Urban/ Built Up	0.02
8	Grass Land To Urban/ Built Up	0.18
9	Swamp Area To Urban/ Built Up	0.14
	Total Area (Km²)	**54.2**

Figure 5. Transition map from all land use/land covers to urban area, 1990–2013.

3.3. Multiple hazard Risk

Pokhara is a disaster-prone area due to its steep terrain, a rugged and fragile geomorphic condition, surrounding high peaks and slopes, volatile tectonic process, variable climatic condition and the frequency of earthquakes, floods and landslides. The likelihood of disaster in Pokhara is therefore high. The Pokharel 2013 report has mentioned Archalboat, Dhad, Chhorepatan, Simla Tunda and the Phusre Khola edge as potential subsidence risk areas, the western edge of Bijayapur Khola and the Firke Khola as edge fall risk areas, Phusre Khola, Dowilla, Simal Tunda, both sides of the Seti River bank and the Chhorepatan area as high flood hazard risk areas, Chhorepatan as a sinkhole risk area and Batule chaur as a land collapse risk area [20]. Mentioned (Figure 6) represented the Seti George and Devis's Fall of Pokhara valley.

Figure 6. (**a**) Seti Gorge 2014 and (**b**) Devis's Fall 2014.

3.3.1. Sinkholes

In the Pokhara sub-metropolis and adjoined areas, a number of sinkholes and sites of land subsidence have been reported (Table 2). Geologically, all of these recorded sinkholes lie in the Ghachok formation. This formation is the debris flow deposits that comprise boulder- to silt-sized sediments with limestone fragments and capped by a 1–3 meter-thick silty layer. The area from Baidam and Rastra Bank to Baglung Buspark, Chhorepatan and its adjoining areas, Pokhara stadium and adjoining areas and the left bank of the Seti River near the Golf palace are also susceptible to land collapse and subsidence when a large earthquake occurs [21,22]. The southern peripheries of the Powerhouse, Gupteshwar cave, Mahendra cave and Chamero cave areas, which are regularly visited by tourists, have a high likelihood of sinkhole development within the Phewa formation ([23]; p.13). The areas of Chipledhunga, Mahendra Pul, Male Patan, Manaswar, Bhagerthan Lama Chaur and Tersa pati are additional sinkhole collapse-prone areas of Pokhara valley [21].

The agricultural land with some settlement in the Jaimure basin of Armala nearby the meeting point of the local Kali Khola and Jumleti Khola is a highly affected region with more than 100 large to small sinkholes that have formed in recent years (Figure 7). People have been moved from the area due to the threat. Its water flow is recharged through various sources. The flow of water is obstructed on the surface due to the poorly-managed land use practice, and water passes through the gravel layer to the dissolvable lime, resulting in the formation of underground ponds (Figure 8). This then creates sinkholes, soil subsidence and the flow of lime-mixed water. Consequently, the core city area, named New Road nearby the Pokhara sub-metropolitan City Office and the Phirke Khola Bridge near Ganesh Tol in Parsyang have encountered land collapse.

Mahendra cave, Bat cave and Gupteshwor cave consist of huge boulders and rocks. Thus, the areas of the caves and the settlements and land around are also considered as hazard-susceptible zones. The Seti River, originating from glaciers in the Annapurna range and flows through deep gullies, has been widened broadly in some places. In some places, including the Pokhara metropolis area, it has created deep gorges, where one can only hear the sound of the river flow. People in the metropolis have settled at the edge of the river in the gorge. The unmanaged dense settlements themselves are at risk of disasters, whereas expanding settlements at the edge of the gorge are at risk from a geological perspective, as well.

Table 2. Sinkhole and land subsidence hazard areas.

SN	VDC/Municipality	Location	Potential Hazard
1	Armala	Settlement and cultivated area near Kali khola and Jumleti khola meeting place	Sinkhole/land subsidence
2	Pokhara S.M	Mahendrapul area, Ranipawa area and settlement around	Gorge
		Phirke Khola Bridge near Ganesh Tol in Parsyang	Sinkhole/land subsidence
		Mahendra cave, Bat cave and Gupteshwor area	Sinkhole/land subsidence
		New road nearby the Pokhara Sub-Municipality Office	Sinkhole/land subsidence

Source: Field visit, 2014, and interaction with local people. Note: 'Khola' is the local Nepali name for river and 'Tol' is the local Nepali name for settlement.

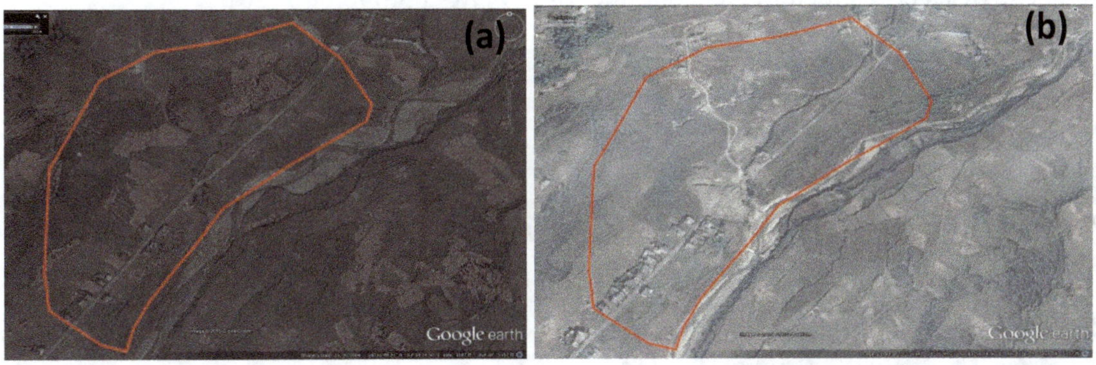

Figure 7. (**a**) Google image of 2004, Armala and (**b**) Google Image of 2014, Armala.

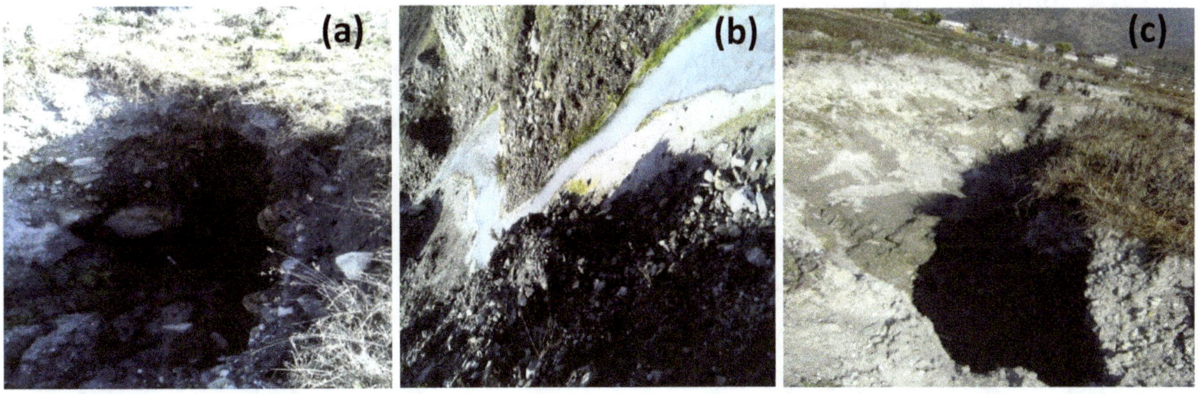

Figure 8. (**a**) Sinkhole in the Armala area; (**b**) Identified watercourse after the sinkhole; (**c**) Sinkhole in the Armala area, 2014.

3.3.2. Floods

The scientific community is well aware that the intensity of flood events, and their potential effects on people, valuables and goods, are directly affected by land use changes in general and increases in the areas of sealed surfaces in particular [24]. Landslides, floods and mass movement disasters are common phenomena in the Seti watershed, associated with extreme weather events and high precipitation, especially during the summer monsoon period, resulting in excessive river discharge.

The upper areas of the watershed have experienced several recent avalanches and debris flows, resulting in lake formation, which have then broken through (Figure 9). When the river carries a large amount of debris, including clay and lime, this challenges the settlements' banks in the lower parts of the catchment area and the workers in the riverbank. A considerable number of lives and the property of urban dwellers residing in the river banks within the Pokhara sub-metropolitan area are likely to be harshly affected by seasonal and flash floods.

Of note was the massive flash flood on 5 May 2012 of the Seti River, which caused serious damage. The flood swept away several settlements and people working along the Seti River, rupturing river banks downstream, destroying tourist spots along the river and damaging the water supply of Pokhara municipality. Altogether 40 people died, an additional 32 were missing, and 31 families were directly affected by the unprecedented flood [25]. A total of 20 houses, one public building and two temples were destroyed, whereas 500 meters of road, two suspension bridges and two water supply systems were smashed. The disaster resulted in an economic loss of NRs 85,000,000 [26].

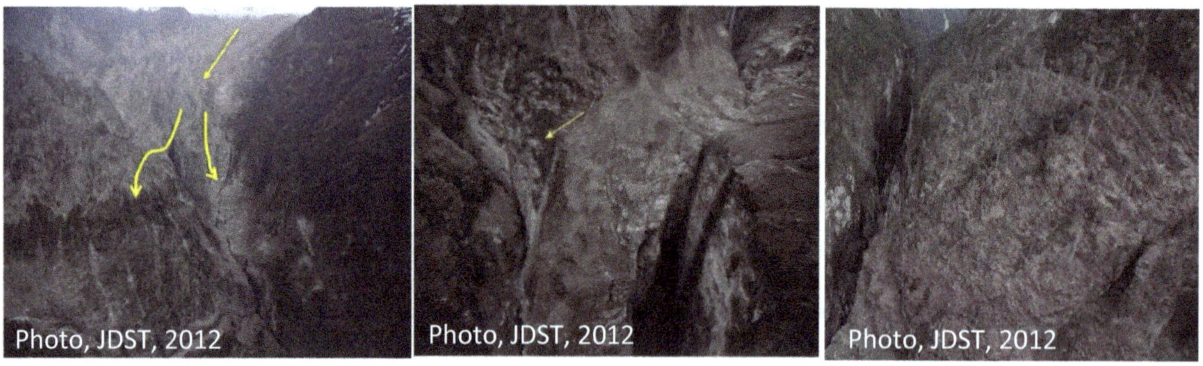

Figure 9. Northern part of Pokhara valley, Seti Gorge (Photos were taken by the Japanese Disaster Survey Team, 2012).

Of the various tributaries of the watershed, it is mainly the Seti River itself that creates the worst effects throughout the year (Figure 10). The flood in the Bhunge Khola has damaged a wide cultivated field and a few houses situated near the confluence with the Phusre Khola (Table 3).

Figure 10. (**a**) Ramghat, 2014; (**b**) Nadipur, 2014; (**c**) Nadipur, 2014.

Table 3. Flood hazard prone areas.

SN	VDC/Municipality	Location	Potential Hazard
1	Arbijaipur	Grass and cultivated land in the bank of Bijaipur khola	Flood
		Kaseri settlement is mostly near the Bijayapur Khola	Flood
2	Machhapuchhre	Grassland at the bank of the Seti River and Mangale Khola	Flood
		Cultivated land with settlement on the bank of the Seti River to the west of Sandal	Flood
		Cultivated land and settlement area near the meeting point of Bhalaudi Khola and the Seti River	Flood
3	Ribhan	Cultivated land near Charym settlement	Flood
		Settlement near Nayapul Dil	Flood
4	Puranchaur	Grassland and cultivated land of the Seti River	Flood
5	Pumdibhumdi	Grass and cultivated land of Chhapswara	Flood
		Cultivated area of Pipaldali on the bank of Phusrekhola	Flood
6	Nirmalpokhari	Grasslands on the bank of Phusrekhola	Flood
		Grassland near the Duwar area	Flood
7	Lahachok	Grassland at the meeting point of Mardi Khola and the Seti River	Flood
		Grassland, cultivated land and some settlements near the bridge and the bank of Mardi Khola	Flood
8	Kristi Nachne Chaur	Cultivated land of Dulekhu settlement	Flood
		Grass and cultivated land to its north	Flood
		Grass and cultivated land on the bank of Phusrekhola	Flood
		Cultivated land on the bank of Mausuli khola, Aambote Khola and Bhunge Khola	Flood
9	Kaskikot	Large settlement and cultivated land on the Kahare Khola bank	Flood
		Wide cultivated land near the meeting point of Laurek khola	Flood
		Large cultivated land near Guhetro and Betani khola near Pame	Flood
		Cultivated land and settlement near Laurek and Harpan Khola	Flood
10	Kalika	Cultivated land at the southern part of the meeting point of Jantre Khola and Madi Khola	Flood
11	Dhital	Grass and cultivated land on the bank of Idikhola	Flood
12	Dhikurpokhari	Cultivated land to the northwest of Bhindabari on the bank of Khahare khola	Flood
		Cultivated land near Dhadkhola near the Lewade and Niware settlements	Flood
13	Bharatpokhari	Cultivated land and grass on the bank of Saraudi khola	Flood
		Grass and settlement of the Chirgadi area	Flood
		Grass, settlement and cultivated land on the bank of Saraudi Khola	Flood
14	Sardikhola	Grass near the Yomu settlement	Flood
		Grass and settlement near Gairakhor of the Sardikhola bank	Flood
		Cultivated land at the meeting point of Bhurujung khola and the Seti River on the border of Sardikhola VDC and Puranchaur VDC	Flood

Table 3. *Cont.*

15	Pokhara sub-metropolitan	Grass, settlement and cultivated land of the Seti bank	Flood
		River bank of Seti at Sitaghat	Flood
		Settlement of Laltinbazar, Upallodip and Tallo dip	Flood
		Damside area grass	Flood
		Grassland on river bank near Manipal Medical College	Flood
		Seti River bank at Ramghat	Flood
		Settlement at the edges of the Seti River at Ramghat	Flood
		Grassland on the bank of the Seti River below Phoolbari and Baraghare	Flood
		Pragati Tol, Lamtara, Dobilla settlements	Flood
		Cultivated land and the grass land at the west of the Seti River bank in Dobilla	Flood
16	Hyangja	Cultivated land and settlement on the bank of Yangdi khola,	Flood
		Cultivated of Yamdi near the Seti dam	Flood
		Grassland, cultivated land and settlement above the Lamachaur settlement on the Seti River bank	Flood
17	Lekhnath municipality	Settlement and grass of the Khudikhola bank and Satmuhane	Flood
		Bijaipur khola bank towards the northeast part from Bijaipur and Seti meeting point	Flood
		Grassland and some settlements in the Musetunda area, Tintara	Flood
		Grassland in the northern and the southern part of Lamiahal area on the Seti River bank	Flood
		Settlement and grass where Khudi and Gaduwa meet	Flood

Source: field visit, 2014, and interaction with local people.

3.3.3. Landslides/Edge Falls

Well-defined erosion fronts are seen in Kabre and below Kuiredanda, both banks of Phusre Khola, between Chhorepatan and Mahatgaunda, Kali Khola, Bhalam Khola, Bijaipur Khola, Mardi Khola, Bhurjung Khola and to the south of Armala and Dihi (Table 4). The largest landslide of the area was observed between Antighari and Bijaipur in a tributary to the Bijaipur Khola. The next largest landslide area in Jumleti had a width of 75 meters, a length of 100 meters and about a 10–15-meter depth. The landslide on the southwestern slope of Thuloswara caused severe damage to the drinking water pipelines and the cultivated land and was about 200 meters long and 50 meters wide. The landslide on the road between Chhorepatan and the Phusre Khola Bridge occurred about 40 years ago and was 800 meters long and 250 meters wide. The landslide at Bamdi (Chapakot) was about 300 meters long, and the one southwest of Harpan on the south facing slope was 120 meters wide. Meanwhile, the prominent 150 meter-high bare rock cliff is located northwest of Jumleti. Other noticeable bare rock cliffs are located at Mukundpur, Kulpidanda, Tamage, along the Handi Khola, Kristi and Matikhan and northeast of Bijaipur [15]. Recent incidents of floods and landslides occurred on 29 July 2015. Thirty seven people were killed, and 19 were injured in Lumle, Bhadaure and Dhikurpokhari of Kaski district. Mentioned (Figure 11) represented the risk area of edge falls.

Table 4. Landslide and edge fall hazard-prone areas.

SN	VDC/Municipality	Location	Potential Hazard
1	Lwanghadel	Bhedabari village (Topo map): surrounded by the rivers Idi, Mardi and Pati Khola from its three sides and remains at risk	Landslide
		Grassland and cultivated areas at the bank of the above 3 rivers	Landslide
2	Sardikhola	Settlement near Gairakhor on the Sardikhola bank	Edge fall
3	Pokhara Sub Metropolitan	Large settlement near Sitaghat	Edge fall
		Settlement of the upper Seti River bank at Nadipur	Landslide/edge fall
		Seti River bank at Nadipur	Edge fall/flood
		Sitaghat settlement on the bank of the Seti River	Flood/edge fall
		Phoolbari and Baraghare area with large grass cover and settlement	Edge fall
		South of Sitapaila, both sides of the main road, Pragati Tol, Lamtara, Dobilla settlements	Flood/edge fall/land slide
		Sahara ground, adjoined settlements, cultivated land and grass land to the west of the Seti River bank	Flood/edge fall/land slide
		Cultivated land to the east of Dobillaghat across the Seti River, Shivalaya area, grassland in Sheepfarm and Bachhi buduwa area	Edge fall
		Gaukharka grass land on the bank of Bijaypur khola, agricultural land of Kalimati	Edge fall
4	Lekhnath Municipality	Grassland on the northern and the southern part of Lamiahal area on the Seti River bank	Flood/edge fall
		Settlement and grass where Khudi and Gaduwa meet	Flood/edge fall
5	Lahachok	Bank of Mardi River towards the northwest of the Seti and Mardi meeting point	Edge fall
6	Bharatpokhari	Pudi khola bank towards the northwest part of the Pudi and Seti meeting point	Edge fall
7	Ghachok	Seti River bank towards the northeast from the meeting of Pallo Khahare and Seti	Edge fall
8	Dhampus	Wall of Dhad khola	Landslide

Source: field visit, 2014, and interaction with local people.

Figure 11. (**a**) Nadipur, 2014; (**b**) Dobilla (both sides), 2014.

3.3.4. Fire Hazards

Another high risk in the study area is fire. The municipal area comprises dense settlements with weak infrastructure, unplanned electrical cables and joined buildings. Petrol pumps are either very close or attached to settlements. The city has a very limited fire brigade, and people lack safety measures. Hence, the fire risks are high due to electrical faults, fuel pump explosions or fires during and following earthquakes, which may result in extensive loss of property and lives (Figure 12).

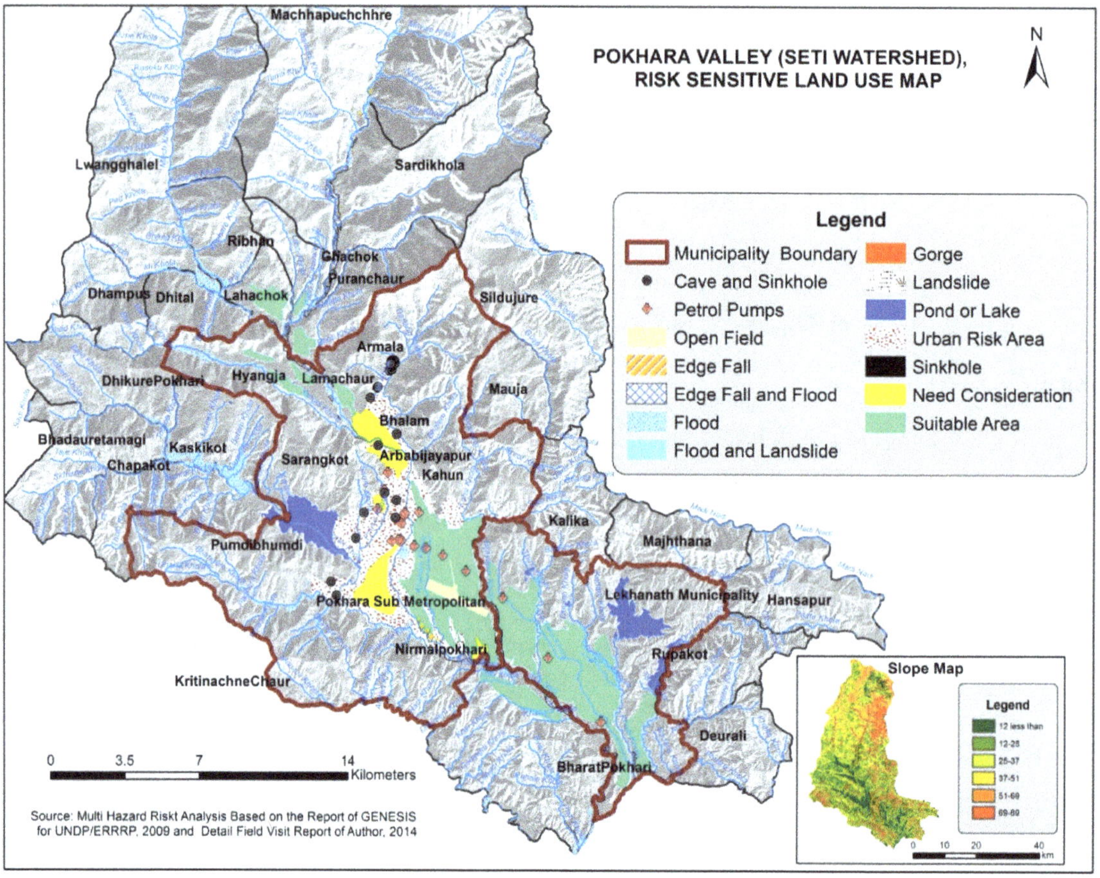

Figure 12. Risk sensitivity map.

3.3.5. Earthquakes

The severity of disasters triggered by seismic events is the outcome of the interaction between the geophysical phenomenon and the characteristics of the built environment [27,28]. Pokhara sub-metropolitan city can be considered to be moderately vulnerable to earthquakes with a probability of intensity of VIII MMI earthquake in the region. In the sub-metropolitan city area, liquefaction-susceptible zones are found on the banks and tributaries of the Seti River, and some patches of liquefaction susceptible zones are found in the KI Singh Bridge, Laltin Bazar and Kamal Pokhari areas [21].

According to the preliminary damage report provided by the Pokhara sub-metropolitan office, at least two deaths and as many as 24 injuries resulted from the tremor of 25 April 2015. Altogether, 517 buildings in 17 wards and more than 300 houses and other infrastructure partially collapsed due to the earthquake and the subsequent aftershocks mainly on 12 May. Arwa, Mijure and Purunchaur are the comparatively more affected VDCs. The educational institutions, tourism industry and the normal daily lives of people were severely affected by the earthquake. This earthquake caused more destruction to the eastern part from the epicenter, Barpak VDC, Gorkha, than Pokhara, located to its west.

The dense settlements of Pokhara sub-metropolitan city and Lekhnath municipality are mostly low buildings, and few are built to withstand earthquakes. The location of petrol pumps close to buildings and unsafe electrical cables increase fire risk if earthquakes occur. All of these factors result in a high threat to life and property.

Our land use/land cover analysis has revealed the increase in settlement and built-up areas in disaster-prone areas. It is essential that strategic planning to combat risks is undertaken.

Increase in the size of human population in the region is one of the main reasons for the change in land use and land cover in the study area. Due to rapid population growth, the city areas are increasing in size, and cultivated land and open fields are being converted to urban areas. This results in increased environmental degradation.

The land sales market and business development have played vital roles in increasing the density of settlements. Land value has been rising rapidly in comparison to past decades. There have been no specific measures put in place to fix the value and price of land. There are no standards for infrastructure development, and this is likely to result in considerable destruction and loss of lives and property from potential disasters, such as earthquake, landslide, flood, epidemic diseases and fire. High rise apartments are being constructed without proper building codes and soil tests. The Land Use Policy, 2012, includes scientifically-based standards, and their implementation is urgently needed.

Development and settlement activities should be restricted in those areas that are identified as sensitive to multiple hazards (Figure 13). To mitigate their vulnerability and the possible damage to life and property, existing settlements at risk should be relocated to safer areas, such as Hemja, Lamachaur, Ghachok, Ghardi khola, Bharatpokhari, Puranuchaur and Lahachok (Figure 13).

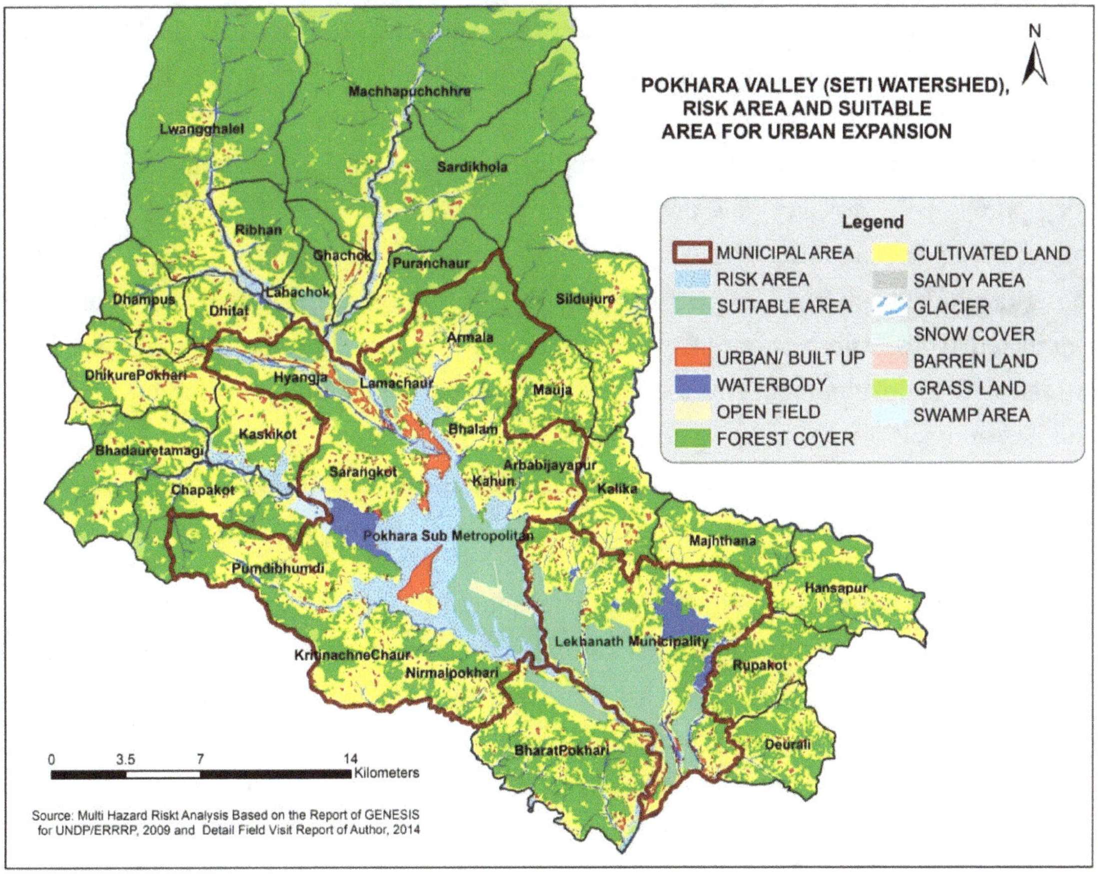

Figure 13. Map for risk areas and suitable places for urban expansion.

4. Conclusion

Our study has assessed trends in urban growth, land use/land cover change and multiple hazard risk in the Seti watershed within Pokhara valley using an integrated approach, including GIS and remote sensing. Land use/land cover maps for the years 1990 and 2013 were derived using an object-oriented approach, which provided information on the spatio-temporal distribution of the region. According to the land use/land cover analysis, the total 24.03 km^2 of urban area in 1990 increased to 54.20 km^2 in 2013. In contrast, 349.63 km^2 of cultivable land decreased to 320.81 km^2 in 2013, whilst forest cover experienced a small increase from 470.80 km^2–479.41 km^2 over the same time period. Consequently, the study area has faced rapid demographic changes in two decades.

Large cultivated areas and some settlements at Armala, some city core areas, such as Mahendrapul and Ranipawa, and the periphery of Mahendra cave, Bat cave and Gupteshwar area, along with the new road area, have already witnessed sinkholes, gorges and land subsidence, respectively. The settlements and the landscapes around these areas are at high risk for future disasters. The settlements, cultivable lands and grasslands near the river banks of the Seti River and its tributaries have a high threat of flooding, and the settlements there need to be relocated to safer areas.

Lessons learned from the Pokhara valley study suggest that rapid urban spread and growth has resulted in an increase in impermeable land use, and these areas of high population growth are at severe risk of multiple hazards, such as floods, landslides, edge falls, fire, earthquakes, sinkholes and land subsidence. Disaster risk management would indicate that these settlements at high risk require scientifically-planned relocation.

Finally, disasters hinder development, and development creates new risks by fostering land use change and environmental degradation. In order to reduce the landslide and edge fall problem, sustainable development practices should be adopted. The Initial Environmental Evaluation (IEE), Environmental Impact Assessment (EIA) and Strategic Environmental Assessment (SEA) should be the compulsory prerequisites for any new development projects. Since the regulations adopted in the Land Use Policy, 2012, are scientifically based, their implementation is urgently needed. Nevertheless, some building codes have been implemented in Pokhara city, but these exclude those areas that have been newly annexed to the sub-metropolis. Fixed scientific measures should be strictly adopted in terms of infrastructure development in these newly-developing urban, fringe and rural areas. Geological studies and soil tests should be conducted prior to construction of infrastructure and road standards maintained for road construction. Trained manpower with certified licenses should be used in the construction of infrastructure.

- Sustainable hazard mitigation and land use plans should be the distinct priority of the local and the central government for long-term social and economic reasons. Disaster risk management has not explicitly been assimilated in the national, regional and local policies, programs and projects. Therefore, revisiting the policies, programs and projects is necessary.
- To minimize development in disaster-prone areas and environmental damage due to urban sprawl, integrated, proactive and strong legislative actions should be implemented.
- Restricting development, structures and settlements in hazard-prone areas, relocation of the people to safer areas, developing coping capacity at the community level with awareness about

frequent and probable natural hazards and wise land utilization practice should be a priority. The government needs to introduce effective resettlement plans with appropriate livelihood schemes in the course of the relocation of the people from high risk-prone areas.

- Since flood risk is the function of hazards, exposure and vulnerability, it is essential to reduce exposure and vulnerability to floods. Predictive watershed management planning identifies the sensitive areas, controls landslides, designates acceptable land uses and specifies overall management targets.
- Integration of flood risk management (structural and non-structural measures) in regular urban planning and identification and mapping of flood plains and prone areas, urban, fringe and rural areas at risk is essential.
- Institutionally, more attention and investment should be paid to hazard prevention; information sharing, raising awareness, reinforcing preparedness and early warning system development are required.
- Retrofitting of existing structures is crucial to make them disaster resilient.
- Road standards should be maintained during road construction. Restrictions to steep slope structures/road constructions is essential to prevent landslide and edge fall.
- A new compact settlement strategy should be adopted to manage settlement spread.
- Soil testing and geological studies are essential before the construction of private and public infrastructure.
- Preparing a large-scale topographic map for detailed analysis of the structure and the preparation of comprehensive land use maps, land capability maps, land suitability maps and land zoning maps are needed.

Since disaster management is a fundamental part of development, our results show that it is essential that policy makers, public officials and concerned authorities include multiple hazard risk mitigation in land use plans and policies. Research establishes the links between urban expansions, escalating demography and multiple hazard risk assessment, which assist in modelling the latent impact of future catastrophes and emergency management. It also provides a solid base for the effective management of disasters when they occur and their mitigation.

Acknowledgments

We acknowledge to contributions of the participants in this study (business persons, school teachers, university lecturers and local people) for their invaluable information for the research.

Author Contributions

Bhagawat Rimal had the original idea for the paper, and was took overall responsibility for the study, including data collection and analysis, preparation of figures and finalization of the manuscript. Himlal Baral contributed materials and expertise concerning the study site and the research theme. Nigel Stork carried out the comprehensive revision of the article. Kiran Paudyal contributed to put the research in the local context. Sushila Rijal facilitated the field work activities and drafted the field report. All authors have reviewed and contributed to finalize the manuscript. In addition, all authors approved the final manuscript.

Conflicts of Interest

The authors declare no conflict of interest.

References

1. Batuk, F.; Şengezer, B.; Emem, O. Relations between disaster management, urban planning and NSDI. *Int. Arch. Photogramm. Remote Sens. Spat. Inf. Sci.* **2008**, *37*, 441–444.
2. GPRB (Government of the People's Republic of Bangladesh). *National Plan for Disaster Management 2010–2015*; Disaster Management Bureau Disaster Management & Relief Division, Government of the People's Republic of Bangladesh: Bangladesh, 2010.
3. UNISDR (United Nations International Strategy for Disaster Reduction). *Disaster Statistic 1995–2005*; UNISDR: Geneva, Switzerland, 2006.
4. Wamsler, C. Managing Urban Disaster Risk, Analysis and Adaptation Frameworks for Integrated Settlement Development Programming for the Urban Poor. Ph.D. Thesis, The Housing Development & Management (HDM), Architecture and Built Environment Lund University, Lund, Sweden, 2007.
5. UNDP (United Nations Development Programme). *Urban Risk Management*; UNDP: New York, NY, USA, 2010.
6. Thapa, R.B. Spatial Process of Urbanization in Kathmandu Valley, Nepal. Ph.D. Thesis, The Graduate School of Life and Environmental Sciences, The University of Tsukuba, Tsukuba, Japan, 2009.
7. Li, X.; Anthony, G. Modeling sustainable urban development by the integration of constrained cellular automata and GIS. *Int. J. Geogr. Inf. Sci.* **2000**, *14*, 131–152.
8. Rimal, B. Urban Development and Land Use Change of Main Nepalese Cities. Ph.D. Thesis, The Faculty of Earth Science and Environmental Management, University of Wroclaw, Wroclaw, Poland, 2011.
9. Rimal, B. *Urban Landscape Dynamics of Nepal*; Everest Geoscience Information Service Center: Kathmandu, Nepal, 2012.
10. FAO/UNEP (Food and Agriculture Organization/United Nations Environment Programme). *Negotiating a Sustainable Future for Land, Structural and Institutional Guidelines for Land Resources Management in the 21st Century*; FAO: Rome, Italy, 1997.
11. MoHA (Ministry of Home Affairs). *Nepal Disaster Report*; Ministry of Home Affairs, Government of Nepal: Nepal, 2011.
12. Yamanaka, H.; Yoshida, M.; Arita, K.; Terrace landform and Quaternary deposit around Pokhara Valley, central Nepal. *J. Nepal Geol. Soc.* **1982**, *2*, 95–112.
13. Gautam, P.; Panta, S.R.; Ando, H. Mapping of subsurface karst structure with gamma ray and electrical resistivity profiles: A case study from Pokhara valley, central Nepal. *J. Appl. Geophys.* **2000**, *45*, 97–110.

14. CBS (Central Bureau of Statistics). *Environment Statistics of Nepal and Nepal Population Report*; National Planning Commission Secretariat; 2011.
15. Nepal Environmental and Scientific Services (NESS). *Environmental Survey for Upgrading Feasibility Study on Upper Seti (Damauli) Storage Hydroelectric Project for Nepal Electricity Authority (NEA)*; NESS: Kathmandu, Nepal, 2001.
16. Anderson, J.R.; Hardy, E.E.; Roach, J.T.; Witmer, R.E. *A Land Use and Land Cover Classification System for Use with Remote Sensor Data*; United States Government Printing Office: Washington, DC, USA, 1976.
17. Gurung, H.B. *Pokhara Valley, Nepal Himalaya: A Field Study in Regional Geography*; University of Edinburgh: Edinburgh, UK, 1965.
18. Poudel, K.R. Urban growth and land use change in the Himalaya Region: A case study of Pokhara sub metropolitan city, Nepal. *GIS Ostrava* **2008**, *27*, 1–11.
19. Rimal, B. Urbanization and the decline of agricultural land in Pokhara sub metropolitan city, Nepal. *J. Agric. Sci.* **2013**, *5*, 54–65.
20. Pokharel, R.P. Vulnerable land in Pokhara city and its use potentiality. *Himal. Geogr.* **2013**, *8*, 51–58.
21. Genesis Consultancy Pvt. Ltd. *Report on Impact of Settlement Pattern, Land-Use Practice and Options in High Risk Areas, Pokhara Sub-Metropolitan City*; Report for UNDP/ERRRP; 2009.
22. Genesis Consultancy Pvt. Ltd. *Report on Impact of Settlement Pattern, Land Use Practice and Options in High Risk Areas, Pokhara Sub-Metropolitan City*; Revised Report for UNDP/ERRRP; 2010.
23. Koirala, A.; Rimal, L.N.; Sikrikar, S.M.; Pradhananga, U.B. Engineering and environmental geological map of Pokhara valley (1:50000). Department of Mines and Geology in cooperation with BGR: Kathmandu, Nepal; Hannover, Germany, 1996.
24. Ebert, A.; Welz, J.; Heinrichs, D.; Krellenberg, K.; HansJürgens, B. Socio-environmental change and flood risks: The case of Santiago de Chile. *Erdkunde* **2010**, *64*, 303–313.
25. MoHA (Ministry of Home Affairs). *Nepal Disaster Report*; Ministry of Home Affairs and DPNet-Nepal: Kathmandu, Nepal, 2013.
26. JDST (The Japanese Disaster Survey Team). *Survey Report on Seti River Flood, Nepal 2012*; Available online: http://www.sff.or.jp/nepal/osirase/ (accessed on 2 Feburary 2015).
27. Hewitt, K. *Regions of Risk: A Geographical Introduction to Disasters*; Addison Wesley Longmann: Essex, UK, 1997.
28. Martins, V.N.; Cabral, P.; Silva, D.S. Urban modelling for seismic prone areas: The case study of Vila Franca do Campo (Azores Archipelago, Portugal). *Nat. Hazards Earth Syst. Sci.* **2012**, *12*, 2731–2741.

4

Changing Livelihoods and Landscapes in the Rural Eastern Cape, South Africa: Past Influences and Future Trajectories

Sheona Shackleton [1,*] and Marty Luckert [2]

[1] Department of Environmental Science, Rhodes University, Grahamstown 6140, South Africa
[2] Department of Resource Economics and Environmental Sociology, University of Alberta, 515 General Services Building, Edmonton, AB T6G 2H1, Canada; E-Mail: marty.luckert@ualberta.ca

* Author to whom correspondence should be addressed; E-Mail: s.shackleton@ru.ac.za

Academic Editors: Claudia A. Radel and Jacqueline M. Vadjunec

Abstract: This paper seeks to understand the drivers and pathways of local livelihood change and the prospects for transformation towards a more sustainable future. Data are used from several studies, and a participatory social learning process, which formed part of a larger project in two sites in the Eastern Cape, South Africa. Secondary information from a wealth of related work is used to place our results within the historic context and more general trends in the country. Findings indicate that livelihoods in the rural Eastern Cape are on new trajectories. Agricultural production has declined markedly, at a time when the need for diversification of livelihoods and food security seems to be at a premium. This decline is driven by a suite of drivers that interact with, and are influenced by, other changes and stresses affecting local livelihoods. We distil out the factors, ranging from historical processes to national policies and local dynamics, that hamper peoples' motivation and ability to respond to locally identified vulnerabilities and, which, when taken together, could drive households into a trap. We end by considering the transformations required to help local people evade traps and progress towards a more promising future in a context of increasing uncertainty.

Keywords: livelihoods; farming; traps; transformation; South Africa; landscapes; poverty; change; multiple stressors; Eastern Cape

1. Introduction

Many parts of the world are facing accelerated change, rising insecurity and vulnerability, and greater uncertainty brought about by a multitude of interacting drivers and shocks ranging from global economic downturns to growing corruption and weak governance, escalating poverty and food insecurity, failing health systems under pressure from epidemics such as HIV/AIDS, and extreme weather events [1,2]. At the same time, the longevity of the goods and services provided by the planet's ecosystems is questioned [3]. These global risks and changes interact with, and may be compounded by, country-level structural, economic and political processes that create and sustain inequities. The effects of these global-and national-level interactions are also felt in localized contextual dynamics, manifesting in institutional breakdown, crime, landscape modification through changing land uses, declining agricultural production, land degradation and biodiversity loss [4].

A case in point is the agrarian change in the Eastern Cape, South Africa that has been taking place over several decades. The present day landscape has been largely shaped by discriminatory apartheid policies that placed most of the land in the hands of white commercial farmers and systematically undermined peasant production amongst black farmers in the Bantustans [5]. However, despite this history, and up until relatively recently, farming continued to provide an important source of livelihood and food security for black rural families. Many people in the rural communal areas of the country identified themselves as "farmers", undertaking both arable and livestock farming. However, more recent work, especially in the Eastern Cape, has suggested that farming, particularly extensive arable production is declining and either shifting to more intensive cultivation of home gardens or being abandoned altogether [5–14]. The trends also suggest a reduction in livestock ownership and numbers, although this change is not as striking [13–15]. Such adjustments in agricultural production are influenced by, and interact with, many other trends at a variety of scales. For instance, under global climate change, the region is projected to become hotter and drier, facing more droughts and, ironically, more severe and frequent floods [16]. These climatic changes are superimposed on a suite of other stresses impacting local livelihoods and human well-being [17,18]. Short-term shocks, such as death and illness in the household, job loss, hunger, violence and crime, often arise as a consequence of longer-term "background" shifts, which, we argue, are fundamentally altering local social–ecological systems. Examples of these "slow variable" background changes include the systemic effects of HIV/AIDS on household structure and assets [19–21], urbanization and changes in demographics, household structure and relations, and values.

For many of the progressive changes observed (such as agricultural decline), no clear trajectories into the future are evident, with short-term stressors working in synchrony with longer-term change to create feedback loops that are potentially leading to increasing vulnerability and food insecurity [22,23]. We, therefore, argue that the legacy of deprivation and neglect in South Africa's rural communal areas, combined with more recent changes that decrease flexibility, diversification and local agency, as well as exposure to increased risk, may lock some households and communities into situations that prevent innovation and response. Such situations are often viewed as being "stuck" or in a trap [24]. It is in these contexts that the possibility of transformation becomes crucial. New policies and actions will need to be put in place to protect and support vulnerable people, and innovative practices or different "ways of doing" need to be developed and implemented. This approach requires recognition and understanding of the changes taking place, the drivers of these changes, and the various factors that may hinder people's ability to accommodate or respond to change.

In this paper we consider changes in ways of life in two marginalized rural areas of the Eastern Cape and investigate how interactions amongst these changes and their drivers emerge and impact on livelihoods and current and future vulnerability. To assemble a holistic and composite picture of the complexity, dynamics and heterogeneity of local livelihoods, and to tease out the multiple factors across scales that drive the observed trends, we synthesize findings from several specific sub-studies conducted in the same sites, and then situate these findings within the relevant literature. Our aim is to investigate changes in livelihoods and landscapes in the Eastern Cape and consider whether some households, through combinations of shocks and long-term stressors, may be trapped in situations that prevent them from responding to vulnerabilities and, thus, to securing resilient livelihoods. We specifically seek to understand how historical processes, national policies and contemporary local dynamics have affected current livelihood portfolios and the capacity to adapt to change, particularly in relation to future climate change. We argue that with an improved understanding of these processes, transformative policies and strategies may be better developed to avoid traps and facilitate more promising futures.

In the next section, we describe our research approach and its theoretical underpinnings. We then present key findings, which identify large changes in the nature and structure of livelihoods and landscapes that have occurred in our study areas, and suggest reasons behind these changes drawing on both our own research and other studies and sources. Next, we consider a number of features of the broader political and socio-economic contexts that have also contributed to, and resulted from, these livelihood changes. We conclude with policy considerations regarding transformations and potential future trajectories that could assist rural Eastern Cape households evade traps and make choices regarding their livelihood strategies, including engagement in arable production.

2. Study Sites

We selected two sites for the project: Gatyana in Mbashe Local Municipality and Lesseyton in Lukanji Local Municipality in the Eastern Cape Province, South Africa (Figure 1). The study sites were chosen to correspond to a gradient of increasing rainfall and decreasing accessibility to urban amenities and markets. Each site consists of multiple villages, with the boundaries of these being less clear-cut in the Gatyana site due to the scattered nature of homesteads. The inland site, Lesseyton, is 20 km from the city of Queenstown. It falls at the intersection of the Grassland and Thicket biomes [25] and has a mean annual precipitation of between 350 and 500 mm. Homesteads are built in a typical village layout and generally have only a small garden space for vegetable cultivation. However, there is access to communal land, which is used primarily for grazing by livestock owners. The coastal site, Gatyana, is more isolated and is about 30 km from the small town of Willowvale. It still lags behind in infrastructure such as tarred roads, and has marginal local markets and poor transport systems. Most homesteads are widely dispersed with large plots of land (half a hectare or more). The site crosses two major biomes, the Indian Ocean Coastal Belt Forests and Savanna, and receives a mean annual rainfall of between 950 and 1100 mm [25].

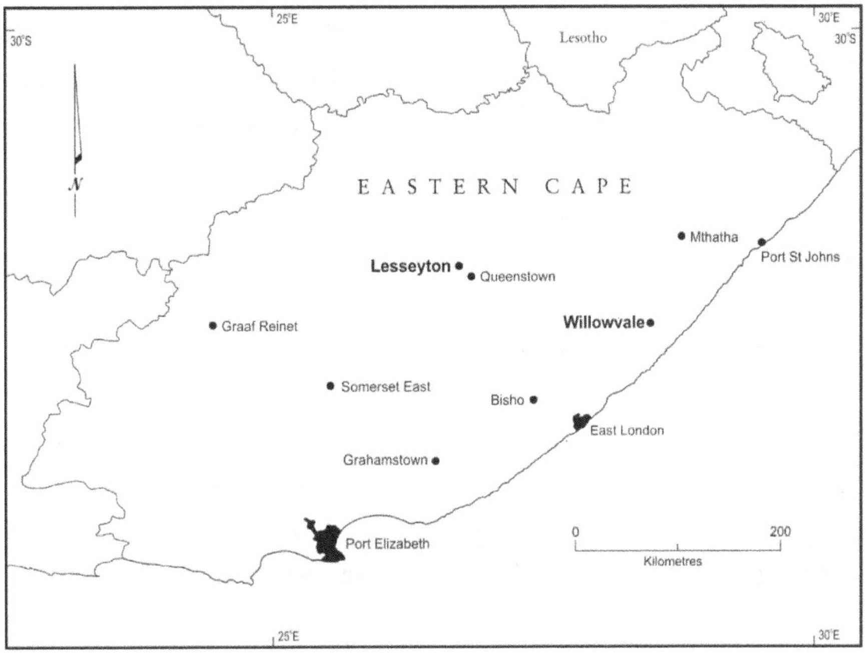

Figure 1. Map of the Eastern Cape, South Africa showing study sites.

3. Approach and Methods

To construct an integrated and cohesive understanding of the complex livelihood and landscape changes in our study sites, and to analyze the drivers behind these changes, we draw on and synthesize findings from a mix of separate sub-studies (five in total from Honors undergraduate to PhD level) conducted as part of a larger project funded by the International Development Research Centre (IDRC) [26–31]. Using livelihoods and assets-based approaches (guided by the sustainable livelihoods framework and the concepts of vulnerability and multiple stressors), this project sought to explore how climate change, together with several other livelihood stressors including HIV/AIDS, affect vulnerability, food security, livelihood strategies and choices, adaptive capacity and coping, adaptive and transformational responses amongst different types of households.

The data presented in this paper are mainly derived from unpublished Masters theses and draw heavily on the baseline household survey used by several students. This survey was administered in 2011 and 2012 to 170 randomly-selected households at each site (total n = 340). Questions related to livelihood and income sources were repeated an additional two times at six monthly intervals to capture seasonal differences. The data presented here are averaged across the three rounds. Data collected included information on household demographics, health and HIV/AIDS proxy indicators, assets, welfare perceptions, shocks and responses, livelihood activities and household income in cash and kind [30]. We interviewed heads of households or the next most senior adult. Men often left the interviews to the women in the household. For some analyses, we disaggregated the data by gender structure (based on the sex of household head and adult members) and wealth categories (see [32]). These data provided a holistic picture of household income shares, livelihood activities and community and household assets.

In addition to this quantitative data, we use qualitative data from a study by Clarke [26] on responses to climate change and HIV/AIDS amongst vulnerable rural households, in which she undertook 20 life histories. Five main "themes" were used to guide the interviews: (1) general household information such as who lived there, and what household members did for a living; (2) the village's history, and any changes that occurred over the period of the household's residence; (3) the respondent's personal life story from growing up with their parents to the present, focusing on the hardships they experienced and the specifics of how they and their family coped with these hardships; (4) the respondent's main concerns for the future; and (5) the types of support the respondent felt were needed in their community. Each interview was primarily conducted and recorded in *isiXhosa* and later transcribed and translated into English through a translator. The findings were systematically analyzed using a software package for textual analysis called MAXQDA. A coding system was created using the following categories: vulnerability (changes, shocks and stressors experienced), responses (responses made towards shocks and stressors), assistance (help received), health (shocks and stressors related to health), climate, farming, and natural-resource use (climate-related changes, shocks and stressors experienced), and future concerns (future shocks and stressors interviewees were concerned about, as well as what support they felt was needed). The text retrieval option allowed specific categories and study sites to be analyzed separately. These broader categories gave rise to more specific themes such as unemployment, social-grant dependency, and crime-related shocks and stressors. Here, we extract representative quotes from the life histories as supporting evidence for the changes and vulnerabilities emerging at the local level. These provide insight into local, lived experiences of vulnerability and change.

Participatory mental maps were constructed at both sites to record the stressors believed by local men and women to be creating vulnerability, and their perceptions of the linkages between these [26]. We held four small focus groups (4–8 participants each) with men and women separately, with each group creating their own mental map. Mental maps are "qualitative representations of a system consisting of variables and the causal relationships between them" [33]. For our study, these took the form of a spider-gram drawn on large sheets of paper, with key stressors being linked via directional lines. HIV/AIDS was suggested to participants as an initial item, to which they added the various cause and effects of this disease. Respondents then added further stressors and drivers of change and began connecting these with what was already on the map, until they felt they had exhausted all the key factors contributing to their vulnerability. All links and decisions were discussed within the focus groups. These maps provide a rich interpretation of the local drivers of vulnerability and their impacts on people's lives. In addition, to obtain the perspectives of the youth on local vulnerabilities and concerns, participatory workshops were held with mixed groups of grade 11 learners in each site (20 learners in Lesseyton and 53 in Gatyana). These scholars discussed and tabulated their viewpoints related to the past, present and future of their area.

For data on agrarian change, we draw on the published results of Shackleton *et al.* [14], which investigated farmers' opinions of landscape change, ecological succession in abandoned fields, and changes in land-cover categories. In addition, we report on unpublished research undertaken in 2013 and 2014 by Shackleton and colleagues on arable production in the Eastern Cape. To assess broad trends in landscape use and land-cover change, 1961 aerial photographs were compared to 2009 photographs by means of randomly selected sample plots (50 m × 50 m). Eight hundred forty-four plots were sampled on the 1961 photographs and 949 on the 2009 photographs, representing approximately 5 percent of the study area. Each plot was categorized by its dominant land use or land cover, *i.e.*, abandoned garden, abandoned field, current field, current garden, forest, woodland, and grassland. In May and June 2012, a household questionnaire was administered to 31 current and 50 former field cultivators to solicit views on changes in land-use practices, including causes and what happens to abandoned fields. A snowball sampling approach was adopted to identify households in each category. To obtain an adequate sample, 38 villages within the wider Gatyana area were surveyed. For the more recent research, we conducted follow-up, in-depth interviews with the same cultivators in October 2013. Some had already ceased cultivating their fields. Furthermore, we undertook a separate survey of 90 randomly-selected households in two villages in the study site regarding cultivation of home gardens. Key data collected related to cultivation and garden details, the reasons for gardening, what garden inputs were used (for example manure, fertilizers, compost and labor), methods of plowing, seasonal determinants, changes/trends in gardening, challenges involved, estimations on yields and sale of crops.

These research results are complemented by the outputs of a parallel, participatory social learning and knowledge exchange process at both study sites, which provided in-depth insights on key research themes [34]. This process, which took place over three years, involved mixed groups of 10–15 people and commenced at the same time as the household survey. We aimed to build problem-solving capacity and strengthen local practices that had evolved in response to multiple stressors. In some of the foundational meetings, we asked participants to share stories of vulnerability and changes in their communities and lives based on their first-hand experiences. Participant conceptualizations of vulnerability and change covered shocks and stressors (drought, illness), "community problems" (truancy amongst the youth, crime, rape of women and girls) and structural aspects such as poverty, poor education and unemployment. The entire process is documented in a handbook [34].

Also important for our analysis are other complementary research findings from the Eastern Cape and various critical commentary papers on higher-level trends and processes in the country. We use this literature to support and contextualize our findings and to explore underlying drivers.

4. Conceptual and Analytical Lens

Our analytical approach to considering past influences and future trajectories is informed by the concepts of transformation and traps that have been developed in the social–ecological systems literature. In this literature, transformation is generally defined as "the fundamental alteration of the nature of a system once the current ecological, social or economic conditions become untenable or undesirable" [35] or a "profound, substantial and irreversible change" [36]. While transformation may be spontaneous or planned, the process often implies conversion to a better or more beneficial state [37,38]. Consequently, it can take on a normative and deliberative dimension. The normative nature of transformation can lead to ethical concerns related to what constitutes a more beneficial state, for whom, at what scale and who determines this. In the climate change literature, the IPCC uses much the same definition, though alterations in various sub-systems are emphasized, including value systems; regulatory, legislative or bureaucratic regimes; and financial institutions and technological or biological systems. Other climate change researchers prefer the term "transformational adaptation", where the distinction between adaptation and transformation is less clear cut, with transformation being related to the scale, intensity and novelty of adaptations made [39–41]. Transformation is said to occur within social–ecological systems when principal members of these systems change core functions through collective action and institutional change (coordinated and uncoordinated action) and from the accumulation of numerous smaller, often behavioral, changes made by individuals. Transformation can also occur through changes in policies and governance, institutions, livelihoods, landscapes, farming systems, locations or identity [36,40]. From this perspective, transformation is, thus, both similar to, and different from, adaptation. In contrast to adaptation, which refers to small, incremental responses to risk and change that keep a system in the same state, transformation is conceptualized as being required when human-environmental systems or components of these are over-whelmed or trapped and a complete shift is required [39]. However, transformation is rarely a discrete event—it often emerges messily over space and time through a series of smaller transformations [36]. We argue in reality the distinctions between these concepts become blurred when faced with the variety and complexity of conditions influencing livelihoods and their development, particularly given the lack of clarity and the uncertainty regarding future development directions and pathways and the normative nature of transformation.

Transformation may be hindered by traps, but, at the same time, transformation is generally required to escape traps. Traps are seen as negative states and are often the outcome of multiple and accumulative barriers to action that prevent or block the ability of people to transform the way they live. Several types of traps are recognized in the socio-economic ecological resilience literature: poverty traps, rigidity traps, and locked-in traps [24]. These traps convey different meanings for different disciplines, but simply put, they imply a situation of "being stuck", often in a dysfunctional or degraded state, and being unable to respond to shocks and changes. In the resilience community, a poverty trap embodies the loss of options to develop or deal with change [24], and represents a condition that inhibits positive responses to a stimulus. Such a trap requires external support to escape [24]. Similarly, rigidity traps can equally prevent a shift towards upwardly mobile trajectories; they represent a system in which the members of organizations and their institutions become highly connected, rigid and inflexible and resist change [24]. This rigidity constrains the ability to respond to new problems and opportunities [24]. Locked-in traps are similar [42]. Walker and Salt [43] argue that to avoid being trapped on an unacceptable trajectory that is difficult to escape, it may be necessary to facilitate almost continual transformational change. This depends on creating new opportunities that, in turn, are dependent on internal sources of resilience and adaptive capacity, and various forms of external support [44]. Various structural (legacy of Apartheid), cognitive (forms of denial), normative and cultural (conformity, traditions, place), and behavioral factors can create path dependence and keep people (or certain groups of people) and systems in these traps. Path dependence occurs when the legacy of past events, states, decisions and actions affect current dynamics and future changes though reinforcing feedback mechanisms and entrenchment of particular self-reproducing institutional arrangements [42,45]. This makes it difficult to divert the system to a new course [46]. In our case, the notion of path dependence helps us to understand the long-term consequences of separate development (under the Apartheid regime) in our study sites.

The starting point for our analysis (below) is to describe the changes in livelihoods and landscapes that have occurred in the rural Eastern Cape. However, surrounding this change, our research identified important past and current contexts that have created several types of interlinked shocks, stressors and other factors that could act as barriers to livelihood improvement, both now and in the future, for certain groups of people. All of these are closely associated and often mutually reinforcing, together creating a situation that resembles aspects of both poverty and locked-in traps. However, some of the changes also have positive dimensions that need to be considered and weighed up against any negative aspects, with social grants being a particular case in point. The primary changes are described below with supporting evidence from both our own research and the literature.

5. Results and Discussion

5.1. Changing Livelihoods and Landscapes

5.1.1. Shifting Livelihood Portfolios

Land-based income (from crops, livestock, and natural resources) makes up, on average, only 15–25 percent of livelihoods in the Lesseyton and Gatyana sites, respectively (see Figure 2). The cash income received from employment (formal, small businesses, casual labor and remittances) is of a similar magnitude, but reversed, with the more rural site receiving 16 percent of their livelihoods from employment activities, while households in Lesseyton receive 29 percent of their income from these employment sectors. The overwhelming source of income for both sites comes from social grants that make up over half of income.

The snapshot of income shares (see Figure 2) is the result of ongoing processes, detailed below, which have caused land-based and employment incomes to decrease, while reliance on social grants has grown (also see [7]). These changes represent a narrowing of livelihood activities/options and a reduction in the use of local landscape resources as a basis for livelihoods and as safety nets needed in response to shocks. In the next three sections, we consider trends within each of these sources of livelihood and the important contexts and underlying causes for such changes through analysis of the broader literature.

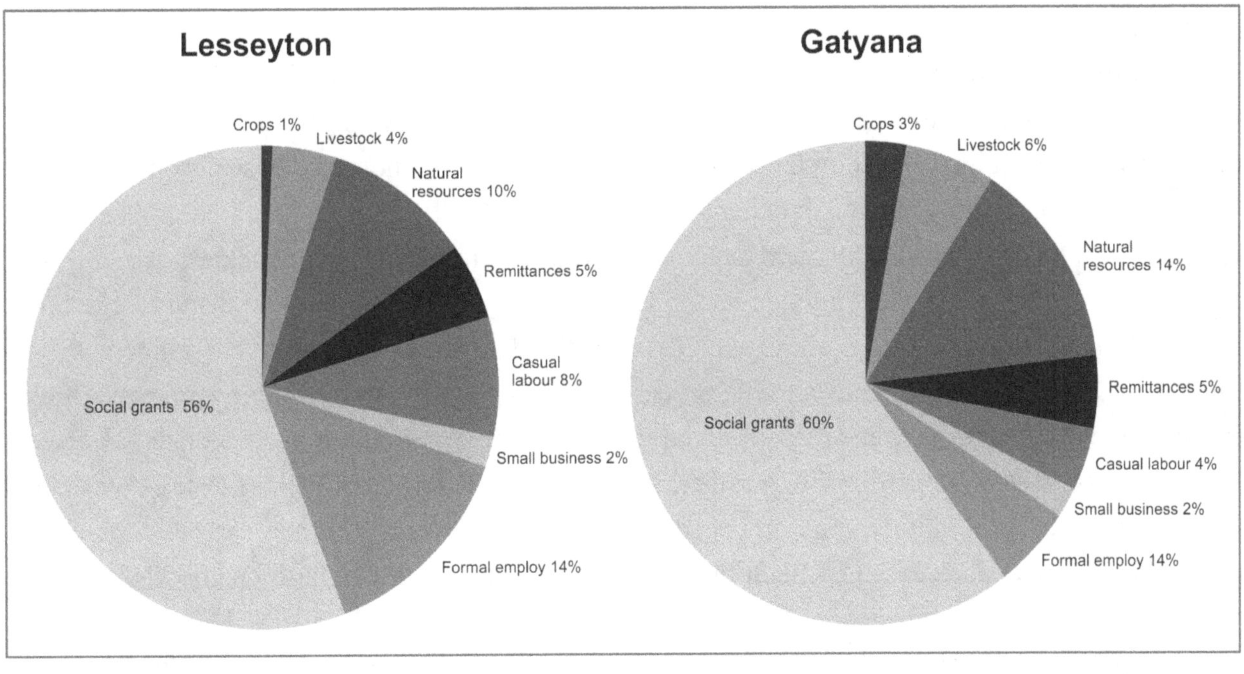

Figure 2. Average income shares from different livelihood sources across all households (n = 340) in Lesseyton and Gatyana (Source: Stadler 2012, unpublished thesis [30]).

5.1.2. Abandonment of Fields and a Decline in Livestock Production

Results from our study sites show that cultivation forms, on average, only between 1 percent and 3 percent of total household income in Lesseyton and Gatyana respectively (Figure 2). This is a substantial decline relative to the past (*i.e.*, from greater than 20 percent of income in 1950) [12], and is reflected in changes in the landscape, especially the increasing area of abandoned fields. This field abandonment and decline in arable farming for both cash and subsistence purposes has been happening over several decades and can be viewed as a social–ecological transformation in itself (Table 1) [12,14,28]. In the Gatyana study site, landscape change over a 50-year period involved mainly a decline in cultivated fields and an increase in woodlands, primarily due to the establishment of the thorny, pioneer species *Acacia karoo* in old lands (Table 1). The encroachment of woody vegetation into these areas acts as a barrier to their future cultivation.

Table 1. Landscape change in Gatyana over a 50-year period.

Aerial Photography Analysis of the Gatyana Landscape over the past 50 Years (1961–2009) has Shown:	
• Reduction in cultivated field cover from 12.5% to 2.7%	• Increase in woodland from 13.9% to 28.8%
• Increase in abandoned field cover from 1.5% to 6.9%	• Increase in forests by 5%
• Reduction in grassland of 22.5%	• Increased abandonment of home gardens, but less than abandonment of fields

Source: Shackleton *et al.* 2013 [14].

Many researchers have argued that this move away from agriculture started in the 1940s (or even earlier as discussed by Hebinck and van Averbeke [47]) with the concentration and relocation of people into the Bantustans, and a deliberative process of "depeasantization". This was expedited through the creation of "labor pools" to supply the mining industry and the marginalization of black farmers, while white farmers were heavily subsidized [8,47]. At the same time, the policy of so called "Betterment" that placed people in villages and zoned land for particular uses, served to separate and distance households from their fields and contribute further to their disuse [7,8,14,47,48]. However, some farming still continued, facilitated by income from jobs in the mines, which in those days could mainly be sent home because accommodation (in hostels) and transport were provided by the mining companies [47,49].

The greatest levels of field abandonment have occurred in recent years, with Shackleton *et al.* [14] finding that most of the past farmers interviewed in Gatyana had stopped farming on distant fields, on average, about 18.5 years ago. This timing coincided with the democratic transition and an increase in social welfare [5,50], as well as changes in the process of "formal" migration of mine workers, which has now ceased [50]. Other drivers and reasons given for the decline in agriculture include: lack of access to markets; exhausted soils; unpredictable weather; lack of labor; poverty and lack of access to credit; infrastructure decline, especially fencing and roads; destruction of crops by cattle and wild animals, related to both the former and to the loss of child labor for herding; a decline in farming knowledge; and institutional factors that prevented the emergence of a land rental market [5,6,14,47]. Regarding the latter, Kepe and Tessaro [5] argue that a lack of land rights in the rural communal areas of South Africa has limited the implementation of agricultural programs devised by the state because *de facto* owners simply do not trust anyone else to cultivate their land in the absence of secure tenure. Livestock production has also declined (attributed locally to disease) and become more asymmetrical, although to a lesser extent than arable farming [6]. Indeed, the lack of cattle for draught has been given as another reason for the decline in cropping [6,14,48,51].

Despite the substantial decline in field cultivation, households are still planting so called "home gardens" (fields attached to homesteads), especially in areas like Gatyana where residences are scattered in the landscape and plots can be up to one hectare in size (in Lesseyton this is not the case, and there is little home gardening). Andrew and Fox [6] contend that in the coastal areas of the former Transkei (known as the Wild Coast and of which Gatyana is part), there has, in fact, been a shift from field to homestead cultivation. Our research into the extent and use of gardens shows that most households have gardens, although they may not cultivate them annually. Women (63 percent) are mainly responsibility for managing these home gardens and for making cultivation decisions. Constraints to gardening mentioned included financial (mentioned by about 40 percent of respondents), biophysical (weather, pests, water, soil fertility, fencing and draught power, with at least one of these being mentioned by most respondents) and lack of labor and disinterest among the youth (46 percent). When asked about trends, the majority of households interviewed felt that garden cultivation was declining, with 46 percent saying the greatest decline had been in the last fifteen years. This was also reflected in the land use change data (Table 1).

Given the declines in extensive arable agriculture and to a lesser extent home gardening, an important question is whether this land use trajectory is narrowing options for the future and placing people in a state of greater vulnerability, including blocking positive transformation in relation to future food security. The current high reliance on purchased food and other goods was perceived by the majority of vulnerable people to be detrimental due to an increased need for scarce cash (Table 2) [26], and could become a more serious problem as food prices rise with the impacts of climate change [52]. Moreover, for households in isolated rural areas such as Gatyana, local supplies are costly and the transport to distant markets expensive. This further stretches meager cash resources. The waning in the perceived value of farming and homestead gardening also results in the potential loss of a set of other related benefits. These include communal ethics relating to farming, and other social and cultural benefits [6], reinforced ownership and entitlement rights [51], the supplementation towards incomes and food security, and provision of a safety net [12,53,54]. Gardening, for instance, has become a very individualistic endeavor, with not a single one of the households interviewed mentioning making use of work parties. Hebinck [5] and Andrew and Fox [6] propose that these trends have led to a disconnect between people and the land and the loss of a farming identity, with the prospects for the future suggesting that the youth are unlikely to be interested in farming [7,14]. Regarding the latter, Hebinck and Van Averbeke [47] describe a scenario of rural residents, especially the youth, being in a mental state of "waiting to leave" the moment an urban opportunity arises, and as drifting between urban and rural settings (also mentioned by Dubbeld [49]). Bank [7] also discusses how new opportunities for women (in state-supported off-farm commodity groups), as well as their increased migration to towns and cities, have severed them from their traditional farming and gardening role. These shifts, together with the woody invasion and reforestation of old lands and grasslands [14], constitute a strong disincentive for future agricultural production in the area.

Despite this downward trend in production, cultivation has not totally disappeared; there are still a few dedicated cultivators (Table 3), and about a third of gardening households in Gatyana mentioned that crop cultivation is an important part of *Xhosa* culture and identity and is likely to continue. Some 15 percent believe cultivation will enter a phase of expansion again in the future. A survey conducted for the Mbhashe Local Municipality IDP (2012–2017) [55] showed that out of 31 wards, only seven did not list an agricultural-related need, suggesting that residents of the region are still interested in continuing to farm. However, even if some of the farming challenges could be overcome, and there was effort to support agricultural development in the region, the next challenge for farming is climate change. It has already been mentioned that the climate future for southern Africa, including South Africa, will be increasingly severe. Work in the study areas has shown that people are perceiving changes in weather patterns, which include shifts in the start of the rains to later in the season and more frequent dry periods.

These changes were attributed to factors such as "God's punishment for the way society is going", and were seen as a risk particularly by more well-off farming and livestock-owning households [30]. Such risk perception could serve as a further disincentive for farming. If agriculture is to be reinvigorated in these areas, it will be necessary to take into account shifts in seasons, rising temperatures (2–4 °C), and more extreme events such as droughts and floods.

Table 2. Reliance on purchased rather than on collected natural resources and cultivated food amongst vulnerable households in Gatyana and Lesseyton and perceptions of the impacts of this reliance (n = 100).

Site	Does Your Household Rely More on Purchased Goods than in the Past?		Has This Made It Harder or Easier for Your Household?		
	% yes	% no	% Easier	% Harder	% Unsure
Gatyana (n = 50)	72	28	16	72	17
Lesseyton (n = 50)	69	29	23	60	12
Some stated reasons			Gardening wastes money Shop gives credit Can't rely on gardening Crops don't grow Too old to garden	Transport expensive High costs in town Shops far Pay for load in taxi No other option	

Source: Clarke 2012, unpublished thesis [26].

Table 3. Narratives from interviews with households still cultivating fields indicating that there is still life in arable farming in the rural Eastern Cape (n = 19).

Why Are You Farming?	What Do You Think about the Future of Farming?
"No one forces me to farm, it is something I enjoy doing, I love it and have been farming since I was a child. I have never worked but in farming." "I don't buy everything, for example I grind maize here at home and don't buy it in town. Farming saves a lot of money". "I feel very good about being a farmer, I wish I would plant more if I had resources and if I was healthy." "I started farming from 1941 to today, it`s something I have been doing all my life." "I`m passionate about farming chickens and crops. Farming guarantees food, you cannot experience hunger when farming."	"There is little interest from the younger generation in farming." "We are still young and now we are negotiating with government to revitalize our fields, we are really pushing for them to come and support us." "My eldest child will definitely continue with farming." "My children and grandchildren will continue with farming because we raised them well regarding the importance of farming. They were raised to respect farming." "There is hope of farming in the future here, people do talk about wanting to plant their fields again, and their hope is on government."

Source: Shackleton and Hebinck (unpublished data) [56].

5.1.3. Unemployment, Declining Remittances and Less Cash

Remittances now only form some five percent of total household income sources in our study areas (Figure 2), whereas they used to be one of the main sources of cash income, especially for farm investment, for rural households [7,47]. Remittances have decreased with declining employment opportunities nationally in several sectors and in the cities, especially for the less-connected and less-educated rural poor [5,57]. Marais [58] estimated that between 1994 and 2000, more than 500,000 jobs were lost in the country. Further factors include de-industrialization in some of the smaller towns in the Eastern Cape [7,58] and the process of more-permanent relocation to urban areas, which leaves little money to send home [47]. The impact of HIV/AIDS on the productive-age classes of the population has also reduced remittances, as ill people return home for care or to die [54]. Bank and Minkley ([59]; p. 24) make the point that "…Evidence suggests that the rates and real value of rural remittances have been declining steadily since the 1980s," and go on to suggest that "The main dynamic in the post-apartheid era is that rural labor—once so eagerly desired—has become a burden to the state and an irrelevance to capital." The decline in remittances has had the effect of squeezing and stretching the cash obtained from grants and can also be linked to the decline in agriculture as mentioned previously, possibly further trapping people. Connected to a decline in remittances is a change in values related to how income from employment should be spent. Kepe and Tessaro ([5]; p. 271) mention that "Even if there are people who still hold jobs, nowadays migrant workers buy cars instead of cattle that should be used for plowing." In this case, a change in one source of cash income has resulted in cascading effects on livelihood diversification and land use in terms of both cultivation and animal husbandry.

5.1.4. Social Grants: Window of Opportunity or a Barrier to Transformation?

Social grants form the largest percentage share of livelihood income in both study sites and are sometimes the only source of income for households (Figure 2, Table 4). This contribution has become increasingly important as other income sources have declined and as social protection has increased. Post 1994, in an effort to alleviate poverty and redistribute wealth, old-age pensions in South Africa were equalized (*i.e.*, black and white pensioners now receive the same amounts) and social grants were extended to include a child-support grant, initially for children of up to the age of seven but increased in 2010 to children up to 18 years old [60]. Furthermore, in 2007 the eligibility age of men for a pension was dropped from 65 to 60, the same as for women. While the pension amount is significant and reflects the minimum wage, the child grants are small. These changes have increased the availability of state support to the poorest South Africans. Some 15.2 million people receive grants.

Table 4. Narratives from life histories indicating dependency on social welfare grants (n = 20).

"I worry about how I will survive if the government stops the old age grant."
"There are six people living in this house. Me, my two daughters, my brother and grandchildren. We live on the pension that I get."
"I cannot afford to build a decent dwelling for my family due to financial constraints. And to raise children on a social grant is very hard for me. It was better when my husband was alive."
"I am not employed; I do not have a job. I live on the grant I get. I live with my grandchildren. Nobody in this house is employed."
"Life is much harder now. There is no employment for the youth. They depend on my pension for survival. Everything they want they expect me to provide."
"If my grandmother were to pass on I would have no-one to support me, as I depend on her old age grant for surviving."

Source: Clarke, 2012, unpublished thesis [26].

Grants tend to be pooled within the household and are critical in contributing to food security, children's education and reducing the negative impacts of HIV/AIDS [27,60] (Table 4), but they have also created high levels of reliance on them (forming more than half of household income, Figure 2) and there has been some diversion of benefits from the intended beneficiaries (something acknowledged by local people themselves). Many life-history interviewees stressed how they depended on grants as their only source of cash, and that these payments often did not stretch beyond supplying everyday needs (Table 4). This dependence on grants, however, cannot be separated from the apartheid history of the country and its disempowering effects on poor people's self-determination, confidence, and self-sufficiency, as well as the marginalization of the areas in which they live.

While various authors have suggested that grants can provide a window of opportunity, this does not seem to have been fully mobilized in our study areas, except by more educated households who have greater means of using social grants to invest in improved livelihoods [27]. Hagen-Zanker *et al.* [60] found that pensions can help adults within the pensioner's household seek employment by providing the resources for job-hunting. However, these results would only apply to individuals who are competitive in the job market, especially those with education (as revealed by Ndlovu [27]). Other research has shown the child grants have increased the likelihood of mothers participating in the labor market [61], but probably not in rural areas where there are few local employment opportunities. Other positive aspects of grants have been to reduce risky livelihood activities such as transactional sex, contribute to health care, and support other livelihood strategies.

Recent in-depth, ethnographic work by Dubbeld [49] in the village of Glendale in Kwazulu Natal, South Africa provides a more nuanced perspective on the role of social grants. His work suggests that not everyone sees social grants as a blessing and seldom as a substitute for wage labor. Young men in particular are not eligible for the current types of grants, but at the same time have limited job opportunities. Thus, the grant system puts money in the hands of some but not others. Interestingly, in our study sites it was only young males, especially those living alone, who were food insecure [32]. Dubbeld [49] further writes about how some of his informants perceived grants as "socially corrosive," destroying family values, morals and traditions such as marriage, absolving men of their parental responsibilities and contributing to disrespect of elders by the youth.

The situation in our study sites is complex and multidimensional, and there are no straight-forward answers regarding the impacts of social welfare and how it is affecting future trajectories. Fay [9], working in a neighboring site to Gatyana, found a correlation between the mean number of social grants per household and the presence or absence of field cultivation, with households with more child grants less likely to practice cultivation. Kepe and Tessaro ([5]; p. 271), also working in the Eastern Cape, explain how informants "unanimously agreed that with the prevalence of these grants, many people would rather use the monthly income from the grant to purchase food items rather than cultivate crops." Certainly, this trend seems to make sense, as purchasing food is much less risky and time consuming than investing in cultivation where the outcomes are unpredictable, especially as people are beginning to perceive and experience greater uncertainty in the climate. At the same time, growing food locally could reduce food insecurity into the future as previously highlighted.

The notion of a "cycle of dependency," described by Kofinas and Chapin [44] as a situation where individuals, households and even communities move away from traditional livelihood practices into systems with disincentives for self-sufficiency, is too simplistic in our context. While dependencies on external support could increase vulnerabilities to change, through depressing innovation and productive activity and narrowing the livelihood portfolio, these negative results may be outweighed by the importance of these grants in providing a safety net, especially for the most vulnerable (women, orphans, the elderly and infirm, and those with low levels of education and opportunity). However, the grants are neither permanent nor secure and are lost when the pensioner passes away or the child grows up, potentially plunging households back into highly vulnerable states. Moreover, our results suggest that, given people's situations and other stressors in their lives such as HIV/AIDS, these grants are able to do little more than keep households within the status quo, preventing them from falling deeper into poverty. They cannot be the basis of investment for the future [49]. Other commentators have also recognized that the grant system alone is unable to change rural people's lives [58] and claim that the South African government's social security system is not designed to lift a significant proportion of individuals out of poverty. For such a result to occur, social grants would likely need to co-exist with other complementary transformations such as in education [27] and/or reductions in the effects of HIV/AIDS, and/or be more universally distributed. Consequently, the role of social grants in facilitating transformation, rather than merely helping people cope and survive, will need further investigation [26,30]. Indeed, at the moment, grants are unlikely to play a role in reinvigorating agriculture.

5.2. Broader Contexts and Changes Influencing Landscape and Livelihood Change

The changes in livelihoods described above have occurred within the milieu of a number of broader contexts, changes and stressors that characterize our study sites. These include the legacy of apartheid, HIV/AIDS impacts, changing values, demographic changes, unsustainable and maladaptive strategies such as crime, changing demographics and the erosion of social capital. The mental maps in Figure 3 illustrate the multiple stressors and changes recognized by men and women from the study sites and the links between these, including to farming activities; whereas Table 5 reflects the perspectives of the youth.

5.2.1. The Legacy of Apartheid and the Immovability of Path Dependence

While apartheid ended two decades ago, inequality and poverty in South Africa have worsened [54,62,63]. This reflects the legacy of the Bantustan system, with former "homelands" in rural areas remaining under-developed in comparison with former "white" South African areas and cities [57]. Consequently, the communal areas of the Eastern Cape continue to represent some of the poorest and most vulnerable regions in the country. The residual effects of poor education, poor service delivery, lack of land rights, the undermining of agriculture and self-sufficiency, and the creation of a labor pool can still be seen and continues to impact on people's livelihood options today. What is more, many of these deprivations are being perpetuated by the current government (Table 5). National transformation has not yet trickled down to the rural communal areas, and they still face neglect at multiple levels: "More than a decade into our successful democracy, the Eastern Cape Province remains trapped in structural poverty that shows in all aspects of its demographic, health and socioeconomic profile" ([64]; p. 15). This sentiment is recognized and reiterated at the local level as captured in Table 5, Figure 3 and the following quote from one of the in-depth interviews from our study: "The so-called freedom is for the educated people. As for the uneducated, it has brought them nothing." The quality of rural education is in a dismal state, as highlighted by students in our study areas (Table 5). Moreover, local government is failing in many of its delivery mandates (Table 5). Land rights in these areas are still in the hands of the state despite many promises to resolve the land question [5].

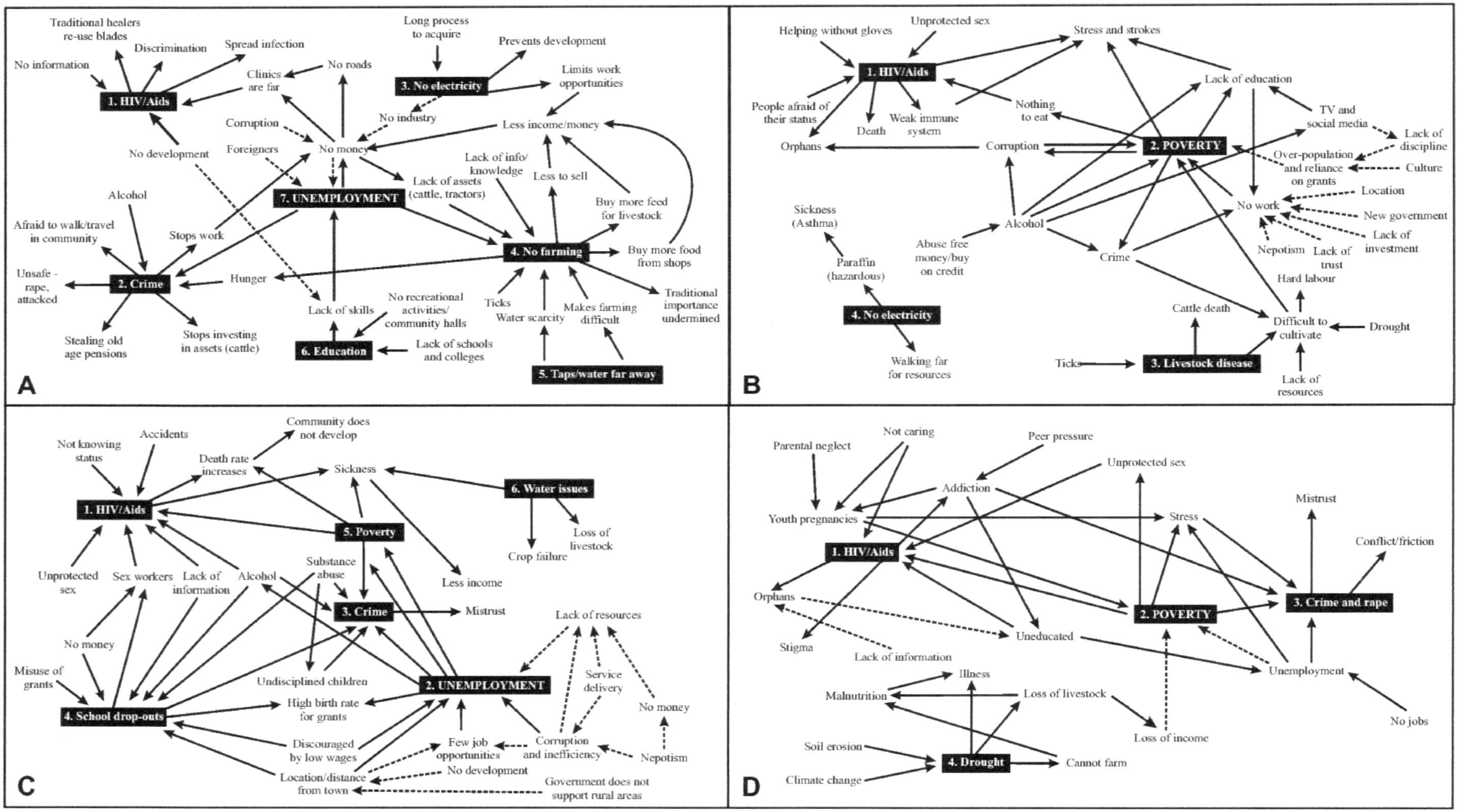

Figure 3. Participatory mental modeling spider-diagrams of perceived stressors in the community and their interconnections; done by (**A**) men in Gatyana, (**B**) women in Gatyana, (**C**) men in Lesseyton and (**D**) women in Lesseyton. Numbered items in the black boxes indicate the order of discussion of the major livelihood challenges/vulnerabilities identified; arrows pointing towards these boxes can be thought of as causes while those pointing away are effects, and dotted arrows indicate relationships revealed from a separate problem tree exercise. (Source: Stadler 2012, unpublished thesis [30]).

Table 5. Past, present and future problems and path dependency as discussed by the youth (Grade 11 learners) in Lesseyton and Gatyana (n = 20).

Lesseyton	Problems	Perceived Causes	Responses
Past	No electricity, not educated, low income	Service delivery, low education level, low income	Look for employment, sell livestock
Present	Not enough water, unemployment, health problems, crime—murder, rape, unhealthy environment (litter, animals), substance abuse, lack of information—distance, teenage pregnancy	Lack of education, theft of taps, dumping in the bush - unclear access rights to dumping site, drugs and alcohol linked to lack of recreation, peer pressure, media, ignorance, unemployment	Rainwater tanks, sell vegetables, sew clothes and other self-employment, prostitution, community punish criminals, contraception, use school library, people clean up area
Future	The same problems as present, but worse	Youth abusing drugs and alcohol, setting bad example for peers, leaders are corrupt and selfish	Knowledge sharing (guest speakers, evening classes, community library, adults should start this), leaders should be more educated so they take issues seriously, recreational activities, more police and police stations, more community skills-based programs, environmental regulations, elder supervision for youth, employ more nurses, street committees
Gatyana	**Problems**	**Perceived Causes**	**Responses**
Past	No education, transport, forced marriage, no rights	Children working, doing domestic duties, no money for transport, wanting money for marriages, Apartheid	Electricity
Present	Inadequate school and recreational facilities, teachers, no electricity, service delivery and roads, walk far for water, livestock theft, violence and crime, farming less, people are judgmental, lack of information, e.g., for bursaries	Corruption in school, no money, corruption in municipality, no water tanks and dams, poverty leads to theft, substance abuse, lack of education	Stronger and more policing No bail for arrests—harsher consequences More trained/qualified teachers Government should provide agricultural support Government should regulate child grants
Future	Violence and crime, no farming, no proper schools, teenage pregnancy increase	Government failing to deliver, lack of discipline and democracy, no consequences for crime	

Source: Clarke 2012, unpublished thesis [26].

The lingering effects of the entrenched policies of apartheid will thus continue to form a barrier to transformation in the previous Bantustans for a long time (see Figure 3), and new policies, for example, that aim to put traditional authority structures back into power have the potential to reinforce the "separateness" of these areas, disempower some groups of people (e.g., women) and heighten social differentiation [65]. Furthermore, education is one of the most important assets to extract households out of poverty [66] but seems to be in a trap of its own in the rural Eastern Cape. Ndlovu's [31] work showed that with education, social grants can be used more effectively to build household resilience. These continued effects of the former apartheid system are supported by Adato *et al.* ([62]; p. 227) who claim that "The legacy of apartheid is an economy in which social exclusion and poverty continue to interact in a mutually self-sustaining fashion" and is likely to persist well into the future. This characteristic, referred to as path dependence [38], we believe it is one of the fundamental aspects to be addressed to avoid traps and support transformation in the rural Eastern Cape. Indeed, if people in our study sites could move beyond living hand to mouth, it is possible that some would invest in revitalizing cultivation, as our research has shown that it is often the more wealthy (a minority) who are continuing to farm [14].

5.2.2. HIV/AIDS: Long-Term, Insidious and Asset-Eroding Impacts

In the rural areas in which we were working about half of households have been impacted in some way by HIV/AIDS (Table 6). This prevalence of HIV is supported by qualitative data, where many informants mentioned the loss of family members to HIV/AIDS and the increasing incidence in the community (see Table 7). With South Africa considered as having one of the highest HIV/AIDS prevalence rates in the world, this disease has had major negative impacts on households and communities across the country; a situation that was further exacerbated by the government's lack of action and slow roll out of antiretroviral drugs. HIV/AIDS is a long-term stressor on livelihoods and not only a shock as frequently framed; its impacts are often gradual, incremental and uneven within communities and regions. Impacted households have higher and more specific dietary needs, and face the dual pressure of less productivity and income, coupled with increasing health-related expenses, often resulting in a downward spiral of increasing stress and food insecurity and the erosion of assets as these are sold to pay for healthcare (Table 7). At the household and community level, the deaths of adults equate to a loss of labor and skills, an upset of knowledge transfer systems, loss of professionals and a rise in orphans. Home-based care also diverts household labor from economically or agriculturally productive activities [67].

The effects of HIV/AIDS tend to be longer lasting than other causes of mortality for two reasons. Firstly, because so many families are affected, traditional coping strategies (such as support from relatives) have been eroded [22,68], which can affect responses to other shocks. Secondly, with the significantly higher death rates of prime-aged adults the number of orphans has grown [22]. These orphans are most often either taken in by relatives, usually grandparents, adding to the strain on household resources, or they form their own child-headed households, where they frequently experience a lack of supervision, stunted growth and poor socialization [69]. Many of these orphans have dropped out of school, and so the effects will last for the rest of their lifetime, if not generations. In the absence of adult caregivers (a major concern in the communities where we worked; Figure 3), these children are particularly vulnerable to exploitation, abuse and HIV infection themselves. Many authors have made reference to the "vicious circle" of HIV/AIDS and poverty, which can operate in mutually re-enforcing cycles, as HIV/AIDS depletes multiple forms of capital, whilst a lack of capital diminishes capacity to recover from other shocks and increases vulnerability to HIV/AIDS [20,67,70]. Therefore, HIV/AIDS in itself can create a trap for some households, possibly putting them in the position where the income they obtain from social grants merely sustains them. HIV/AIDS can thus impact the transfer of knowledge on land-based livelihood strategies to the youth, erode culture and traditional ways of life (such as farming) and impact labor availability for farming activities.

Table 6. HIV/AIDS experiences and presence of proxy indicators in households by site*.

Type of Impact	Lesseyton (n = 170)	Gatyana (n = 170)
Non affected %	42.4	34.7
Chronic illness and receiving free care %	45.9	47.6
Illness-related death in previous 10 years %	18.2	22.9
Presence of *de facto* orphans %	24.1	17.6

* Percentages do not total 100 as households may have multiple experiences. Source: Stadler 2012, unpublished thesis [30].

Table 7. Narratives from the life histories (n = 20) suggesting the long-term impacts of HIV/AIDS, crime, and the erosion of social capital, but also how family still tries to assist in times of stress.

Contextual Factor	Representative Sample Quotes from Life Histories
HIV/AIDS	"In the past, the people were healthy and fit and working. Funerals were not an everyday thing as it is the case now. Every weekend there is a funeral, it has become the norm. HIV/AIDS is rife. Some disclose their status and others do not." "The death rate due to HIV is very high." "Sickness is worse now. There was no HIV and no asthma." "Health is not good at all. People could reach 80 years and above; but now we die at age 40. We are a cursed generation, we think."
Crime	"We got a phone call on a Saturday at about 12 midnight to say Grandma was being attacked by thugs. They stabbed her 4 times. Grandma stayed with a 14 year old girl. They ran away with the girl. We searched for the girl and when we arrived we found the girl dead; they had raped her and broke her neck. We are disappointed in the police in the way the case was handled. In fact the case was dismissed. Justice had failed us." "Crime is rife. I cannot leave my chickens in the fowl run. When it is bed time, I fetch the chickens and they sleep with us in the house." "It was a quiet place when I first arrived here but now things have changed. There was no crime here and no murders."
Social capital and networks	"People in the community are unreliable. You cannot count on them." "It is not easy; very few people are willing to assist. If someone assists you with anything, they expect you to pay back." "I seek help from my sisters. The community does not help. They gossip about other peoples' problems so I do not bother asking them for help." "I borrow money from my relatives; they do give when in a position to."

Source: Clarke 2012, unpublished thesis [26].

5.2.3. Changing Perspectives and Values

Interviews with different actors in the two communities, including the youth, the elderly and mixed groups, suggested that the rural areas of the Eastern Cape are undergoing a process of social and cultural change as new values, goals and ambitions take root. As the above sections suggest, the youth do not appear to have a strong interest in systems of local production and prefer the idea of working in formal employment (Table 5). This has also been observed by other authors with respect to food gardening in other areas of the Eastern Cape [50,71]. Moreover, adult members of the communities felt that the younger generation has lost respect for traditional ways of life and the authority of their elders and is flaunting discipline, family and community values, rules and norms (Table 7) [26,30]. One of the biggest issues that emerged in our participatory social learning meetings in both sites was concern about the future of the youth (Figure 3). A lack of parental care and discipline were mentioned repeatedly in both sites as major factors contributing to community vulnerability. This can be partially linked to the HIV/AIDS epidemic as described above. At the same time, interviews with the youth suggested that they feel misunderstood and neglected. All groups, however, commented on the serious problems of substance abuse, truancy, crime and teenage pregnancy amongst the younger generation and saw this as a barrier to development, farming and lower dependence on grants (Table 5).

These trends, together with the general perception amongst the youth of a hopeless future, their own general lack of empowerment and few adult role models, are in line with findings of youth in marginalized communities elsewhere in the country [72], and emphasize the need to incorporate youth in community development and adaptation planning. Reflecting the modernization that is happening, discussions with youth groups in both sites indicated a desire for a strong emphasis on education, schooling and knowledge (Table 5), whereas these themes were largely absent from discussions with elderly groups. Knowledge and agency are important to counter-act the negative behaviors arising from not being able to imagine a bright, hopeful future [72].

In addition to changes in how the youth see their lives and futures, and the intergenerational tensions this causes, Bank [7] asserts that other tensions in household dynamics relate to the increasing influence and decision-making power of women, encouraged partly through state projects designed to support women's development through (off-farm) commodity groups. He contends that households have "crumbled under the pressure of change," and that this social fragmentation has serious consequences for any efforts to re-institute household-based activities such as family farming.

5.2.4. Urbanization and Shifts in Demographics

Migration has always been an important process shaping rural demographics in the Eastern Cape, but more recently much of this migration of both men and women has become permanent in that migrants rarely return to their rural homes on either a part-time or full-time basis [7]. Population statistics from the province show highly skewed, concave population pyramids for those municipalities that consist mainly of rural, communal areas. In these areas, the proportion of young and middle-aged adults who make-up the economically active sector of the population (the peak age for out-migration is between 25 and 39 years) is substantively less than in urban areas [7,64,73]. The Eastern Cape has the highest rate of permanent net out-migration of any province in South Africa, with most of these migrants going to the Western Cape, followed by Gauteng and Kwazulu-Natal. Some 73 percent of black Africans living in the Western Cape were born in the Eastern Cape [74]. As described above, the permanence of this migration means that there is little spare cash to send remittances back to the rural family [7]. The dominance of children, looked after by the elderly, in the rural Eastern Cape population and in our study sites is another factor affecting the ability of households to use their land productivity. Lack of labor as pensioners become less mobile was a common reason provided for withdrawal from arable farming in our study sites, particularly in the Gatyana site. Over the four years of our study, we observed several elderly households give up farming their fields due to their inability to continue with the physical work. The poor socio-economic conditions in the province described above is reflected in this pattern of outward migration.

5.2.5. Crime and Other Maladaptive Behavior

Crime was an issue that surfaced repeatedly in our interviews and workshops (Table 7, Figure 3). This is unsurprising as South Africa's incidences of criminal activity are one of the highest in the world [57]. Amongst men, women and youth in the two study sites, crime was seen as a major factor contributing to vulnerability, preventing productive activity including farming and small business establishment, and leading to mistrust and conflict in the community (Table 5, Figure 3, Table 7). The Lesseyton youth mentioned how the theft of communal taps was affecting water security, and women in Gatyana cited how it has become unsafe for them to walk far distances to collect natural resources or to cultivate fields. Women and girls, in particular, are vulnerable to rape and violence.

Crime was linked to a host of problems, including: unemployment, poverty, a failing education system and school drop-outs, lack of parental care and discipline amongst the youth, and alcohol abuse, amongst others. The absence of a parent/parents within a household can impact emotionally on a child, and may increase the likelihood of children (primarily boys) becoming involved in crime [75]. In both sites, crime was described as an inhibiting factor preventing people from investing in assets, notably cattle, fences and other farming equipment for fear that these would be stolen. Theft was also shown to be one of the main constraints to farming in a study by Perret [76]. In addition, the youth in both sites associated crime with the lack of recreational activities, peer pressure and social media. The youth believed that there were no real consequences for crimes committed and therefore people had no fear of repercussions. Crime was often described in terms of cyclic knock-on effects: crime leading to limited livelihood options, in turn driving people to substance abuse and other desperate crimes. The effects crime has on a community's social capital and livelihood options, not to mention the health and well-being of the individual members affected, quite clearly indicates how crime can be viewed as a persistent barrier to action and transformation and a contributor to the trap people are in. Bhorat and Kanbur [57] argue that there is a fair degree of policy inertia in terms of dealing with crime, and this is evident in some of the narratives obtained from the life histories (Table 7). Corruption (outside and within the community) and nepotism were also mentioned as important factors affecting access to services, jobs and assets, as well as impacting on community development. Crime and corruption thus form major barriers to agriculture, innovation, motivation and livelihood improvement (also see Table 5).

Alcohol abuse was also mentioned in several discussions as a key factor that kept people in poverty (Table 5, Figure 3). About a quarter of all households, regardless of income group, were purchasing alcohol every month (Table 8), although this was much lower in female-only households. It has been estimated that about one in four men and one in 10 women in South Africa experience symptoms of alcohol-related problems [77]. Passing time in the taverns was also mentioned as a serious problem amongst young men that prevented them from getting involved in productive and healthy activities. Alcohol is recognized as a driver of risk-taking sexual behavior in South Africa, with HIV intervention strategies that focus on responsible drinking showing positive results in reducing the risk of infection [78]. Alcohol abuse also increases a person's vulnerability if they experience social marginalization and discrimination as a result of their addiction [79].

Table 8. Differences in percentages of households purchasing alcohol.

		Lesseyton (n = 168)	Gatyana (n = 170)
Gender of adults in the household	Male only	39.1	32.4
	Adult male head with adult female/s	26.7	34.1
	Adult female head with adult male/s	19.2	25.6
	Female only	4.0	4.2
	P Value (Pearson Chi-Square)	0.008	0.002
Income category	Lowest income	19.5	20.9
	Low income	33.3	17.4
	Moderate income	25.0	28.2
	High income	20.5	26.8
	P Value (Pearson Chi-Square)	0.458	0.604

Source: Stadler 2012, unpublished thesis [30].

5.2.6. Erosion of Social Capital

Seeking assistance from family and friends was one of the most common responses to shocks revealed in our sites [26]. However, there was also evidence that drawing on others for assistance is not as easy as it was in the past (Table 7). Adato *et al.* [62] explain that the erosion of social capital amongst black South Africans started with the social exclusion policies of the apartheid government, which barred people from opportunities to earn income, access to assets, decision-making, social services and community and family support. However, this loss of social capital seems to have accelerated in recent years. In both our study sites, community members in various discussions (Figure 3) mentioned how crime and corruption were undermining community trust and cohesion, and how HIV/AIDS was resulting in a strain on social networks. Generally, family members may assist in preventing destitution, but usually cannot do much more because they also lack resources (Table 8) [62]. A general decline in social capital was identified in our study sites [30], although female-headed households seemed to have more social capital than those headed by men [32].

This situation is not unique to the Eastern Cape. Hebinck and Bourdillon [80] also observed a decline in kinship ties in rural areas of Zimbabwe, while Adato *et al.* [62] found that one third of the households they had surveyed in KwaZulu-Natal reported problems of conflict and distrust between families. Linked to the phenomenon of declining cultivation is the disappearance of communal labor-sharing work parties and after feasts, which were essential to plow and plant large areas [14,28,48]. Bank [7] and Dubbeld [49] also write about the fragmentation and loss of social capital within rural households as a consequence of several of the factors discussed above, which in turn affect collective family activities such as farming.

6. Conclusions: Traps, Transformations and Possible Future Trajectories

There are multiple factors impeding or making it difficult for people in the rural communal areas of the Eastern Cape, in particular female-headed households, HIV-impacted households, and those with low levels of education, to escape their current situation of poverty and vulnerability and to make choices regarding their livelihoods. The various stressors and changes identified, combined with the highly polarized legacy of the country, effectively hamper local people's ability to take action to better their situation and respond to new or future shocks and stresses. It thus could be argued that the majority of households are in a poverty trap, affected to some extent by a locked-in trap (with path dependency) at a higher level. In particular, a key consequence of the changes is that poor rural households in the Eastern Cape are becoming increasingly dependent on cash income, more specifically social grants (which in themselves are not secure), for their livelihoods and food security. Moreover, residents of these areas see these changes as placing a strain on their livelihoods, with consequences for savings and coping with future shocks. The decline in cultivation and the high dependence on purchased food can put poor people at the mercy of markets; something to be viewed with concern in the context of growing food demands and the projected effects of climate change on food prices [81,82]. Maintenance of small-scale farming systems, or at least some form of food self-reliance, is therefore seen by many to be a priority for food security, health, livelihood resilience, and adaptation to future climate change [52,83]. Given the multiple drivers of the decline in cultivation in the Eastern Cape, and the barriers to reviving it, as identified in this paper, new approaches that cater to different groups with different aspirations will be needed. These approaches could range from boosting household food security to small-scale farming, but not before the fundamental issue of land rights is addressed. As pointed out by Kepe and Tessaro [5], without transformation in communal area tenure systems, it is unlikely that much progress will be made with efforts to reinvigorate agricultural production. At the same time, it will be necessary to consider the potential impacts of changing weather patterns and future climate trends on production and the consequences of the increasing woody invasion of the landscape. Participatory scenario development could be used as a tool to explore different pathways to multiple futures [24].

The notion of people in the rural Eastern Cape being "stuck" is supported by work by Adato *et al.* [62], who argue that large numbers of South Africans are trapped in poverty without a pathway to upward mobility. Similarly, Bank and Minkely [59] talk about people in the rural Eastern Cape being caught-up in a process of "involution"; a situation in which, "Economic marginalization and disconnection inhibit social change because they lock people into a system of shared poverty, where the poor reinvent communitarian traditions of mutual dependence and risk-spreading in an effort to protect themselves from the effects of deepening poverty" ([59]; p. 26). They further go on to identify the rural Eastern Cape as a "case of the entrenchment of a livelihood system which has failed to either stabilize or transform itself into a new pattern under changing conditions." Such arguments are reiterated in the work of Cundill and Fabricius [84]. They show how the Eastern Cape's history, together with more recent experiences of poverty and mobility, has resulted in a situation characterized by high levels of vulnerability, low levels of institutional capacity and the reduced production of, particularly, provisioning ecosystem services through agriculture. They contend that these conditions represent a trap that will require a great deal of energy to escape, and that transformability must be at the focus of discussions about future trajectories in the region.

The complexity of livelihood change and the multiplicity of factors interacting at an immediate level in the study communities—some common and others uncommon, some emerging from within the community and others arising from wider spheres of influence—together with the mixture of economic and social drivers, highlight the importance of understanding local perspectives of change and stressors to plan for effective transformation. Like Levine *et al.* [85], we recognize that "There is no neat distinction between risk, shock, stressors, vulnerability, barriers and responses (whether coping, adaptive and transformative), because in the real world they can be one and the same" ([85]; p. 2). As our results have shown, a stressor or change can become a barrier, just as can a maladaptive response or an unintended transformation (e.g., decline in agriculture). Moreover, research from our study sites shows that escaping traps that result from multiple stressors and change are likely to require multiple transformations. Trying to address the complexity of these traps with simple, single transformations may perpetuate, or even exacerbate, these. For example, Ndlovu [27] indicates that the potentially transformative policy of initiating social grants for large segments of the poor population has led to reduced labor supply in productive activities including farming. However, results also suggest that this negative effect could be overcome with improved education. Conditions at our study site thus imply, in addition to clearer land rights, addressing problems related to health and HIV/AIDS, crime and other social welfare issues may also be pre-requisites for transformative change, as will addressing basic structural issues related to continued neglect. Though the focus of our work in understanding multiple stressors is a fundamental step in better comprehending, and reacting to, poverty and vulnerability in the Eastern Cape, more work is needed on assessing the impacts of multiple responses, such as social grants, education, land tenure reform, health programs, and crime prevention. Moreover, the dynamic context of vulnerability and transformation is an area that livelihood investigations, such as ours, rarely have high quality, longitudinal data to address. For example, in our work, most of our data were collected within three years, requiring our interpretations to be supplemented with secondary findings from other researchers that may not be directly comparable. Collecting panel data over decades would greatly help in developing a more nuanced understanding of the time dimension to multiple stressors and responses. This was highlighted in Chapter 13 of the 2014/15 IPPC report, where it is stated that the majority of studies "Do not focus on continuous struggles and trajectories but only offer snapshots. An explicit analysis of livelihood dynamics would reveal more clearly how people respond to series of climatic stressors and shocks over time" ([86]; p. 818).

Within this complexity of traps and changes in our study sites, we argue that, while some of the changes observed could be thought of as unplanned transformations, their future trajectories remain unclear and their desirability ambiguous given the uncertainty of the future. Within this context, introducing specific, as opposed to enabling, transformations could become risky. For example, despite the decline of cultivation, and the problems associated with this trend, we believe, as mentioned above, that to introduce policies that seek to re-establish agrarian livelihoods in these rural areas will need careful consideration and a suite of approaches so as to prevent these being counterproductive. Further, we do not know whether the resources that households would have to divert to agriculture would be worth the increased food security and reductions in cash expenditures on food that such changes may bring. Instead, if an enabling environment can be created, where people are educated, healthy, live in communities with minimal crime, and have tenure security and the rights to self-determination, then they will be in a better position to consider pros and cons of alternative livelihood options themselves, including reinvigoration of agricultural production as a livelihood diversification activity. Such an approach also helps to avoid the normative nature of specific pathways and aligns with Leach *et al.*'s [87] notion of multiple pathways to sustainability, where promoting numerous solutions through diversity helps to recognize the values, needs and interests of different people and places. Such an approach also keeps options open and allows people to change.

However, creating an amenable environment for making choices is not easy, and we agree with Bank and Minkley ([59]; p. 21) that to transform these areas requires:

> "...more detailed analysis and understanding of the social, political and economic forces that have deepened and entrenched poverty in rural areas over the past decade. We need to know how existing inequalities are being maintained and reproduced. It is not enough to assume that the removal of 'obstacles' and the introduction of localized incentive packages will be enough to re-orientate an entire region with a long and deep history of poverty and under-development."

Indeed, deep political commitment to structural and discursive transformations that take into account aspects of distributional justice, that recognize the shifts in rural livelihoods and household dynamics, and that build on local visions, strengths and agency are needed. In the words of Matiwane and Chimera-Dan ([64]; p. 16), this entails "extraordinary measures by the government and all development partners ...... in order to break the shackles of structural poverty and its consequences among the population of the province."

Acknowledgments

We are grateful to the International Development Research Centre (IDRC) Ecohealth Programme and the National Research Foundation (NRF), South Africa for funding. We acknowledge all members of the project team whose work contributed to this synthesis paper. In particular, we recognize the essential contributions of the following postgraduate students: Caryn Clarke, Leigh Stadler and Ross Shackleton from Rhodes University, and Patrick Ndlovu and Amy Trefry from University of Alberta (their theses are available online at their respective universities). Georgina Cundill, as the postdoctoral leader of the social learning process, made a substantive intellectual contribution to the project and the synthesis papers derived from it, while Monde Ntshudu expertly facilitated our close relationship with the communities of Lesseyton and Gatyana, whom we also thank most sincerely for their time and cooperation. Paul Hebinck was a collaborator for the follow-up work with field farmers, and we are working on a publication based on this research.

Author Contributions

Sheona Shackleton and Marty Luckert were co-leaders of the project and conceived and wrote this synthesis paper drawing on the work of other team members.

Conflicts of Interest

The authors declare no conflict of interest.

References

1. Faber, D.F.; McCarthy, D. Neo-liberalism, globalization and the struggle for ecological democracy: Linking sustainability and environmental justice. In *Just Sustainabilities: Development in an Unequal World*; Agyeman, J., Ed.; MIT Press: London, UK, 2003; pp. 38–63.

2. Schröter, D. Our vulnerability to changes in ecosystem services. In *Assessing Vulnerability to Global Environmental Change: Making Research Useful for Adaptation and Decision Making and Policy*; Patt, G., Schröter, D., Klein, R.J.T., De la Vega-Leinert, A.C., Eds.; Earthscan: London, UK, 2009; pp. 97–114.
3. Rockström, J.; Steffen, W.; Noone, K.; Persson, Å.; Chapin, F.; Lambin, E.; Lenton, T.; Scheffer, M.; Folke, C.; Schellnhuber, H.; *et al.* A safe operating space for humanity. *Nature* **2009**, *461*, 472–475.
4. Fraser, E.D.G.; Dougill, A.J.; Hubacek, K.; Quinn, C.H.; Sendzimir, J.; Termansen, M. Assessing vulnerability to climate change in dryland livelihood systems: Conceptual challenges and interdisciplinary solutions. *Ecol. Soc.* **2011**, *16*, 3.
5. Kepe, T.; Tessaro, D. Trading off: Rural food security and land rights in South Africa. *Land Use Policy* **2014**, *36*, 267–274.
6. Andrew, M.; Fox, R. "Undercultivation" and intensification in the Transkei: A case study of historical changes in the use of arable land in Nompa, Shixini. *Dev. S. Afr.* **2004**, *21*, 687–706.
7. Bank, L. On family farms and commodity groups: Rural livelihoods, households and development policy in the Eastern Cape. *Soc. Dyn.* **2005**, *31*, 157–181.
8. Fay, D. The trust is over! We want to plough! Land, livelihoods and reverse resettlement in South Africa's Transkei. Ph.D. Thesis, Boston University, Boston, MA, USA, 2003.
9. Fay, D. Cultivators in action, Siyazondla inaction? Trends and potentials in homestead cultivation. In *In the Shadow of Policy: Everyday Practices in South African Land and Agrarian Reform;* Hebinck, P., Cousins, B., Eds; Wits University Press: Johannesburg, South Africa, 2013; pp. 247–262.
10. MacAllister, P.A. *Building the Homestead: Agriculture, Labour and Beer in South Africa's Transkei*; Research Series/African Studies Centre No. 16; Ashgate: Aldershot, UK, 2001.
11. Hajdu, F. Relying on jobs instead of the environment? Patterns of local securities in the rural Eastern Cape, South Africa. *Soc. Dyn.* **2005**, *31*, 235–260.
12. Hebinck, P.; Lent, P.C. *Livelihoods and Landscapes: The people of Guquka and Koloni and their Resources*; Brill Press: Leiden, The Netherlands; Boston, MA, USA, 2007.
13. Hebinck, P.; Van Averbeke, W. What constitutes "the agrarian" in rural Eastern Cape settlements? In *In the Shadow of Policy: Everyday Practices in South African Land and Agrarian Reform;* Hebinck, P., Cousins, B., Eds.; Wits University Press: Johannesburg, South Africa, 2013; pp. 189–204.
14. Shackleton, R.; Shackleton, C.; Shackleton, S.; Gambiza, J. Deagrarianization and forest succession in abandoned fields in a biodiversity hotspot on the Wild Coast, South Africa. *PLoS ONE* **2013**, doi:10.1371/journal.pone.0076939.
15. Beinhart, W.; Brown, K. *African Local Knowledge and Livestock Health: Diseases and Treatments in South Africa*; Boydell & Brewer Ltd: Suffolk, UK, 2013.
16. Christensen, J.H.; Hewitson, B.; Busuioc, A.; Chen, A.; Gao, X.; Hold, I.; Jones, R.; Kolli, R.K.; Know, W-T.; Lpaise, R.; *et al. Climate Change 2007: The Physical Science Basis*; Contribution of Working Group 1 to the Fourth Assessment Report to the Intergovernmental Panel on Climate Change; Cambridge University Press: Cambridge, UK; New York, NY, USA, 2007.
17. Shackleton, C.M.; Shackleton, S.E.; Gambiza, J.; Nel, E.; Rowntree, K.; Urquhart, P.; Fabricius, C.; Ainslie, A. *Linking Ecosystem Services and Poverty Alleviation in the Arid and Semi-Arid Lands of Southern Africa*; Nova Publishers: New York, NY, USA, 2010.

18. Reid, P.; Vogel, C. Living and responding to multiple stressors in South Africa—Glimpses from KwaZulu-Natal. *Glob. Environ. Change* **2006**, *16*, 195–206.
19. Loevinsohn, M.; Gillespie, S. *HIV/AIDS, Food Security and Rural Livelihoods: Understanding and Responding*; Discussion Paper No. 157; IFPRI: Washington, DC, USA, 2003. Available online: http://ebrary.ifpri.org/cdm/compoundobject/collection/p15738coll2/id/71905/rec/2 (accessed on 28 June 2015).
20. Piot, P.; Greener, R.; Russell, S. Squaring the circle: AIDS, poverty, and human development. *PLoS Med.* **2007**, *4*, 1571–1575.
21. Marks, S. An epidemic waiting to happen? The spread of HIV and AIDS in South Africa in a social and historical perspective. *Afr. Stud.* **2002**, *61*, 13–26.
22. Misslehorn, A.A. What drives food insecurity in southern Africa? A meta-analysis of household economy studies. *Glob. Environ. Change* **2005**, *15*, 33–43.
23. Shackleton, S.E.; Shackleton, C.M. Linking poverty, HIV/Aids and climate change to human and ecosystem vulnerability in southern Africa: Consequences for livelihoods and sustainable ecosystem management. *Int. J. Sust. Dev. Wld. Ecol.* **2012**, *19*, 275–286.
24. Folke, C.; Chapin, S.F.; Olsson, P. Transformations in ecosystem stewardship. In *Principles of Ecosystem Stewardship: Resilience Based Natural Resource Management in a Changing World*; Chapin, F.S., Kofinas, G.P., Folke, C., Eds.; Springer: Amsterdam, The Netherlands, 2009; pp. 103–125.
25. Mucina, L.; Rutherford, M.C.; Powrie, L.W. *Vegetation Map of South Africa, Lesotho and Swaziland*, 2nd ed.; South African National Biodiversity Institute: Pretoria, South Africa, 2007.
26. Clarke, C. Responses to the Linked Stressors of Climate Change and HIV/AIDS amongst Vulnerable Rural Households in the Eastern Cape, South Africa. Master's Thesis, Rhodes University, Grahamstown, South Africa, 2012.
27. Ndlovu, P. Effects of Social Grants on Labour Supply and Food Security of South African Households: Is There a Disincentive Effect? Master's Thesis, University of Alberta, Edmonton, AB, Canada, 2012.
28. Shackleton, R.T. Deagrarianization and Forest Succession in Abandoned Fields over 65 Years: Gatyana, South Africa. Honors Dissertation, Rhodes University, Grahamstown, South Africa, 2012.
29. Shackleton, S.; Luckert, M.; Cundill, G.; Cobban, L.; Clarke, C.; Shackleton, R.; Ndlovu, P. Transformation and barriers in the context of multiple stressors: Understandings from two rural sites in the Eastern Cape, South Africa. In Proceedings of the Transformation in a Changing Climate, Oslo, Norway,19–21 June 2013; pp. 136–148.
30. Stadler, L.T. Assessing Household Assets to Understand Vulnerability to HIV/Aids and Climate Change in the Eastern Cape, South Africa. Master's Thesis, Rhodes University, Grahamstown, South Africa, 2012.
31. Trefry, A.; Parkins, J.; Cundill, G. Culture and food security: A case study of homestead food production in South Africa. *Food Secur.* **2014**, *6*, 555–565.
32. Shackleton, S.; Cobban, L.; Cundill, G. A gendered perspective of vulnerability to multiple stressors, including climate change, in the rural Eastern Cape, South Africa. *Agenda* **2014**, *28*, 73–89.
33. Bunce, M.; Rosendo, S.; Brown, K. Perceptions of climate change, multiple stressors and livelihoods on marginal African coasts. *Environ. Dev. Sustain.* **2010**, *12*, 407–440.

34. Cundill, G.; Shackleton, S.E.; Sisitka, L.; Ntshudu, M.; Lotz-Sisitka, H.; Kulundu, I.; Hamer, N. *A Social Learning Approach to Building Capacity to Adapt to an Uncertain Future: A Descriptive Guide for Practitioners and Action Researchers*; Rhodes University/IDRC/Ruliv: Grahamstown, South Africa, 2014.
35. Walker, B.; Holling, C.S.; Carpenter, S.R.; Kinzig, A.P. Resilience, adaptability and transformability in social–ecological systems. *Ecol. Soc.* **2004**, *9*, 5. Available online: http://www.ecologyandsociety.org/vol9/iss2/art5/ (accessed on 10 April 2015).
36. Brown, K.; O'Neill, S.; Fabricius, C. Social science understandings of transformation. In *World Social Science Report 2013. Changing Global Environments*; ISSC and UNESCO: Paris, France, 2013; Chapter 9, pp. 100–106.
37. Chapin, F.S., III. Managing ecosystems sustainably: The key role of resilience. In *Principles of Ecosystem Stewardship. Resilience Based Natural Resource Management in a Changing World*; Chapin, F.S., Kofinas, G.P., Folke, C., Eds.; Springer: Amsterdam, The Netherlands, 2009; pp. 29–53.
38. Chapin, F.S., III; Folke, C.; Kofinas, G.P. A framework for understanding change. In *Principles of Ecosystem Stewardship. Resilience Based Natural Resource Management in a Changing World*; Chapin, F.S., Kofinas, G.P., Folke, C., Eds.; Springer: Amsterdam, The Netherlands, 2009; pp. 3–28.
39. Kates, R.W.; Travis, W.R.; Wilbanks, T.J. Transformational adaptation when incremental adaptations to climate change are insufficient. *Proc. Natl. Acad. Sci. USA* **2012**, *109*, 7156–7161.
40. Marshall, N.A.; Park, S.E.; Adger, W.N.; Brown, K.; Howden, S.M. Transformational capacity and the influence of place and identity. *Environ. Res. Lett.* **2012**, doi:10.1088/1748-9326/7/3/034022.
41. Stafford Smith, M.; Horrocks, L.; Harvey, A.; Hamilton, C. Rethinking adaptation for a 4°C world. *Phil. Trans. R. Soc. A* **2011**, *369*, 196–216.
42. Carpenter, S.R.; Brock, W.A. Adaptive capacity and traps. *Ecol. Soc.* **2008**, *13*, 40. Available online: http://www.ecologyandsociety.org/vol13/iss2/art40/ (accessed on 29 September 2015).
43. Walker, B.; Salt, D. *Resilience Practice. Building Capacity to Absorb Disturbance and Maintain Function;* Island Press: Washington, DC, USA, 2012.
44. Kofinas, G.P.; Chapin, S. Sustaining livelihoods and human well-being during social–ecological change. In *Principles of Ecosystem Stewardship. Resilience Based Natural Resource Management in a Changing World*; Chapin, F.S., Kofinas, G.P., Folke, C., Eds.; Springer: Amsterdam, The Netherlands, 2009; pp. 55–75.
45. Page, S.E. Path dependence. *Quart. J. Pol. Sci.* **2006**, *1*, 87–115.
46. Greener, I. The potential of path dependence in political studies. *Politics* **2005**, *25*, 62–72.
47. Hebinck, P.; van Averbeke, W. Rural transformation in the Eastern Cape. In *Livelihoods and Landscapes: The People of Guquka and Koloni and their Resources*; Hebinck, P., Lent, P.C., Eds.; Brill Press: Leiden, The Netherlands; Boston, MA, USA, 2007; pp. 33–66.
48. Schmidhuber, J.; Tubiello, F.N. Global food security under climate change. *Proc. Natl. Acad. Sci. USA* **2007**, *104*, 19703–19708.
49. De Klerk, H. The Mutual Embodiment of Landscape and Livelihoods: An Environmental History of Nqabara. Master's Thesis, Rhodes University, Grahamstown, South Africa, 2007.

50. Dubbeld, B. How social security becomes social insecurity: Unsettled households, crisis talk and the value of grants in a KwaZulu-Natal village. *Acta Juridica: Marriage Land Custom* **2013**, *2013*, 197–217.
51. Aliber, M.; Hart, T. Should subsistence agriculture be supported as a strategy to address rural food insecurity? *Agrekon* **2009**, *48*, 434–458.
52. Timmermans, H.G. Rural Livelihoods at Dwesa/Cwebe: Poverty, Development and Natural Resource Use on the Wild Coast, South Africa. Master's Thesis, Rhodes University, Grahamstown, South Africa, 2004.
53. Hendriks, S. The potential for nutritional benefits from increased agricultural production in rural KwaZulu-Natal. *South Afr. J. Agric. Ext.* **2003**, *32*, 28–44.
54. Hebinck, P. Investigating rural livelihoods and landscapes in Guquka and Koloni: An introduction. In *Livelihoods and Landscapes: the People of Guquka and Koloni and their Resources*; Hebinck, P.; Lent, P.C., Eds.; Brill Press: Leiden, The Netherlands; Boston, MA, USA, 2007; pp. 1–32.
55. Integrated Development Plan: 2012–2017; Mbhashe Local Muncipality: Dutywa, South Africa, 2012.
56. Shackleton, S.E.; Hebinck, P. Rhodes University, Grahamstown, South Africa. Unpublished data from a re-survey of households farming fields in Gatyana, Wild Coast, 2014.
57. Bhorat, H.; Kanbur, R. *Poverty and Well-Being in Post-Apartheid South Africa: An Overview of Data, Outcomes and Policy*; Working Paper 05/101; Development Policy Research Unit, University of Cape Town: Cape Town, South Africa, 2005.
58. Marais, H. *South Africa Pushed to the limit: The Political Economy of Change*; UCT Press: Cape Town, South Africa, 2011.
59. Bank, L.; Minkley, G. Going nowhere slowly? Land, livelihoods and rural development in the Eastern Cape. *Soc. Dyn.* **2005**, *31*, 1–38.
60. Hagen-Zanker, J.; Morgan, J.; Meth, C. *South Africa's Cash Social Security Grants: Progress in Increasing Coverage*; ODI Working Paper; ODI: London, UK, 2011.
61. Eyal, K.; Woolard, I. Female labor force participation and South Africa's child support grant. In Proceedings of the CSAE 25th Anniversary Conference: Economic Development in Africa, Oxford, UK, 20–22 March 2011. Available online: http://www.csae.ox.ac.uk/conferences/2011-EdiA/papers/467-Eyal.pdf (accessed on 18 June 2015).
62. Adato, M.; Carter, M.R.; May, J. Exploring poverty traps and social exclusion in South Africa using qualitative and quantitative data. *J. Dev. Stud.* **2006**, *42*, 226–247.
63. Özler, B. Not separate, not equal: Poverty and inequality in post-apartheid South Africa. *Econ. Dev. Cult. Change* **2007**, *5*, 487–529.
64. Matiwane, M; Chimere-Dan, D. *The People Matter: the State of the Population in the Eastern Cape. Research and Population Unit, Eastern Cape Department of Social Development*; South African Government: East London, South Africa, 2010.
65. Claassens, A. Law, land and custom 1913–2013: What is at stake? In *Land Divided: Land and South African Society in 2013, in Comparative Perspective*; Proceedings of the Land Divided Conference 2013, Cape Town, South Africa, 24–27 March 2013; p. 17.
66. May, J. Persistent poverty, asset accumulation and shocks in South Africa: Evidence from KwaZulu-Natal. In *Poverty and Policy in Post-Apartheid South Africa*; Bhorat H., Kanbur R., Eds.; HSRC Press: Pretoria, South Africa, 2006.

67. Wiegers, E.; Curry, J.; Alesandra, G.; Hourihan, J. Patterns of vulnerability to AIDS impacts in Zambian households'. *Dev. Change* **2006**, *37*, 1073–1092.

68. Shackleton, S. *Forests as Safety Nets for Mitigating the Impacts of HIV/AIDS in Southern Africa*; Forest Livelihood Briefs, 4th ed.; CIFOR: Bogor, Indonesia, 2006.

69. Barnett, T.; Whiteside, A. *AIDS in the Twenty-first Century: Disease and Globalization*; 2nd ed.; Palgrave MacMillan: New York, NY, USA, 2002.

70. Gillepsie, S.; Haddad, L.; Jackson, R. HIV/AIDS, food and nutrition security: Impacts and actions. In Proceedings of the 28th Session of the ACC/SCN Symposium on Nutrition and HIV/AIDS, Nairobi, Kenya, 3–4 April 2001. Available online: http://www.ifpri.org/themes/HIV/accscn2001en.pdf (accessed on 26 June 2015).

71. Møller, V. Attitudes to food gardening from a generational perspective: A South African case study. *J. Int. Relat.* **2005**, *3*, 63–80.

72. Campbell, C.; MacPhail, C. Peer education, gender and the development of critical consciousness: participatory HIV prevention by South African youth. *Soc. Sci. Med.* **2002**, *55*, 331–345.

73. Hamann, M.; Tuinder, V. *Introducing the Eastern Cape: A Quick Guide to its History, Diversity and Future Challenges*; Stockholm Resilience Centre: Stockholm, Sweden, 2012.

74. Stats S.A. *Census 2011 Municipal Report—Eastern Cape/Statistics South Africa*; Report No.: 03-01-50; Statistics South Africa: Pretoria, South Africa, 2012.

75. Coovadia, H.; Jewkes, R.; Barron, P.; Sanders, D.; McIntyre, D. The health and health system of South Africa: Historical roots of current public health challenges. *The Lancet* **2009**, *374*, 817–834.

76. Perret, S. *Livelihood Strategies in Rural Transkei (Eastern Cape Province): How does Wool Production Fit in?* Working Paper 2002–20; University of Pretoria: Pretoria, South Africa, 2002.

77. Parry, C. South Africa: alcohol today. *Addiction* **2005**, *100*, 426–429.

78. Kalichman, S.; Simbayi, L.; Vermaak, R.; Cain, D.; Smith, G.; Mthebu, J.; Jooste, S. Randomized trial of a community-based alcohol-related HIV risk-reduction intervention for men and women in Cape Town South Africa. *Ann. Behav. Med.* **2008**, *36*, 270–279.

79. Vermeulen, S.J. *Climate Change, Food Security and Small-Scale Producers*; CCAFS InfoBrief; CCAFS: Copenhagen, Denmark, 2014.

80. Khanal, N.R.; Watanabe, T. Abandonment of land and its consequences. *Mountain Res. Dev.* **2006**, *26*, 32–40.

81. Baiphethi, M.N.; Jacobs, P.T. The contribution of subsistence farming to food security in South Africa. *Agrekon* **2009**, *48*, 459–482.

82. London, L. The "dop" system, alcohol abuse and social control amongst farm workers in South Africa: A public health challenge. *Soc. Sci. Med.* **2009**, *48*, 1407–1414.

83. Hebinck, P.; Bourdillon, M.F.C. Analysis of livelihoods. In *Women, Men and Work: Rural Livelihoods in Central-eastern Zimbabwe*; Bourdillon, M.F.C., Hebinck, P., Eds.; Weaver Press: Harare, Zimbabwe, 2001; pp. 1–13.

84. Cundill, G.; Fabricius, C. Adaptive co-management under resource-poor conditions: Lessons from South Africa. In *Exploring Sustainability Science: A Southern African Perspective*; Burns, S., Weaver, A., Eds.; Sun Press: Stellenbosch, South Africa, 2008; pp. 537–568

85. Levine, S.; Pain, A.; Bailey, S.; Fan, L. *The Relevance of Resilience*; HPG Policy Brief 48; ODI: London, UK, 2012.

86. Olsson, L.; Opondo, M.; Tschakert, P; Agrawal, A., Eriksen, S.H.; Ma,S.; Perch, L.N.; Zakieldeen, S.A. Livelihoods and poverty. In *Climate Change 2014: Impacts, Adaptation, and Vulnerability. Part A: Global and Sectoral Aspects. Contribution of Working Group II to the Fifth Assessment Report of the Intergovernmental Panel on Climate Change*; Field, C.B., Barros, V.R., Dokken, D.J., Mach, K.J., Mastrandrea, M.D., Bilir, T.E., Chatterjee, M., Ebi, K.L., Estrada, Y.O., Genova, R.C., *et al.*, Eds.; Cambridge University Press: Cambridge, UK; New York, NY, USA, 2014; pp. 793–832.
87. Leach, M.; Raworth, K.; Rockstrom, J. Between social and planetary boundaries: Navigating pathways in the safe and just space for humanity. In *World Social Science Report. 2013. Changing Global Environments*; ISSC and UNESCO, Eds.; OECD Publishing and UNESCO Publishing: Paris, France, 2013; pp. 84–89.

5

Forest Transition in Madagascar's Highlands: Initial Evidence and Implications

William J. McConnell [1,*], Andrés Viña [2,3], Christian Kull [4] and Clayton Batko [5]

[1] Center for Global Change and Earth Observations, Michigan State University, East Lansing, MI 48823, USA

[2] Center for Systems Integration and Sustainability, Michigan State University, East Lansing, MI 48823, USA; E-Mail: vina@msu.edu

[3] Department of Geography, University of North Carolina, Chapel Hill, NC 27599, USA

[4] Institut de Géographie et Durabilité, Université de Lausanne, 1015 Lausanne, Switzerland; E-Mail: Christian.Kull@unil.ch

[5] Department of Anthropology, Michigan State University, East Lansing, MI 48823, USA; E-Mail: clayton.batko@gmail.com

* Author to whom correspondence should be addressed; E-Mail: mcconn64@msu.edu

Academic Editors: Claudia A. Radel and Jacqueline M. Vadjunec

---

Abstract: Madagascar is renowned for the loss of the forested habitat of lemurs and other species endemic to the island. Less well known is that in the highlands, a region often described as an environmental "basket-case" of fire-degraded, eroded grasslands, woody cover has been increasing for decades. Using information derived from publically available high- and medium-resolution satellites, this study characterizes tree cover dynamics in the highlands of Madagascar over the past two decades. Our results reveal heterogeneous patterns of increased tree cover on smallholder farms and village lands, spurred by a mix of endogenous and exogenous forces. The new trees play important roles in rural livelihoods, providing renewable supplies of firewood, charcoal, timber and other products and services, as well as defensible claims to land tenure in the context of a decline in the use of hillside commons for grazing. This study documents this nascent forest transition through Land Change Science techniques, and provides a prologue to political ecological analysis by setting these changes in their social and environmental context and interrogating the costs and benefits of the shift in rural livelihood strategies.

Keywords: afforestation; forest transition; Landsat; rural livelihoods

1. Introduction

Madagascar is still losing its natural forest cover, despite decades of conservation efforts [1]. Many temperate countries and a few tropical countries have undergone a "forest transition," whereby net deforestation is reversed, but Madagascar as a whole does not yet fit this pattern [2]. However, at the regional scale, the island nation exhibits large pockets where tree cover is expanding. In the highlands, for instance, which a century ago were largely grasslands devoid of tree cover, introduced eucalypts, pines, acacias, and diverse fruit trees have become ubiquitous.

While widespread and seemingly driven by rural people together with state foresters, this phenomenon presents several challenges for researchers and environmental managers. For one, methodologically, it has been difficult to map and quantify these tree cover dynamics. Traditional remote sensing techniques struggle with these sparse land covers, which are also very patchy and distributed across a landscape mosaic (unlike large blocks of closed canopy forest). Second, such tree cover—largely dominated by exotic species—is often ignored by conservationists for its lesser biological value, yet it turns out to be central to livelihoods and rural development, as well as being relevant to soil and water conservation, carbon budgets, invasion biology, and other aspects of environmental management.

In this paper, we seek to determine the degree to which the highlands of Madagascar are experiencing a sub-national forest transition, and to identify the causes and livelihood consequences of changes in tree cover in a region widely regarded as the scene of ecological and economic devastation. The specific objectives of the study were to document the dynamics of low-density woody cover in the highlands region, and to explain these dynamics and their livelihood implications. We use a simple, two-step mixed Land Change Science – Political Ecology approach [3], in which we first characterize the nature of tree cover change using remote sensing, and then subsequently paint a picture of the diverse factors—livelihood strategies, forestry and land tenure policies, struggles for resource control, ideas about proper environmental management, and plant invasiveness—that combine to cause these changes. More specifically, we applied novel remote sensing techniques involving "continuous" measures of tree cover at the pixel level, rather than delimiting strict forest cover categories. This allows us to characterize the extent and dynamics of low-density woody vegetation in the highlands over the past two decades. We then explain these changes, situating them in the context of land use and livelihoods in the highlands, setting the stage for further political ecological investigation. This study seeks to contribute to the rapidly maturing theoretical literature on the forest transition, which is briefly reviewed below, before an exposition of the novel remote sensing methods we bring to bear on the debate over forest change in Madagascar.

1.1. Theoretical Motivations

The realization that large parts of the world had undergone net increases in forest cover during the 20th Century has sparked one of the most important lines of inquiry in human-environment research over the past ten years, owing in part to growing interest in questions of carbon sequestration and other ecosystem services. The process of slowing deforestation and a subsequent shift towards net forestation in Western Europe and eastern North America as those economies shifted from agricultural to industrial bases has been coined the "forest transition" [4,5]. Detection of similar phenomena in East and South Asian countries [6,7] has fueled interest in the possibility that this phenomenon might be reproduced elsewhere—particularly in the tropics where deforestation continues to be a dominant and still growing trend [8]. This has become one of the most promising avenues for theoretical advancement in the emerging science of land change [9–12].

It was initially theorized that forest transitions mark the end of the process of clearing forest land to make way for agriculture. According to this explanation, industrialization leads to the concentration of population in urban centers, and consequent rural depopulation, spurring a spatial contraction of increasingly intensive agricultural production. It also leads to levels of affluence at which forests are increasingly valued for aesthetic purposes [4,5]. Cross-national analyses have provided some support for this explanation, but have also suggested a second—so-called "forest-scarcity"—path to the forest transition in developing countries, *i.e.*, the planting of trees in response to perceived shortage of forest products [13].

The development of forest plantations around town centers has been a central tenet of land-use theory since its inception; von Thünen [14] theorized the development of a zone of tree production in proximity to towns, just outside a more proximate zone dedicated to the production of perishable products. Initial attempts have been made to link the von Thünen model with forest transition theory under idealized conditions [15], but a great need exists for empirical examination of the actual patterns and processes of the suggested "forest scarcity" pathway [13].

As it begins to mature, forest transition theory has attracted critical review on several counts. In terms of establishing the facts of forest cover change, there is concern that the "dependent variable"—spatio-temporal forest cover patterns—requires more systematic treatment. Generally, the procedures used to assess forest dynamics from remote sensing at global and continental scales suffer from lack of standardization, sometimes failing to differentiate natural forest from exotic monoculture plantations [16]. This issue is particularly acute in sub-humid landscapes, where trees are very important elements of agroecosystems even at relatively low densities [17].

In terms of attributing changes in forest cover to particular causes, forest transition theory is cited for lack of attention to scalar dynamics—the recognition that different processes tend to be more strongly associated with land change at different levels of observation [18–21]. Finally, critics caution proponents of forest transition theory against a "universalist approach and a structuralist-functionalist account," calling instead for research emphasizing "historical contingencies, variable causes among cases, and other contextual specificities" ([16]; pp. 109, 111).

Place-based research on tropical forest transitions has so far focused largely on Latin America [5,22–25] and South and Southeast Asia [26–28] (see also [29,30]). Meanwhile, studies of the role of trees in African landscapes have not yet, to our knowledge, been framed within the forest transition debate. However, some of the foundational contributions to the fields of Political Ecology [31,32] and Environmental History [33] set out to challenge the dominant received wisdom of peasant-driven environmental degradation in Africa [34–37] (see also [38]). In one rather famous case, the landscape of Machakos, Kenya, was observed to have undergone significant intensification under conditions of rapid population growth, upending the conventional population-poverty spiral view [39].

African smallholders have traditionally retained beneficial trees within their fields, and are now experimenting with the incorporation of other woody species into their production systems, often in collaboration with government and NGO efforts [40–42]. These outside interventions, along with differential access to land and other resources constitute a crucial set of institutional factors shaping the distribution and consequences of trees in African landscapes [43,44]. Understanding such institutional factors is a key challenge for the development of forest transition theory [45], and the governance of forests and other resources more generally [46,47].

1.2. Forest Change Studies in Madagascar and Advances in Remote Sensing

A vast literature exists documenting the loss of forest in Madagascar, almost exclusively focusing on what is considered "natural" or pristine formations, under the assumption that greater height, greater canopy closure and higher species diversity are indicators of habitat quality for endangered wildlife, such as lemurs. These assumptions may generally hold for the purposes of conserving biological diversity. However, from the standpoint of local livelihoods, woody species play important roles at much lower density and diversity.

Detecting the presence of woody cover at low densities using traditional remote sensing techniques presents methodological challenges and this, along with the conservation motives of most land-cover change studies in places like Madagascar, means that changes involving such land covers are typically not captured. In particular, traditional satellite image processing techniques reflect cartographic technology favoring choropleth maps, in which the landscape is divided into units of internally homogenous land-cover types, with the number of categories limited to ease the readability of the map. In practice, however, it is quite rare for studies to actually specify what they consider to constitute forest cover. For example, among nearly two dozen studies of land change in Madagascar (See Table S01 in Supplementary Materials), we were able to identify only one that provided an explicit definition of forest. Harper and colleagues, in their review of half a century of deforestation and forest fragmentation in Madagascar, defined forest as "areas of primary vegetation dominated by tree cover at least seven meters in height, with neighboring tree crowns touching or overlapping when in full leaf" ([1]; p. 2). The issue of conflicting definitions of forest is treated explicitly by Dufils [48], and McConnell and Kull [49].

The past decade has witnessed rapidly growing interest in the development of techniques for the classification of remotely sensed imagery to obtain continuous depictions of the landscape, along gradients such as "percent tree cover" [50]. Such techniques allow the detection of trees and other woody vegetation at a range of densities, including those with important livelihood implications that are overlooked in most studies. At the same time, the rapid increase in the availability of recent high resolution commercial satellite imagery, through platforms such as Google Earth®, provides an unprecedented ability to collect the kind of information needed to "train" an automated classification of multispectral imagery. Of course, understanding the causes of the changes revealed through these remote sensing and image classification techniques still requires "boots on the ground." The present study brings together decades of such concerted fieldwork with novel remote sensing techniques to document and explain changes in the open woodlands, woodlots and orchards of highland Madagascar.

2. Study Area

Madagascar is classified as a low-income country by the World Bank, with 88 percent of the population subsisting on $1.25 per day, and four-fifths of the population consists of families operating farms averaging roughly 1 ha with manual labor and animal traction. Large, mechanized farms are the exception. The smallholder farms face high transport costs due to poor infrastructure, usually combining subsistence and market-oriented production. Most market production is destined for local or regional (within the island) consumers, though certain regions have found specialized export niches, such as green beans in the central highlands around Antananarivo [51–54].

Like elsewhere in Africa, farmers seek to diversify their livelihoods as a risk-reduction strategy. If they can, households combine crop production, livestock rearing, unskilled wage labor, seasonal or periodic migration (as labor on farms or in mining), small-scale artisanal or trading activities, urban employment, opportunistic resource exploitation, and other activities [53,55]. In this way, households respond to opportunities and constraints, such as the expansion of economic opportunities linked to preferential trade agreements [53], or the contraction of the livestock sector due to increased dangers of cattle rustling [56]. Natural resource extraction helps as a safety net in the case of crises, whether drought, cyclone, market shifts, or illness or death of household labor [57,58]. Woody tree species—whether native, planted, or invasive—often serve such a role, offering rural people either regular income or a kind of backup option to earn supplemental cash in times of difficulty [59,60].

The central highlands (defined by the current administrative regions of Bongolava, Analamanga, Itasy, Vakinankaratra, Amoron'i Mania and Haute Matsiatra; Figure 1) cover an area of approximately 96,000 km^2, contain three of the island's largest cities (including the capital), and support nearly half (*ca.* 7.5 million) of the island's human population. Total annual rainfall decreases westward, with the eastern portion of the highlands hosting tropical rainforests, while the western portion is mostly composed of low productivity rangeland. The human imprint on Madagascar's highlands is characterized by a mix of irrigation infrastructure and extensive cattle grazing, illustrating the blending of African and Asian cultural heritages [61,62]. Farmers cultivate irrigated rice and a number of upland crops, such as manioc, sweet potatoes, beans, and potatoes. All farms keep at least some animals, and the region's grasslands are typically burned nearly every year to provoke late dry season forage for cattle, and to prevent bush encroachment [63]. These fires, combined with the region's ubiquitous erosion gullies, known as *lavaka*, have given the highlands a reputation as a site of environmental degradation [64–66].

3. Methods

The application of forest transition theory in landscapes dominated by smallholder plots requires a multi-level approach that enables the consistent detection of woody cover dynamics across a broad area, combined with highly contextualized explanations of processes occurring in specific landscapes. We therefore begin with a broad scale assessment of land-cover change using innovative techniques for the classification of moderate resolution (*i.e.*, 30m horizontal resolution) satellite imagery. The procedure yields change maps that highlight particularly dynamic areas, from which we select some for closer examination using field observations in combination with ancillary information.

3.1. Remotely Sensed Data, Digital Image Processing and Change Detection

Woodland dynamics in the highlands were evaluated using three multi-spectral scenes (path 159, rows 73 to 75 of the World Reference System-2 (WRS-2)) acquired by the Thematic Mapper (TM) sensor onboard the Landsat-5 satellite system and by the Operational Land Imager (OLI) sensor onboard the Landsat-8 satellite system on September 25, 1994 and September 16, 2014, respectively. The use of anniversary data from the dry season (*i.e.*, September) was preferred because it enabled the use of mostly cloud-free scenes. These scenes account for *ca.* 80 percent of the central highlands (Figure 1).

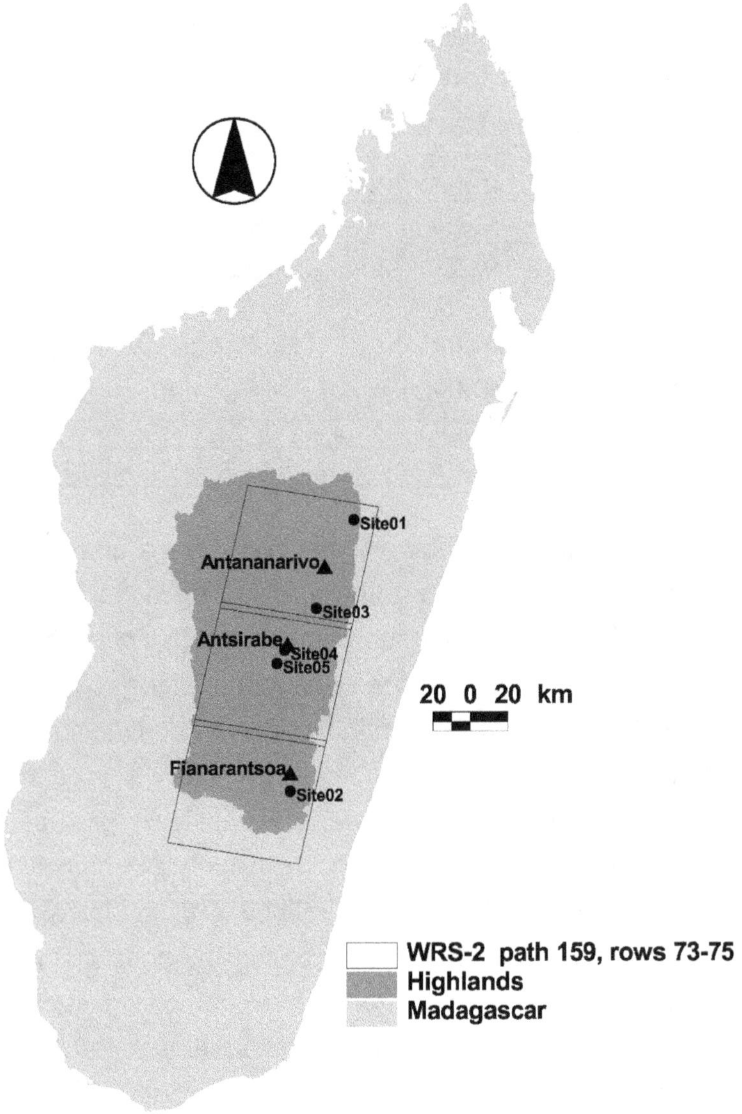

Figure 1. The study region was defined by the area within three Landsat scenes (path 159, rows 73 to 75 of the World Reference System-2 (WRS-2)), comprising *ca.* 80 percent of Madagascar's Highlands. Also shown are the locations of three main cities and five field sites visited between 2003 and 2006 whose land-cover dynamics are explained in more detail in the text.

The three scenes per date were mosaicked, and a relative radiometric normalization was applied to render the radiometric structure of the scenes from the two anniversary dates (*i.e.*, 25 September 1994 and 16 September 2014) comparable. The normalization consisted of the development of linear regression models for each of six optical bands (*i.e.*, bands 1–5 and 7 of the Landsat TM, and bands 2–7 of the Landsat OLI), to predict the brightness values of the 2014 imagery that would have been obtained had they been acquired under the same atmospheric conditions as the 1994 imagery. A total of 40 pseudo-invariant targets (*i.e.*, wet and dry targets present in both scenes assumed to be constant reflectors through time so that any changes in their brightness values could be attributed to sensor calibration, astronomic, atmospheric and phase angle differences) were used to develop the linear regression models. Table 1 shows the coefficients of these models.

Table 1. Coefficients of linear regression models (one for each spectral band) developed using pseudo-invariant targets. These linear models were constructed in order to match the radiometric structure of the six optical bands of the 16 September 2014 Landsat Operational Land Imager (OLI) imagery to that of the September 25, 1994 Thematic Mapper (TM) imagery.

Band	Slope	Intercept	R^2
Blue	0.013	−51.128	0.928
Green	0.007	−28.163	0.910
Red	0.008	−33.207	0.949
Near Infrared	0.006	−20.427	0.971
Shortwave Infrared 1	0.010	−44.685	0.986
Shortwave Infrared 2	0.007	−32.121	0.978

We estimated the probability of an area to exhibit woody cover using a fuzzy classification algorithm based on the principle of maximum entropy [67] using MaxENT, a general purpose machine-learning method for making predictions from incomplete information [68]. To calibrate and validate the maximum entropy classification algorithm, we developed a dataset of 280 "ground truth" polygons of the same area as a Landsat pixel (*ca.* 900 m^2) semi-randomly distributed throughout the study area, in which the interpretation of an initial set of 250 randomly distributed polygons was used to oversample in areas exhibiting some woody cover. Within each of these polygons we randomly distributed 10 points. Using high spatial resolution imagery available in Google Earth®, we visually ascertained the number of points per polygon coinciding with a tree canopy. To assess the reliability of the interpretation of imagery in Google Earth®, two image interpreters independently performed the point counts. The average point count between the two interpreters was obtained, and only those polygons that exhibited a point count difference of 10 percent or less between the two interpreters were used.

Following the classification of forested areas established by the United Nations Food and Agriculture Organization [69], we employed a threshold of 10 percent tree cover as a definition of woodland (encompassing closed canopy forest, plantations, treed savanna, riparian growth and sometimes dense shrubland), and considered a polygon to be at least 10 percent covered by woody vegetation if one or more points per polygon coincided with such cover. Out of the 280 ground truth polygons, 97 (*i.e.*, 34.6 percent) had 10 percent or greater woody cover. Using these data together with the MaxEnt algorithm, we created a woody cover probability (*i.e.*, the probability of having tree cover of 10 percent or higher) map from the multi-spectral bands (*i.e.*, bands 2–7) of the 2014 Landsat imagery. The output probability map was validated using a cross-validation procedure in which 2/3 of the ground truth polygons were used for calibration and the remaining 1/3 used for validation. To reduce dependence on a single random partition into calibration and validation, we generated 10 different random partitions to be used in 10 different classifications, which were then averaged. The 10 output probability maps were validated by means of a receiver operating characteristic (ROC) curve [70]. The ROC curve is a plot of the sensitivity values (*i.e.*, true positive fraction) *vs.* their equivalent 1-specificity values (*i.e.*, false positive fraction) for all possible probability thresholds. The area under the ROC curve (AUC) is a measure of model accuracy, with AUC values ranging from 0 to 1, where a score of 1 indicates perfect classification, a score of 0.5 implies a classification that is not better than random, and values lower than 0.5 imply a worse than random classification. The average AUC value was 0.872, which was significantly ($p < 0.0001$) different from 0.5 and denotes high classification accuracy. The average coefficients of the validated classification of the multi-spectral bands of the 2014 Landsat OLI imagery were then applied to classify the multi-spectral bands (bands 1–5 and 7) of the 1994 Landsat TM imagery.

A simple image differencing technique (in essence, a post-classification comparison [71]), in which the probability of woody cover in 1994 was subtracted from the 2014 probability, was performed on a per-pixel basis yielding a continuous measure of change over 20 years from the 1994 probability, which was used as the baseline. This procedure yields an output map that depicts spatial patterns of more subtle dynamics than is captured by more traditional image "hard" classification approaches in which each pixel is assigned to one of a fixed set of categories. The dynamics revealed by the multi-date Landsat imagery include: no significant change in probability, which we interpret as stability; an increase in probability, which we interpret as gain in woody cover; and a decrease in probability, which we interpret as loss in woody cover. These dynamics were then further explored using the high resolution imagery from Google Earth®, in conjunction with aerial photographs from the early/mid 1990s, along with ground level photos and field visits conducted midway through the study period (*i.e.*, 2003 and 2006).

3.2. Explaining the Observed Woodland Dynamics

Interpretation of the documented land change patterns was based on field observations and previous studies over the past two and a half decades, in conjunction with the published literature. In particular, we relied on two dozen case study sites (selected using a spatially stratified random sample) previously investigated using 1950s and 1990s air photos as well as field visits [72]. Then, in this two-step Land Change Science–Political Ecology study, we move from the more descriptive remote sensing and case study site discussion of tree cover changes to an investigation of the various factors lying behind these changes. Following a political ecological framework, attention was accorded to not only market incentives, demographic pressure, and official policies, but also to access to resources and to the livelihood strategies of rural residents including land tenure arrangements, de-facto rules around trees, and environmental ideologies, as well as the invasive characteristics of some of the trees themselves [55,73,74].

4. Results

4.1. Remote Sensing and Case Study Analysis

The results of the probability map differencing procedure are shown in Figure 2. Increases and decreases in probability between 1994 and 2014 (expressed in the map as gains and losses, respectively) represent increases and decreases in woody cover. Increases in woody cover are indicated in shades of green, and decreases in shades of red, while the initial cover (*i.e.*, baseline probability value in 1994) is given by shades of blue. The combinations of these colors yield gradations among six trends depicting woody cover dynamics between 1994 and 2014 over the study area. What is notable is the wide extent of increasing tree cover, particularly in interior highland areas often treated in broad scale assessments as essentially devoid of trees. Below, we describe in detail the three main categories of change shown in Figure 2 (landscape stability, decreases in woody cover, and increases in woody cover), linking the general patterns to specific dynamics revealed in drill-down analyses of the specific case study sites.

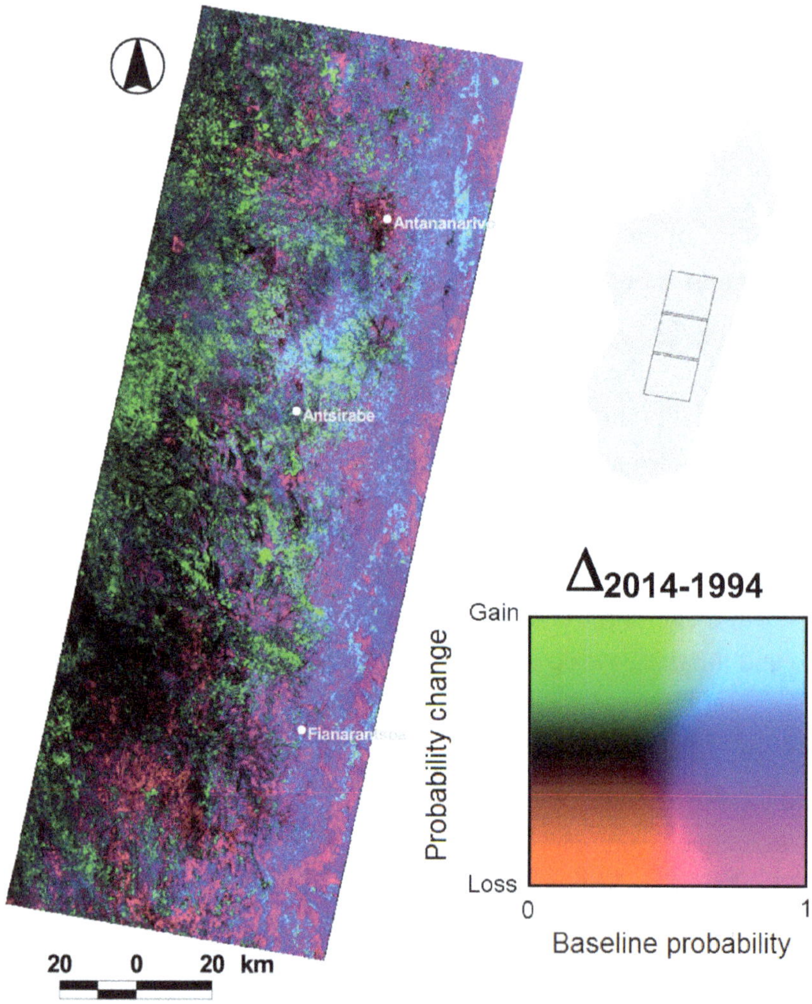

Figure 2. Spatial distribution of the temporal dynamics (*i.e.*, gain, loss, no change) of woody cover in the study area over a 20-year period (*i.e.*, 1994–2014), as a function of the baseline probability of woody cover (*i.e.*, probability of woody cover in 1994). Also shown are the locations of the three major cities (including the capital) in the Highlands of Madagascar.

First, stable landscapes are represented in Figure 2 with black and blue tones, representing herbaceous and forest landscapes, respectively. In the case of more dense (higher probability) woody cover in 1994, this stasis is rendered in dark blue, which one would expect from undisturbed forest. This dynamic is seen in the eastern portion of the study area, which contains the region's remaining dense forests, along the eastern escarpment of the highlands, as well as in some of the larger and longer-term zones of woodlot forests (e.g., east of Antananarivo).

The darkest tones in the map represent areas that had a low probability of woody cover at either date, and these are concentrated in the western, and especially southern, portion of the study area, which are also drier and less populated (Figure 2). These areas are often characterized by lateritic soils with minimal organic layers. Typically, herbaceous vegetation covers the uplands, with woody vegetation restricted to slope hollows and the edges of streams and seasonal drainages.

A ***second*** category of change is decrease in woody cover. A decrease in woody cover is unlikely when the initial probability was low, explaining the paucity of deep red in Figure 2. This dynamic could result, though, from the burning of grasslands containing shrubs and occasional trees, but occurs quite infrequently in our results. Much more common are decreases from a more dense initial cover, shown in magenta. This dynamic is particularly visible along the far eastern edge of the study area, and corresponds to significant degradation of woody cover, as would be expected from clearing of forest for agriculture (Figure 2). In this region, it is likely that some—or even most—of these forests had been subject to some degree of disturbance prior to 1994, but our study period only comprises the dynamics after 1994, which prevents us from distinguishing such prior disturbance. Elsewhere, however, our fieldwork and the literature allow us to identify the conversion of dense forest to other land-cover types. We recognize two distinct processes, one involving cyclical loss of canopy, known as a coppice. This entails removing the stems of tree species capable of re-sprouting from their stump, a property that has made several species of eucalyptus very popular in Madagascar. The large eucalyptus plantations around the capital (Antananarivo), including the Manjakandriana Massif, have been well-described [75]. Such coppicing appears in the change map as large magenta blocks around the black mass of the city in the northeast portion of the study area (Figure 2), interspersed with increases in woody cover (described below).

Smaller, more dispersed stands of eucalyptus have heretofore received less attention in the peer-reviewed literature, but are revealed in our results. A good example is our field site at Ambongabe, north of Anjozorobe (Site 01; Figure 3), where dense woody cover visible in the 1992 aerial photograph was much more open in a 2014 Digital Globe image, resulting in bright magenta tones in the change map. Zooming in on the high resolution imagery reveals this to be a well-organized hilltop plantation with large crowns in 1992, and much smaller crowns in 2014 (Ellipse 1 in Figure 3). Our fieldwork confirms this to be a eucalyptus plantation, which was probably coppiced twice during the intervening 20 years.

Another process explaining decreased woody cover, particularly in plantations, involves more complete harvesting. The study area includes several large government and private plantations subject to regular clearcutting and replanting, much as in other parts of the world. Some of our field sites, however, are characterized by a spectrally similar, yet institutionally different, process. The Landsat change map is dominated by bright magenta tones (representing decreased woody cover) at our Ambozontany field site (Site 02, Figure 4) where portions of the government's Haute Matsiatra pine plantations—shown in the 1991 aerial photos to be closed canopy plantations—had been cleared for upland cultivation when Digital Globe's sensor recorded an image in October of 2014. This process was well under way during a field visit in 2006, which revealed that this conversion represented the unsanctioned actions of communities who had previously been dispossessed of the land for the creation of the state plantations during the first decade following Independence. The village lands in the eastern third of the image have their usual eucalyptus woodlots, but the pine stands were rather threadbare (see Figure S1 in Supplementary Materials).

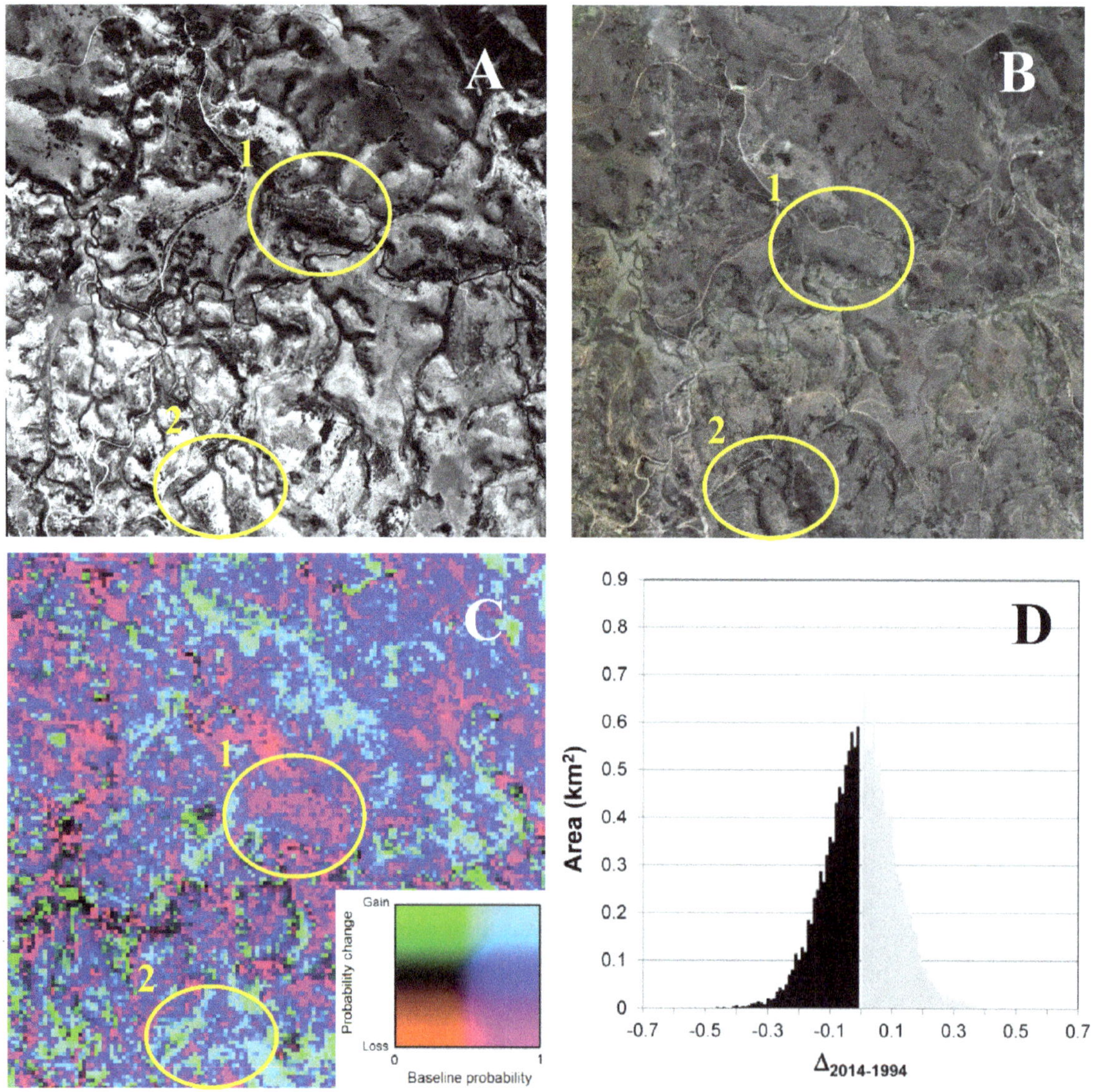

Figure 3. Dynamics of woody cover at Site 01, Ambongabe, near the town of Anjozorobe north of Antananarivo (18°19'S, 47°54'E). (**A**) Aerial photograph taken in 1992. (**B**) Digital Globe image taken in 2014. (**C**) Change map based on Landsat imagery (see Figure 2). (**D**) Histogram depicting the frequency distribution of the delta of woody cover probability values between 2014 and 1994 based on Landsat imagery. Woody cover loss is represented in black, while gain is represented in gray. Ellipses in (A–C) relate to particular woody cover dynamics described in the text. See Figure 1 for the geographic location of Site 01 within the study area.

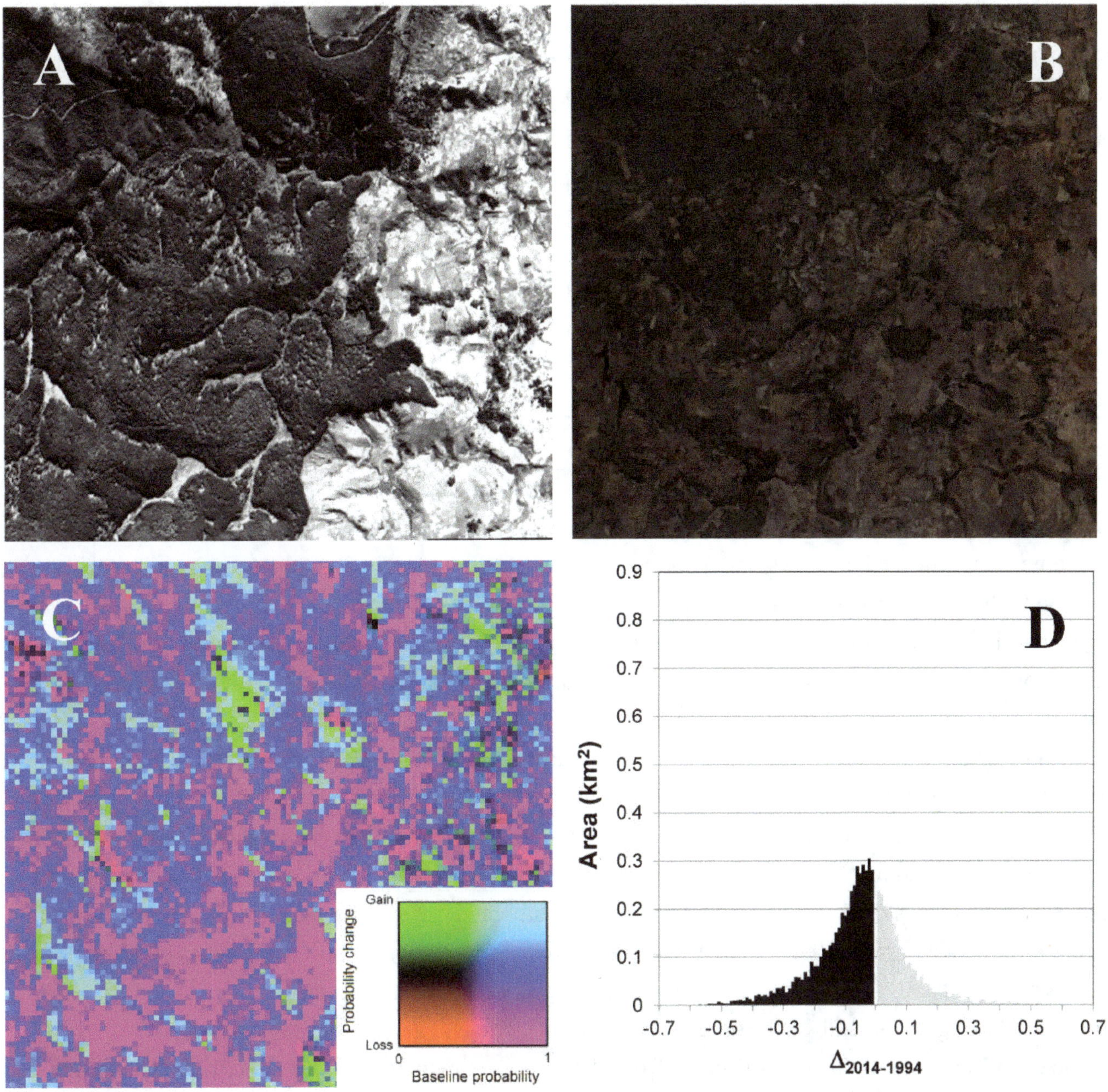

Figure 4. Dynamics of woody cover at Site 02, Ambozontany, south of Fianaratsoa (21°48'S, 47°6'E). (**A**) Aerial photograph taken in 1991. (**B**) Digital Globe image taken in 2014. (**C**) Change map based on Landsat imagery (see Figure 2). (**D**) Histogram depicting the frequency distribution of the delta of woody cover probability values between 2014 and 1994 based on Landsat imagery. Woody cover loss is represented in black, while gain is represented in gray. See Figure 1 for the geographic location of Site 02 within the study area.

The ***third*** and final set of dynamics, representing increases in woody cover, is concentrated in very different parts of the landscape. In the Landsat change map cyan tones represent an increase from an already dense initial cover. These are predominantly found in the eastern area, where secondary growth often occupies land previously cleared for agriculture (Figure 2). They are also particularly visible north of Antsirabe, in a region with a long history of woodlot plantations that continues to intensify. These frequently co-occur with magenta tones, in landscapes where fallow periods are long enough for woody vegetation to reestablish itself following cultivation. This dynamic was also observed, however, farther west, where forests were rare or absent in 1994. For example, in the same field site at Ambongabe described above (Ellipse 2 in Figure 3) as showing evidence of eucalyptus coppices, we see evidence of hilltops covered in brush in 1992 that later showed tree canopies.

Of great interest to the present study are areas appearing in the Landsat change map in bright green tones, representing substantial increases in woody cover over the 20 year period. This can only occur when initial woodland probability was low, and is concentrated in the central and western portions of the study area. Three of our field sites exhibit these dynamics.

Near Ambatolampy (Site 03; Figure 5) well-organized plantations were already evident in the 1991 aerial photographs, and the 2013 CNES Astrium image shows these had expanded significantly, resulting in bright green tones in the Landsat change map (Ellipse 1 in Figure 5). Field work confirms these to be eucalyptus plantations. Farther west at the same site is a dense, well-managed pine forest, with some araucaria, sweet gum, and silver wattle. This is a private 48 ha land holding of a family residing in the capital (Ellipse 2 in Figure 5). A ground-level panorama shot in 2003 (Figure S2 in Supplementary Materials) shows the widespread, low-density eucalyptus plantations, as well as a foreground "scrub" of *Helichrysum* spp. and silver wattle.

Meanwhile Ambohijatovo (Site 04; Figure 6) is a typical intensively cultivated central highland site, marked by eucalyptus woodlots, rice fields, and market vegetables, as shown in a panorama photo taken in 2003 (Figure S3 in Supplementary Materials) from the valley bottom. Comparing the 1991 aerial photograph and the 2013 CNES Astrium image reveals the expansion of a hilltop plantation down to the rice paddies in the valley bottom, once again shown in the Landsat change map in shades of green (Ellipse in Figure 6). Likewise, in the western portion of our Soamahazina site (Site 05; Figure 7), an expansion of tree cover is evident between 1991 and 2013, showing up in the Landsat change map in shades of green. Fieldwork in 2003 documented the development of village woodlots in these cool uplands (1700-2000m) where transport is relatively difficult, but the site is close enough to the town of Antsirabe that charcoal production is economically viable (Ellipse in Figure 7). The haphazard arrangement and sizes of the trees at this site suggest that these woodlots are being expanded in incremental fashion, different from the more regimented rows of even-aged trees seen in other sites.

Finally, it should be noted that there are other "new" tree covers than the spatially dominant eucalyptus and pine woodlots, also responsible for the green tones in the Landsat change map. In particular, Kull [72,73,76] documented a dynamic involving fruit orchards, hedge trees, ornamentals, and other trees in and around settlements. Given their spectral similarities at low stem densities, these cannot easily be distinguished from other woodlots in our remote sensing analysis, but result from different land management processes. Such a landscape element is shown in the panorama photo (Figure S3 in Supplementary Materials) of the Ambohijatovo case study (Site 04), notably on the right hand side (location indicated in Figure 6).

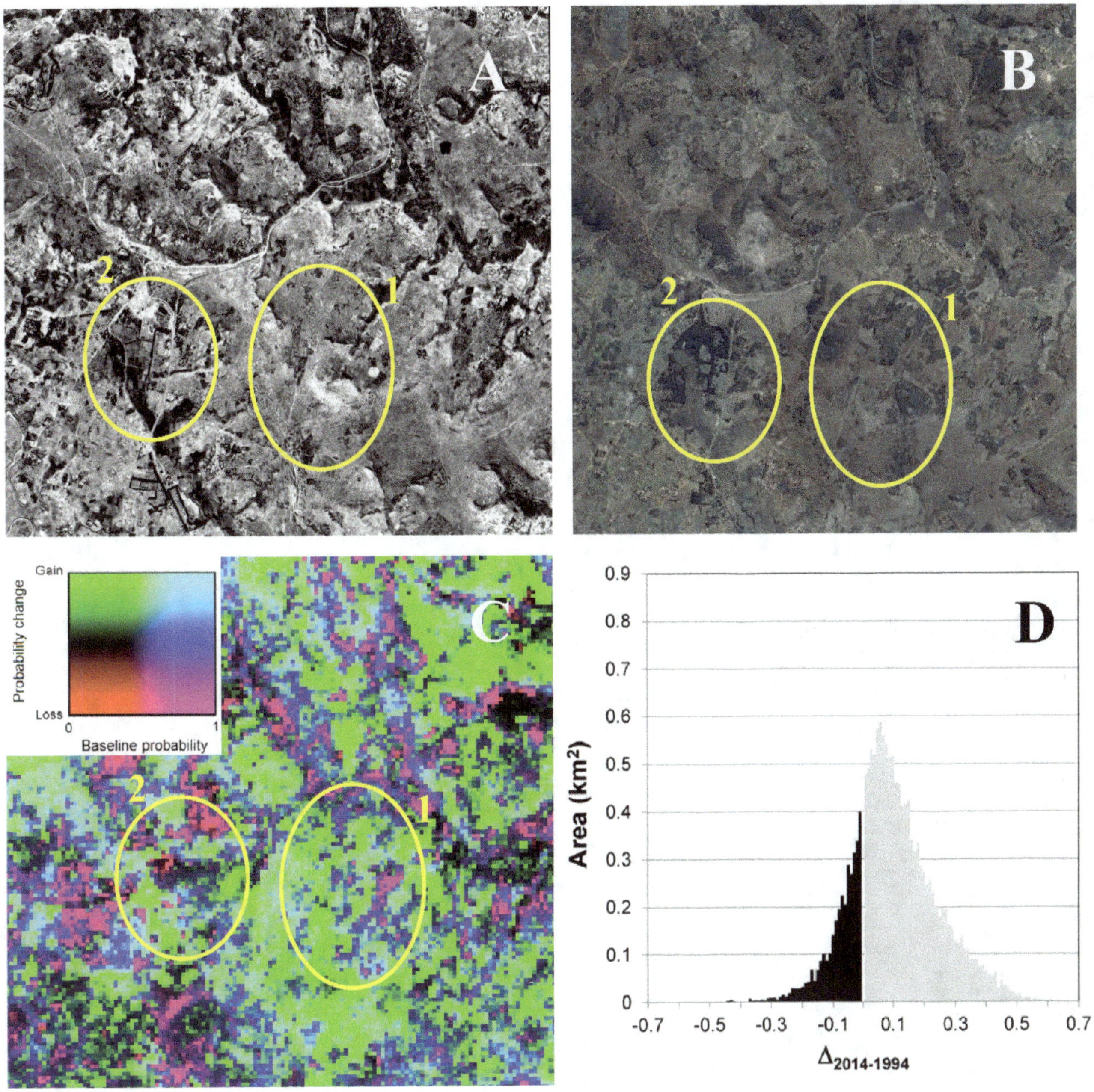

Figure 5. Dynamics of woody cover at Site 03, south of the town of Ambatolampy (19°25'S, 47°25'E). (**A**) Aerial photograph taken in 1991. (**B**) CNES Astrium image taken in 2013. (**C**) Change map based on Landsat imagery (see Figure 2). (**D**) Histogram depicting the frequency distribution of the delta of woody cover probability values between 2014 and 1994 based on Landsat imagery. Woody cover loss is represented in black, while gain is represented in gray. Ellipses in (A–C) relate to particular woody cover dynamics described in the text. See Figure 1 for the geographic location of Site 03 within the study area.

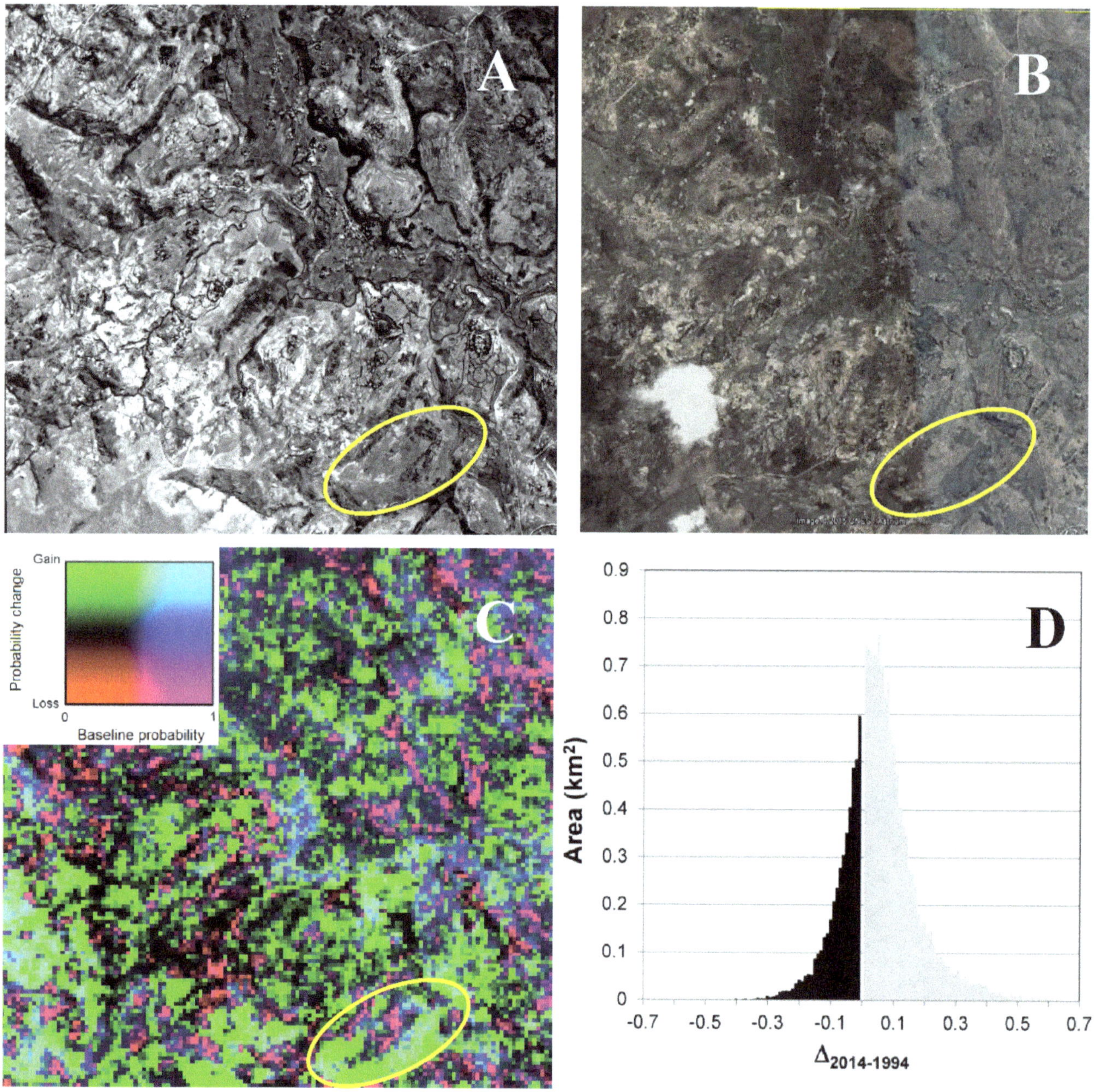

Figure 6. Dynamics of woody cover at Site 04, Ambohijatovo, centered on a rice cultivating valley just south of Anstirabe (19°56'S, 47°0'E). (**A**) Aerial photograph taken in 1991. (**B**) CNES Astrium imagery taken in 2013. (**C**) Change map based on Landsat imagery (see Figure 2). (**D**) Histogram depicting the frequency distribution of the delta of woody cover probability values between 2014 and 1994 based on Landsat imagery. Woody cover loss is represented in black, while gain is represented in gray. Ellipses in (A–C) relate to a particular woody cover dynamic described in the text. See Figure 1 for the geographic location of Site 04 within the study area.

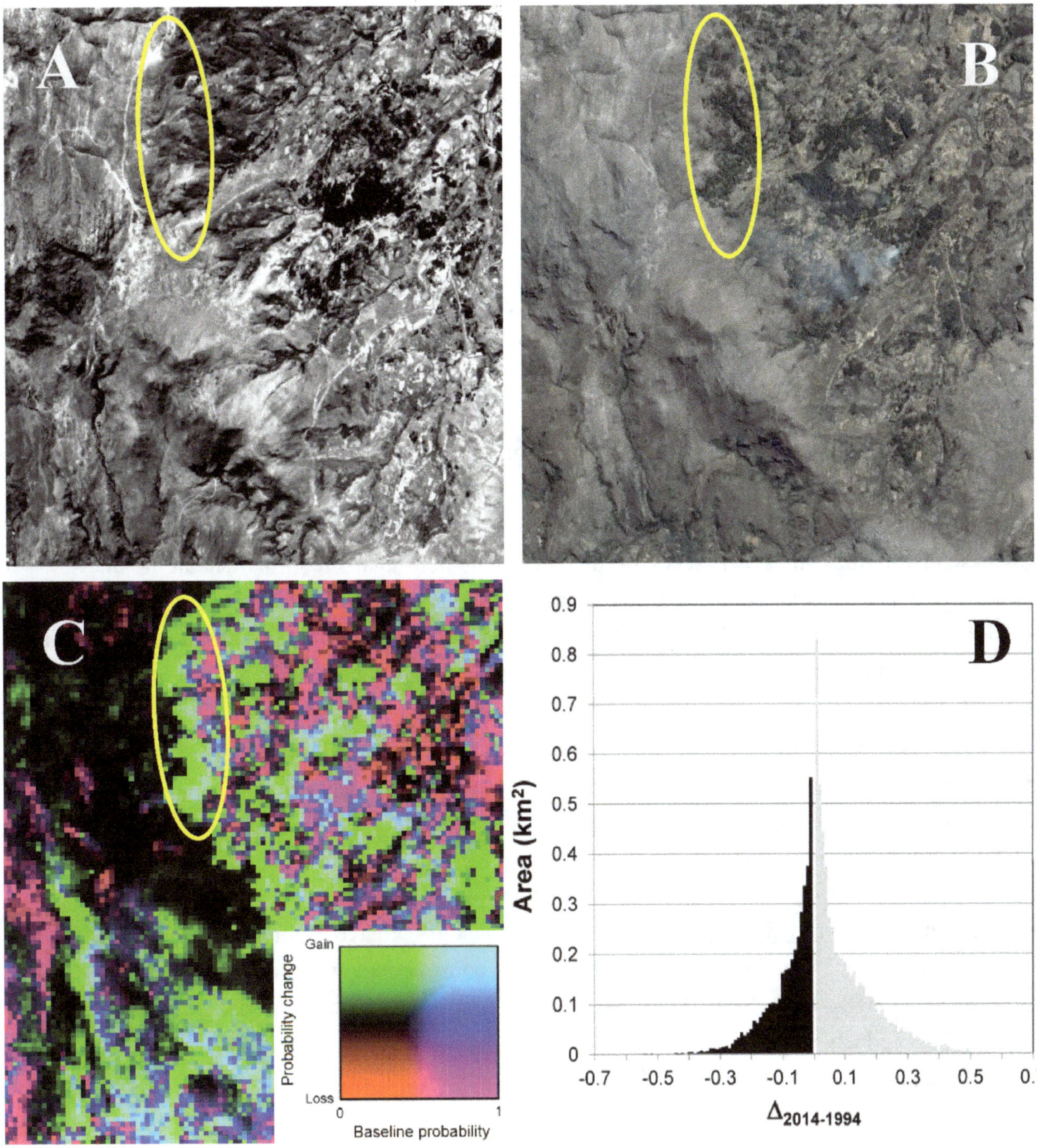

Figure 7. Dynamics of woody cover at Site 05, Soamahazina, an upland area southwest of Antsirabe (20°06'S, 46°54'E). (**A**) Aerial photograph taken in 1991. (**B**) CNES Astrium taken in 2013. (**C**) Change map based on Landsat imagery (see Figure 2). (**D**) Histogram depicting the frequency distribution of the delta of woody cover probability values between 2014 and 1994 based on Landsat imagery. Woody cover loss is represented in black, while gain is represented in gray. Ellipses in (A–C) relate to a particular woody cover dynamic described in the text. See Figure 1 for the geographic location of Site 05 within the study area.

4.2. Political Ecological Context of Tree Cover Change

A variety of intertwined endogenous and exogenous factors can be shown to lie behind the tree cover dynamics of the highlands. Six factors emerged from the analysis of the empirical case study sites and the literature: household livelihood strategies, urban market demand, land tenure struggles, dominant environmental ideas and resulting policy, government forestry interventions, and the invasiveness of some of the trees themselves. Each of these factors was already operative at the beginning of the study period. We argue that the conjuncture of these processes spurred the tree cover increase (following [77] and [78]). Further research could determine the precedence of particular factors in different places. We review each factor in turn.

First, a vast array of household needs is met by the trees increasingly covering the highlands. They are utilized in diverse ways, sometimes to meet subsistence needs, sometimes for cash income. Domestic fuel in rural areas comes almost purely from locally grown woody cover. Construction uses include floors, roofs, balconies, railings, stairs, cattle pens, pig sheds, and fence posts. Carpentry uses include beds, chairs, tables, cooking utensils, oxcarts, toys, and much more. Fruit trees and shrubs—from oranges to mangos, from vines to peaches—provide obvious consumable and marketable products, as well as secondary uses of dead branches as fuel wood, and mulberry leaves as silk worm fodder, among many other uses. Other benefits derived from trees include the production of honey based on eucalyptus flowers, the improvement of soil fertility based on acacia nitrogen fixation, and the sale of eucalyptus essential oils [79–81]. It is important to note that population growth in Madagascar is still quite rapid, with the country's population doubling between 1990 and 2013. Meanwhile, economic stagnation in urban areas has actually led to highly unusual urban-to-rural migration [82]. At the same time, the benefits previously derived from the cultivation of uplands have decreased with declining soil quality, while the benefits of grazing cattle have been undermined by increasing cattle rustling [56], which may in turn be interpreted as symptomatic of a weak, neopatrimonial state [83].

Second, urban market demand for wood fuel—particularly charcoal for urban consumers, or even to run industrial installations—as well as for construction wood, drives investment in woodlots [55,75,79,80,84–87]. Several areas of the highlands contain landscapes particularly devoted to the production of charcoal, firewood, and wood products. The most notable are the eucalyptus-dominated hills east and northeast of the capital, in the districts of Manjakandriana and Anjozorobe, where over 70 percent of the land—mostly former grassland—is now treed. At a smaller scale other similar landscapes are found outside other highland cities, such as the high-elevation pine and wattle groves between Antsirabe and Faratsiho [80].

Third, land tenure issues play a contributing role. Since around 1930 and continuing to the present, the plantation of woodlots was recognized by the state as evidence of land appropriation and development, a prerequisite to formal title registration (it should be noted that, in contrast, extant native forests were long ago declared state domain). Thus, urban and rural investors not only benefit from a growing urban market demand for the products from their woodlots, but they can also use the woodlots as part of a strategy to claim title to lands otherwise considered state-owned [55,84,88] or to lay out claims to common lands.

Fourth, it is important to understand the context of prominent environmental discourses in shaping perceptions of the highlands and policy responses. From early colonial days through to the present, scientists, foresters, and government agents have perceived the highlands as "barren" and deforested, and in need of "re-greening." This discourse is based both in a concern over soil degradation and biodiversity loss, as well as more pragmatic concerns with the supply of fuel and construction material [64,72]. Understanding the context of prominent environmental discourses is crucial to understanding government interventions, the next factor.

Fifth, both the colonial and independent governments, together with non-governmental organizations (NGOs), have repeatedly promoted the afforestation of the highlands. Tree planting to create communal forest reserves was at times obligatory and at other times a punishment for crimes [64,89]. The government's *Direction des Eaux et Forêts* (the Forest Service), founded at the dawn of French colonial control, still oversees a number of tree nurseries, forest stations, and forest plantations. Forestry stations, such as Manankazo (near Ankazobe), Manjakatompo and Antsampandrano (on the Ankaratra massif) and Angavokely (east of the capital) oversaw important reforestation projects and are at the center of regions with smallholder woodlots. The Forest Service also embarked on large, industrial plantations of *Pinus* spp, in particular in the *Haute Matsiatra*, northeast and southeast of Fianarantsoa as well as the *Haute Mangoro*, north of Moramanga (outside our study area). In the past two decades, however, the approaches have been more voluntary, sponsored for instance by projects run through local and foreign NGOs, aid agencies, universities, and conservationists. As just one example among many, the overseas wing of the French *Office National des Forêts* sponsors the work of the Malagasy NGO *Tany Meva* to plant trees across a few hundred hectares in a municipality some 30 km outside the capital [90]. At the same time, the state's inability to provide rural security has created disincentives to traditional livestock practices and surplus grain production, thus encouraging a shift in land use favoring trees, while its regulation of the charcoal trade has had mixed results [91] (see also [92] on the issue of corruption in law enforcement in Madagascar generally, and [93] on the specific question of conservation enforcement).

Sixth, and finally, the trees themselves have played a role in expanded tree cover in the highlands, taking advantage of suitably invasible landscapes. In particular, both pines and acacias have been relatively successful in reproducing themselves and dispersing from initial plantations [94], mirroring their behavior elsewhere [95,96].

5. Discussion and Conclusions

5.1. On Pathways and Processes of Forest Transitions and Landscape Change

While the rainforests of the eastern side of Madagascar's highlands continue to be exploited, a dynamic of woodlot rotations and tree planting is transforming the central and western highlands. Based on the state-of-the-art remote sensing evidence, we argue that this represents an incipient, sub-national forest transition. Building on the analysis of factors facilitating this trend, we furthermore argue that this transition follows a "scarcity" pathway pushed and pulled by endogenous and exogenous factors [97].

The land-use transition underway in the highlands of Madagascar shares little with other cases prominent in the forest transition literature. The economic development pathway is not relevant, since neither rural depopulation driven by industrialization nor rising affluence are operative in Madagascar, where the opposite trends are at work [82]. As for the traditional forest scarcity pathway, the few large-scale government-sponsored plantations account for a modest portion of the increased woody cover detected, and many of those were seen to be reverting to agricultural uses. Meanwhile, there are no large-scale private operations like the rubber and palm oil plantations of Southeast Asia. The situation in Madagascar bears some resemblance to the "smallholder stewardship" dynamic in the Ecuadorian Amazon described by [98], in which farmers with abundant land and labor are increasingly allowing certain trees to grow in their pastures. In the Malagasy highlands, though, pasture clearing is typically less labor-intensive than in Amazonia and, while farmers allowing spontaneous regeneration is certainly a part of the story, smallholders are also purposely planting trees that would not otherwise occur. Perhaps more instructive is the Vietnam case presented by Lambin and Meyfroidt [99] and [100], in which smallholders incorporate trees in their landscapes in response to a mix of endogenous and exogenous factors. In that case, the key exogenous factors included market liberalization and land tenure reform, which increased the relative profitability of lowland cultivation, leading farmers to reallocate labor from shifting cultivation, and planting the uplands with trees. In the case of highland Madagascar, market liberalization has not spurred a similar intensification of paddy land, despite massive investment by the United States Agency for International Development and other donors. Rather, it appears that continued uncertainty in rice prices, in concert with the increasing risk of theft—both of grain stores and cattle purchased with the proceeds—have created incentives for farmers to use former grazing land to plant trees. Thus, external factors, including the provision of material, knowledge and incentives to plant trees at the micro level, seem to have combined with the state's inability to provide functioning markets and rural security to spur farmers to plant trees. To our knowledge, this pathway to the forest transition has not been documented elsewhere.

Outside the forest transition literature, the predominant way to understand the long-term evolution of the highlands landscapes has been the Boserupian model of agricultural intensification linked to intertwined growth of population, markets, and property structures [101,102]. For instance, Rakoto Ramiarantsoa [55] has called the style of population-driven agricultural intensification and expansion found at the margins of the central highlands "*mérinisation*," by which he refers to a progressive transformation of open land to an intensively-cultivated cultural landscape typical of the Merina ethnic group in its homeland of Imerina, centered around the capital city. His concept imbues the Boserupian process with a cultural symbolic role. Yet his use of the term does not capture the conquest of the highlands by woodlots of exotic forestry tree species. Indeed, Rakoto Ramiarantsoa calls these eucalyptus-draped landscapes a departure from the Merina model ([55]; p. 299). Given the widespread adoption of these trees into peoples' livelihoods and landscapes over the past century (as he himself documents: e.g., [103]), these trees arguably deserve to be considered as an integral part of current day highland landscapes, whether Imerina or other cultural and regional divisions, like Vakinankaratra or Betsileo.

5.2. On Resilient Livelihoods and Sustainable Environments

What are the implications of the typically patchy, often low density, largely exotic tree cover that we have documented to be continuing to expand over significant parts of the highlands? We have demonstrated that it is central to rural livelihoods, and meets growing urban needs. Few other places in Africa can boast such a supply of locally produced items, instead being reliant on imported materials (sometimes for local manufacture) such as metal doors and window frames, cooking utensils, *etc.* The supply of such household goods perhaps leaves the Malagasy economy on a much more sustainable footing than it would otherwise enjoy. One could argue that this is a renewable wood supply, particularly in contrast with the rest of sub-Saharan Africa, where biomass for fuel (whether wood or charcoal) comes predominantly from natural forests. The extent and importance of private, and especially smallholder woodlots in meeting—and exceeding—subsistence needs, is a question of great importance. Madagascar, along with Rwanda, appears to be at the leading edge in the proportion of urban energy needs being met by such small-scale woodlots [104] (although doubt has recently been cast on the sustainability of Antananarivo's peri-urban eucalyptus plantations, given observations of a shortening of the coppice cycle below the six years recommended for maximum productivity [87]).The present study contributes to our understanding of this important phenomenon.

On the policy front, the pattern of state support for, and indeed investment in, forestry plantations corroborates the findings of Chidumayo and Gumbo [105] and Zulu and Richardson [106] that the provision of woodfuel supplies from plantations has been a long-standing policy goal in many countries, but the success of large, state-sponsored plantations has been mixed. Highland farmers have long been encouraged in their afforestation efforts by the colonial and independent governments of Madagascar, despite the fact that Ghilardi and colleagues [107] report that charcoal supply is often absent from the energy policies of countries that rely heavily on this source of energy.

In some ways, this is a tale of a Machakos Miracle writ large [39]. It highlights the role of endogenous factors, with government support, as opposed to the alternative approach of large government plantations. However, caution is warranted on several counts. First, closer examination of the Machakos case has revealed that, while it may represent a laudable example of environmental rehabilitation in the face of population pressure, not all people involved benefited equally. In that Kenyan case, elite retirees returned to their home areas, bringing with them intellectual and financial capital not available to others who ended up displaced to marginal lands, with negative environmental and social consequences [108]. In the Malagasy highlands, the tree-based dynamics of land control and connections to urban markets likewise are part and parcel of the dynamics of social differentiation between local or urban elites and marginalized members of society, though perhaps without the obvious displacement to marginal lands seen in Machakos [103].

Second, the ecological consequences of expanded woody cover in the highlands remain to be investigated. The trees affect soil chemistry, soil biology, and soil erosion, as well as hydrology, often in complex ways [109,110]. Several of the species common in the new tree cover (*Acacia* spp., *Pinus* spp.) are widely known to be invasive elsewhere, which is one reason for their spread here, and a reason some farmers like them or at least see them as an opportunity. But this also constitutes a potential challenge for the future, since, for instance, pines have invaded endemic *Uapaca* woodlands and montane heaths.

On the more positive side, this new tree cover is unsurprisingly not of central interest to biodiversity conservation planners, yet there are indications that such landscapes are not irrelevant. For instance, some plantation forests or hedgerows may serve as corridors or nursery landscapes, particularly in the more humid forest edge areas [111–114]. In addition, the new forests have implications for Madagascar's contributions to carbon sequestration [115].

5.3. Future Directions

We can only go so far in quantifying and explaining the changes we observe, and in understanding the social and environmental consequences, until further and more detailed field studies are undertaken. Seeking to balance and assess the complex implications of these environmental transformations in the highlands of Madagascar is a challenge; the most important and relevant assessment is probably that of the residents of the highlands themselves, for which these new woody resources constitute a treasure not to be squandered.

Acknowledgments

Air photo analyses and fieldwork by CK were funded by the Province of Quebec (FCAR 2002-SC-79002), the Australian Research Council (DP0666131), the Anne U. White Fund of the AAG, and Monash University.

Author Contributions

WM and AV conceived the research; WM, AV, CK, and CB collected and analyzed the data; WM, AV, and CK wrote the paper.

Conflicts of Interest

The authors declare no conflicts of interest.

References

1. Harper, G.J.; Steininger, M.K.; Tucker, C.J.; Juhn, D.; Hawkins, F. Fifty years of deforestation and forest fragmentation in Madagascar. *Environ. Conserv.* **2007**, *34*, 325–333.
2. Kull, C.A.; Ibrahim, C.K.; Meredith, T.C. Tropical forest transitions and globalization: NEO-liberalism, migration, tourism, and international conservation agendas. *Soc. Nat. Resour.* **2007**, *20*, 723–737.
3. Brannstrom, C.; Vadjunec, J.M. Notes for avoiding a missed opportunity in sustainability science: Integrating land change science and political ecology. In *Land Change Science, Political Ecology, and Sustainability Synergies and Divergences*; Brannstrom, C., Vadjunec, J.M., Eds.; Routledge: New York, NY, USA, 2013; pp. 1–23.
4. Mather, A.S.; Needle, C.L. The forest transition: A theoretrical basis. *Area* **1998**, *30*, 117–124.
5. Rudel, T.K. Is there a forest transition? Deforestation, reforestation, and development. *Rural Sociol.* **1998**, *63*, 533–552.
6. Foster, A.D.; Rosenzweig, M.R. Economic growth and the rise of forests. *Q. J. Econ.* **2003**, *118*, 601–637.
7. Kauppi, P.E.; Ausubel, J.H.; Fang, J.; Mather, A.S.; Sedjo, R.A.; Waggoner, P.E. Returning forests analyzed with the forest identity. *Proc. Natl. Acad. Sci. USA* **2006**, *103*, 17574–17579.

8. Hansen, M.C.; Potapov, P.V.; Moore, R.; Hancher, M.; Turubanova, S.; Tyukavina, A.; Thau, D.; Stehman, S.; Goetz, S.; Loveland, T. High-resolution global maps of 21st-century forest cover change. *Science* **2013**, *342*, 850–853.
9. Caldas, M.; Walker, R.; Arima, E.; Perz, S.; Aldrich, S.; Simmons, C. Theorizing land cover and land use change: The peasant economy of Amazonian deforestation. *Ann. Assoc. Am. Geogr.* **2007**, *97*, 86–110.
10. Gutman, G.; Janetos, A.C.; Justice, C.O.; Moran, E.F.; Mustard, J.F.; Rindfuss, R.R.; Skole, D.; Turner, B.L., II; Cochrane, M.A. *Land Change Science: Observing, Monitoring and Understanding Trajectories of Change on the Earth's Surface*; Kluwer Academic Publishers: Dordrecht, The Netherlands, 2004.
11. Lambin, E.; Geist, H.; Rindfuss, R.R. Introduction: Local processes with global impacts. In *Land Use and Cover Change*; Lambin, E., Geist, H., Eds.; Springer: Berlin, Germany, 2006.
12. Rindfuss, R.R.; Walsh, S.J.; Turner, B.L.; Fox, J.; Mishra, V. Developing a science of land change: Challenges and methodological issues. *Proc. Natl Acad. Sci. USA* **2004**, *101*, 13976–13981.
13. Rudel, T.K.; Coomes, O.T.; Moran, E.; Achard, F.; Angelsen, A.; Xu, J.C.; Lambin, E. Forest transitions: Towards a global understanding of land use change. *Glob. Environ. Change-Hum. Policy Dimens.* **2005**, *15*, 23–31.
14. Von Thünen, J.H. *Die isolierte Staat in Beziehung auf Landwirtshaft und Nationalökonomie*; Wartenberg, C.M., Translator; Pergamon Press: New York, NY, USA, 1826.
15. Angelsen, A. *Forest Cover Change in Space and Time: Combining the von Thünen and Forest Transition Theories*; World Bank Policy Research Working Paper 4117; World Bank: Washington, DC, USA, 2007.
16. Perz, S.G. Grand theory and context-specificity in the study of forest dynamics: Forest transition theory and other directions. *Prof. Geogr.* **2007**, *59*, 105–114.
17. Whittaker, R.J.; Grainger, A. Tropical open woodlands special issue—Introduction. *Glob. Ecol. Biogeogr.* **1999**, *8*, 177–178.
18. Aspinall, R. Modelling land use change with generalized linear models—A multi-model analysis of change between 1860 and 2000 in Gallatin Valley, Montana. *J. Environ. Manag.* **2004**, *72*, 91–103.
19. Evans, T.P.; Kelley, H. Multi-scale analysis of a household level agent-based model of landcover change. *J. Environ. Manag.* **2004**, *72*, 57–72.
20. Kok, K. The role of population in understanding Honduran land use patterns. *J. Environ. Manag.* **2004**, *72*, 73–89.
21. Walsh, S.J.; Evans, T.P.; Welsh, W.F.; Entwisle, B.; Rindfuss, R.R. Scale-dependent relationships between population and environment in northeastern Thailand. *Photogramm. Eng. Remote Sens.* **1999**, *65*, 97–105.
22. Perz, S.G.; Skole, D.L. Secondary forest expansion in the Brazilian Amazon and the refinement of forest transition theory. *Soc. Nat. Resour.* **2003**, *16*, 277–294.
23. Klooster, D. Forest transitions in Mexico: institutions and forests in a globalized countryside. *Prof. Geogr.* **2003**, *55*, 227–237.
24. Walker, R. Deforestation and economic development. *Can. J. Reg. Sci.* **1993**, *16*, 481–497.

25. Farley, K.A. Grasslands to tree plantations: forest transition in the andes of Ecuador. *Ann. Assoc. Am. Geogr.* **2007**, *97*, 755–771.
26. Southworth, J.; Nagendra, H.; Cassidy, L. Forest transition pathways in Asia—Studies from Nepal, India, Thailand, and Cambodia. *J. Land Use Sci.* **2012**, *7*, 51–65.
27. Meyfroidt, P.; Lambin, E.F. Forest transition in Vietnam and Bhutan: Causes and environmental impacts. In *Reforesting Landscapes*; Nagendra, H., Southworth, J., Eds.; Springer: Dordrecht, The Netherlands, 2010; pp. 315–339.
28. Xu, J.; Yang, Y.; Fox, J.; Yang, X. Forest transition, its causes and environmental consequences: Empirical evidence from Yunnan of Southwest China. *Trop. Ecol.* **2007**, *48*, 137.
29. Hecht, S.B.; Morrison, K.D.; Padoch, C. *The Social Lives of Forests: Past, Present, and Future of Woodland Resurgence*; University of Chicago Press: Chicago, IL, USA, 2014.
30. Nagendra, H.; Southworth, J. *Reforesting Landscapes: Linking Pattern and Process*; Springer Science & Business Media: Dordrecht, The Netherlands, 2009.
31. Peet, R.; Watts, M. *Liberation Ecologies*; Routledge: London, UK, 2004.
32. Robbins, P. *Political Ecology: A Critical Introduction*; Blackwell: Oxford, UK, 2004.
33. Batterbury, S.P.J.; Bebbington, A.J. Environmental histories, access to resources and landscape change: An introduction. *Land Degrad. Dev.* **1999**, *10*, 279–289.
34. Tiffen, M.; Mortimore, M.; Gichuki, F. *More People, Less Erosion*; John Wiley and Sons: Chichester, UK, 1994.
35. Fairhead, J.; Leach, M. *Misreading the African Landscape*; Cambridge University Press: Cambridge, UK, 1996.
36. Leach, M.; Mearns, R. *Lie of the Land: Challenging Received Wisdom on the African Environment*; James Currey: Oxford, UK, 1996.
37. McCann, J.C. The plow and the forest—Narratives of deforestation in Ethiopia, 1840–1992. *Environ. Hist.* **1997**, *2*, 138–159.
38. Kull, C.A. Deforestation, erosion, and fire: Degradation myths in the environmental history of Madagascar. *Environ. Hist.* **2000**, *6*, 423–450.
39. Mortimore, M.; Tiffen, M. *Population and Environment in Time Perspective: The Machakos Story*; John Wiley and Sons: Chichester, UK, 1995.
40. Arnold, J.E.M.; DeWees, P. *Farms, Trees and Farmers*; Earthscan: London, UK, 1997.
41. Clinecole, R.A. Manufacturing Forests & Farming Trees in Nigeria. In *The Lie of the Land: Challenging Received Wisdom on the African Environment*; Leach, M., Mearns, R., Eds.; Heineman: Portsmouth, UK, 1996; pp. 122–139.
42. Weston, P.; Hong, R.; Kaboré, C.; Kull, C.A. Farmer-managed natural regeneration enhances rural livelihoods in dryland west Africa. *Environ. Manag.* **2015**, *55*, 1402–1417.
43. Fortmann, L.; Bruce, J.W. *Whose Trees? Proprietary Dimensions of Forestry*; Westview Press: Boulder, CO, USA, 1998.
44. Rocheleau, D.; Edmunds, D. Women, men and trees: Gender, power and property in forest and agrarian landscapes. *World Dev.* **1997**, *25*, 1351–1371.
45. Mather, A.S. Recent Asian forest transitions in relation to forest-transition theory. *Int. For. Rev.* **2007**, *9*, 491–502.

46. Dietz, T.; Ostrom, E.; Stern, P.C. The struggle to govern the commons. *Science* **2003**, *302*, 1907–1912.
47. Moran, E.; Ostrom, E. *Seeing the Forest and the Trees: Human-Environment Interactions in Forest Ecosystems*; MIT Press: Cambridge, MA, USA, 2005.
48. Dufils, J.M. Remaining forest cover. In *The Natural History of Madagascar*; Goodman, S., Benstead, J., Eds.; The University of Chicago Press: Chicago, IL, USA, 2003; pp. 88–96.
49. McConnell, W.; Kull, C.A. Deforestation in Madagascar: Debates over the island's forest cover and challenges of measuring forest change. In *Conservation and Environmental Management in Madagascar*; Scales, I.R. Ed.; Earthscan Conservation and Development Series; Routledge: London, UK, 2014; pp. 67–104.
50. Hansen, M.; DeFries, R.; Townshend, J.; Carroll, M.; Dimiceli, C.; Sohlberg, R. Global percent tree cover at a spatial resolution of 500 meters: First results of the MODIS vegetation continuous fields algorithm. *Earth Interact.* **2003**, *7*, 1–15.
51. Barrett, C.B. Food marketing liberalization and trader entry: evidence from Madagascar. *World Dev.* **1997**, *25*, 763–777.
52. Fraslin, J.H. Quel avenir pour les paysans de Madagascar? *Afr. Contemp.* **2002**, *202–203*, 93–110.
53. Minten, B.; Randrianarison, L.; Swinnen, J.F. Global retail chains and poor farmers: Evidence from Madagascar. *World Dev.* **2009**, *37*, 1728–1741.
54. Minten, B.; Barrett, C.B. Agricultural technology, productivity, and poverty in Madagascar. *World Dev.* **2008**, *36*, 797–822.
55. Rakoto Ramiarantsoa, H. *Chair de la Terre, Oeil de l'Eau... Paysanneries et Recompositions de Campagnes en Imerina (Madagascar)*; Éditions de l'Orstom: Paris, France, 1995.
56. Fafchamps, M.; Moser, C. Crime, Isolation, and Law Enforcement. *J. Afr. Econ.* **2003**, *12*, 625–671.
57. Olson, S. Environments as shock absorbers, examples from Madagascar. *Environ. Hist. Rev.* **1988**, *12*, 61–80.
58. Shackleton, C.; Shackleton, S. The importance of non-timber forest products in rural livelihood security and as safety nets: A review of evidence from South Africa. *S. Afr. J. Sci.* **2004**, *100*, 658–664.
59. Montagne, P.; Bertrand, A. *Kajiala, Tattali, Djekabaara: Valoriser les produits pour mieux conserver les forêts. Tome 1: les expériences nationales Volume 1: Madagascar*; Projet GESFORCOM. Antananarivo, Madagascar, 2012.
60. Gardner, C.J.; Gabriel, F.U.; St John, F.A.; Davies, Z.G. Changing livelihoods and protected area management: A case study of charcoal production in south-west Madagascar. *Oryx* **2014**, doi:10.1017/S0030605315000071.
61. Raison, J.P. *Les hautes terres de Madagascar et leurs confins occidentaux: enracinement et mobilité des sociétés rurales*; Karthala Editions: Paris, France, 1984.
62. République. *Regions et Developpement: Programmes regionaux et projects locaux (Fianarantsoa)*; République de Madagascar, Ministere de l'Economie at du Plan; Programme des Nations Unies pour le Developpement Projet PNUD/OPS MAG/89/018; Dirasset: Tunis, Tunisia, 1991.

63. Kull, C.A. *Isle of Fire: the Political Ecology of Landscape Burning in Madagascar*; University of Chicago Press: Chicago, IL, USA, 2004.
64. Gade, D.W. Deforestation and its effects in highland Madagascar. *Mt. Res. Dev.* **1996**, *16*, 101–116.
65. Jolly, A. Madagascar: A world apart. *Natl. Geogr.* **1987**, *171*, 148–183.
66. Unruh, J.D.; McConnell, W.; Rodman, J. Environmental change and adaptation in degraded agro-ecosystems: the case of highland Madagascar. *Area* **2010**, *42*, 359–368.
67. Jaynes, E.T. Information theory and statistical mechanics. *Phys. Rev.* **1957**, *106*, 620–630.
68. Phillips, S.J.; Anderson, R.P.; Schapire, R.E. Maximum entropy modeling of species geographic distributions. *Ecol. model.* **2006**, *190*, 231–259.
69. FAO. *Global Forest Resources Assessment 2010*; Food and Agriculture Organization of the United Nations: Rome, Italy, 2010.
70. Hanley, J.A.; Mcneil, B.J. The meaning and use of the area under a receiver operating characteristic (ROC) curve. *Radiology* **1982**, *143*, 29–36.
71. Coppin, P.; Jonckheere, I.; Nackaerts, K.; Muys, B.; Lambin, E. Digital change detection methods in ecosystem monitoring: a review. *Int. J. Remote Sens.* **2004**, *25*, 1565–1596.
72. Kull, C.A. Air photo evidence of historical land cover change in the highlands: wetlands and grasslands give way to crops and woodlots. *Madag. Conserv. Dev.* **2012**, *7*, 144–152.
73. Kull, C.A. Leimavo revisited: Agrarian land-use change in the highlands of Madagascar. *Prof. Geogr.* **1998**, *50*, 163–176.
74. McConnell, W.J.; Sweeney, S.P. Challenges of forest governance in Madagascar. *Geogr. J.* **2005**, *171*, 223–238.
75. Verhaegen, D.; Randrianjafy, H.; Montagne, P.; Danthu, P.; Rabevohitra, R.; Tassin, J.; Bouvet, J.-M. Historique de l'introduction du genre Eucalyptus à Madagascar. *Bois et Forêt des Tropiques* **2011**, 17–25.
76. Kull, C.A. Historical landscape repeat photography as a tool for land use change research. *Norsk Geografisk Tidsskrift* **2005**, *59*, 253–268.
77. Ragin, C.C. *The Comparative Method: Moving beyond Qualitative and Quantitative Strategies*; University of California Press: Berkeley, CA, USA, 2014.
78. Rudel, T.K. *Tropical Forests: Paths of Destruction and Regeneration*; Columbia University Press: New York, NY, USA, 2013.
79. Carrière, S.M.; Randriambanona, H. Biodiversité introduite et autochtone: antagonisme ou complémentarité? Le cas de l'eucalyptus à Madagascar. *Bois et Forêts des Tropiques* **2007**, *292*, 5–21.
80. Kull, C.A.; Tassin, J.; Rangan, H. Multifunctional, scrubby, and invasive forests? Wattles in the highlands of Madagascar. *Mt. Res. Dev.* **2007**, *27*, 224–231.
81. Tassin, J.; Rangan, H.; Kull, C.A. Hybrid improved tree fallows: harnessing invasive woody legumes for agroforestry. *Agrofor. Syst.* **2012**, *84*, 417–428.
82. The World Bank Group. *Madagascar: Systemic Country Diagnostic*; Report #99197; The World Bank: Washington, DC, USA, 2015.
83. Van der Walle, N. *African Economies and the Politics of Permanent Crisis*; Cambridge University Press: Cambridge, UK, 2001.

84. Rakoto Ramiarantsoa, H. Les boisements d'eucalyptus dans l'est de l'Imerina (Madagascar). In *Terre, Terroir, Territoire: Les Tensions Fonciers,* Blanc-Pamard, C., Cambrézy, L., Eds.; OSTROM: Paris, France, 1995; pp. 83–103.
85. Bertrand, A. La Dynamique Seculaire des Plantations Paysannes d'Eucalyptus sur les Hautes Terres Malgaches. *Afr. Stud. Q.* **1999**, *3*, 61–68.
86. Carrière, S.M.; Andrianotahiananahary, H.; Ranaivoarivelo, N.; Randriamalala, J. Savoirs et Usages des Recrus Post-Agricoles du Pays Betsileo: Valorisation d'une Biodiversite Oubliee a Madagascar. *VertigO – La revue en sciences de l'environnement* **2005**, *16*, 1–15.
87. Verhaegen, D.; Randrianjafy, H.; Rakotondraoelina, H.; Trendelenburg Rakotonirina, M.-C.; Andriamampianina, N.; Montagne, P.; Rasamindisa, A.; Chaix, G.; Bouillet, J.-P.; Bouvet, J.-M. Eucalyptus robusta pour une production durable de bois énergie à Madagascar: Bilan des connaissances et perspectives. *Bois et Forêts des Tropiques* **2014**, *320*, 15–30.
88. Bertrand, A. Le boisement, le bail, et la législation environnemental à Madagascar: trois articles courts. *Afr. Stud. Q.* **1999**, *3*, 47–59.
89. Gade, D.W.; Perkins-Belgram, A. Woodfuels, reforestation, and ecodevelopment in highland Madagascar. *GeoJournal* **1986**, *12*, 365–374.
90. Office National de Forets. *Reboisement Communautaire des Hautes Terres, Madagascar*; 2010. Available online: http://www.developpement-durable.gouv.fr/IMG/pdf/Madagascar_ONF_Reboisement.pdf (accessed on 13 November 2015).
91. Minten, B.; Sander, K.; Stifel, D. Forest management and economic rents: Evidence from the charcoal trade in Madagascar. *Energy Sustain. Dev.* **2013**, *17*, 106–115.
92. Root, H. *Environment for Investment in Madagascar: Institutional Reform for Market Economy*; Center for Institutional Reform and the Informal Sector, University of Maryland at College Park: College Park, MD, USA, 1993.
93. Gore, M.L.; Ratsimbazafy, J.; Lute, M.L. Rethinking corruption in conservation crime: Insights from Madagascar. *Conserv. Lett.* **2013**, *6*, 430–438.
94. Kull, C.A.; Tassin, J.; Carriere, S.M. Approaching invasive species in Madagascar. *Madag. Conserv. Dev.* **2014**, *9*, 60–70.
95. Richardson, D.M.; Rejmánek, M. Trees and shrubs as invasive alien species—A global review. *Divers. Distrib.* **2011**, *17*, 788–809.
96. Van Wilgen, B.W.; Richardson, D.M. Challenges and trade-offs in the management of invasive alien trees. *Biol. Invasions* **2014**, *16*, 721–734.
97. Laney, R.M. Disaggregating induced intensification for land-change analysis: A case study from Madagascar. *Ann. Assoc. Am. Geogr.* **2002**, *92*, 702–726.
98. Lerner, A.M.; Rudel, T.K.; Schneider, L.C.; McGroddy, M.; Burbano, D.V.; Mena, C.F. The spontaneous emergence of silvo-pastoral landscapes in the Ecuadorian Amazon: Patterns and processes. *Reg. Environ. Change* **2014**, *15*, 1–11.
99. Lambin, E.F.; Meyfroidt, P. Land use transitions: Socio-ecological feedback *versus* socio-economic change. *Land Use Policy* **2010**, *27*, 108–118.
100. Meyfroidt, P. Environmental cognitions, land change and social-ecological feedbacks: Local case studies of forest transition in Vietnam. *Hum. Ecol.* **2013**, *41*, 367–392.

101. Boserup, E. *The Conditions of Agricultural Progress*; Aldine Publishing Company: Chicago, IL, USA, 1965.
102. Kull, C.A. Saving the land with spades: human landscape transformations in the central highlands. In *Greening the Great Red Island: Madagascar in Nature and Culture*; Kaufmann, J.C., Ed.; Africa Institute of South Africa: Pretoria, South Africa, 2008; pp. 113–135.
103. Blanc-Pamard, C.; Rakoto Ramiarantsoa, H. *Le Terroir et son Double: Tsarahonenana 1966–1992 Madagascar*; Editions de l'Institut de Recherches pour le Developpement: Paris, France, 2000.
104. Owen, M.; van der Plas, R.; Sepp, S. Can there be energy policy in Sub-Saharan Africa without biomass? *Energy Sustain. Dev.* **2013**, *17*, 146–152.
105. Chidumayo, E.N.; Gumbo, D.J. The environmental impacts of charcoal production in tropical ecosystems of the world: A synthesis. *Energy Sustain. Dev.* **2013**, *17*, 86–94.
106. Zulu, L.C.; Richardson, R.B. Charcoal, livelihoods, and poverty reduction: Evidence from sub-Saharan Africa. *Energy Sustain. Dev.* **2013**, *17*, 127–137.
107. Ghilardi, A.; Mwampamba, T.; Dutt, G. What role will charcoal play in the coming decades? Insights from up-to-date findings and reviews. *Energy Sustain. Dev.* **2013**, *2*, 73–74.
108. Rocheleau, D. More on Machakos. *Environment* **1995**, *37*, 3–5.
109. Bruijnzeel, L.A. Hydrological functions of tropical forests: not seeing the soil for the trees? *Agric. Ecosyst. Environ.* **2004**, *104*, 185–228.
110. Baohanta, R.; Thioulouse, J.; Ramanankierana, H.; Prin, Y.; Rasolomampianina, R.; Baudoin, E.; Rakotoarimanga, N.; Galiana, A.; Randriambanona, H.; Lebrun, M. Restoring native forest ecosystems after exotic tree plantation in Madagascar: Combination of the local ectotrophic species Leptolena bojeriana and Uapaca bojeri mitigates the negative influence of the exotic species Eucalyptus camaldulensis and Pinus patula. *Biol. Invasions* **2012**, *14*, 2407–2421.
111. Ganzhorn, J.U. A possible role of plantations for primate conservation in Madagascar. *Am. J. Primatol.* **1987**, *12*, 205–215.
112. Martin, E.A.; Viano, M.; Ratsimisetra, L.; Laloë, F.; Carrière, S.M. Maintenance of bird functional diversity in a traditional agroecosystem of Madagascar. *Agric. Ecosyst. Environ.* **2012**, *149*, 1–9.
113. Kull, C.A.; Carrière, S.M.; Moreau, S.; Ramiarantsoa, H.R.; Blanc-Pamard, C.; Tassin, J. Melting pots of biodiversity: Tropical smallholder farm landscapes as guarantors of sustainability. *Environ.: Sci. Policy Sustain. Dev.* **2013**, *55*, 6–16.
114. Gérard, A.; Ganzhorn, J.U.; Kull, C.A.; Carriere, S.M. Possible roles of introduced plants for native vertebrate conservation: The case of Madagascar. *Restor. Ecol.* **2015**, 23, 768–775.
115. Razakamanarivo, R.H.; Grinand, C.; Razafindrakoto, M.A.; Bernoux, M.; Albrecht, A. Mapping organic carbon stocks in eucalyptus plantations of the central highlands of Madagascar: A multiple regression approach. *Geoderma* **2011**, *162*, 335–346.

6

One Shared Region and Two Different Change Patterns: Land Use Change in the Binational Californian Mediterranean Region

Ricardo Eaton-Gonzalez * and Eric Mellink

Centro de Investigación Cientifica y de Educacion Superior de Ensenada (CICESE),
Carretera Ensenada-Tijuana 3918, Zona Playitas, 22860 Ensenada, B.C., Mexico;
E-Mail: emellink@cicese.mx

* Author to whom correspondence should be addressed; E-Mail: beaton@cicese.edu.mx

Academic Editor: Kenneth R. Young

Abstract: The Californian Floristic Province, ranging from Northern California, USA, to the northwestern portion of the state of Baja California, Mexico, is a region of great biological richness that has a high risk of loss of species due to the effect of human activities. The main stressor that threatens its biodiversity is the change in land use and vegetation cover, which severely impacts the environmental and socio-economic systems' functioning, affecting the provision of environmental services including the maintenance of biodiversity. The Tijuana River Watershed (TRW) is located within this floristic province. It has experienced rapid population growth during the last 50 years, demanding development of infrastructure in areas where native vegetation existed. As a binational watershed, it is an ideal area to study the processes involved in fragmentation and connectivity of natural environments, since both countries, while sharing the same environment, contrast greatly in their economic and social systems, which impose different pressures to these shared natural resources. Our research addresses change in vegetation cover and land use in the TRW, analyzing the changes and differences between Mexico and the United States. This analysis will be a basis to propose future management strategies for the conservation of ecological processes and biodiversity, according to the policies and actions for land management and conservation in both countries.

Keywords: land use and vegetation cover; Tijuana River Watershed; California floristic province; Tijuana; San Diego; binational

---

1. Introduction

Conservation efforts have identified areas of greater biological richness and those that present high risk of loss of species due to human activities. These regions are known as hotspots [1] and of the 34 currently defined around the world, one is the Californian Floristic Province, which is shared between the northern portion of Baja California in Mexico and the states of California and Oregon in the United States (U.S.) (Figure 1). This province is characterized by a Mediterranean climate, a high diversity of communities, and many endemic species of flora and fauna, many of which are threatened by the expansion of urban areas, pollution, and roads construction [1–3].

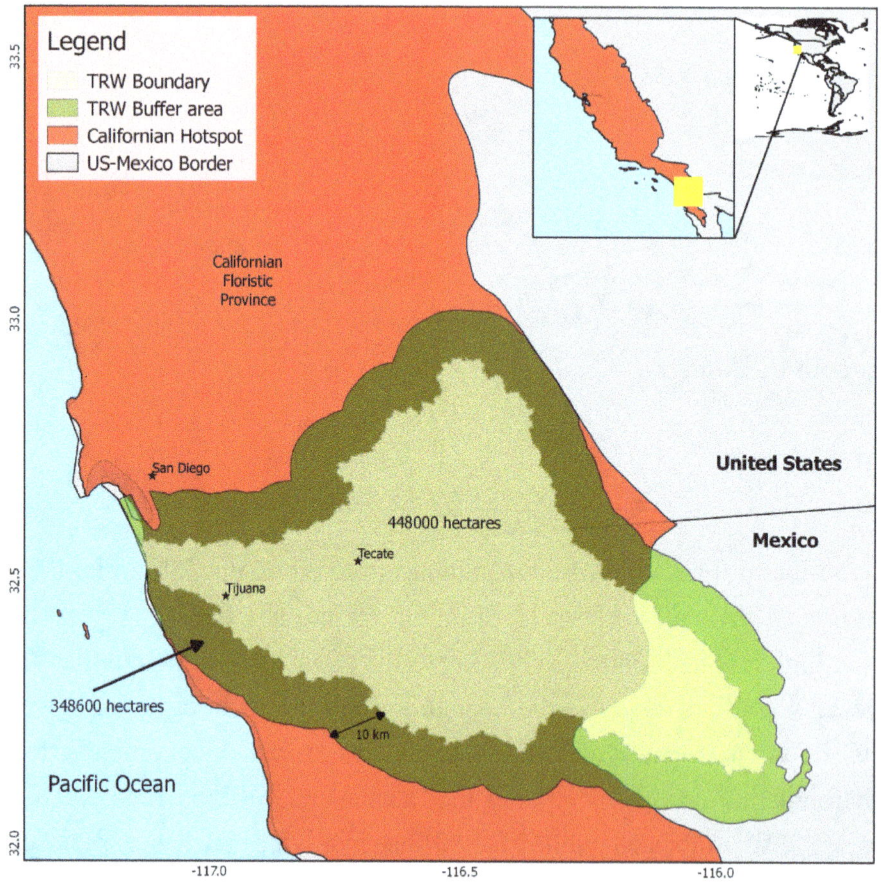

Figure 1. Binational and local settlement of Tijuana River Watershed. Source: Produced with information from http://www.cec.org/ and http://trw.sdsu.edu (accessed on 1 April 2013).

One of the main threats to biodiversity in Hotspots is the change in land use and cover, which severely impacts environmental and socio-economic systems, and affects sustainability, provision of food and supplies, maintenance of biodiversity and vulnerability of populations and ecosystems [4]. These changes in land use and vegetation cover occur through five driving forces: natural, socio-economic, policy-legislation, technology and cultural [5]. Due to their importance in conservation biology, the study of land use and cover change is one of the most important issues in current ecological and environmental research [6].

Since 1940, Northwestern Mexico and Southern California have exhibited accelerated population, urban and economic growth [7]. Three of the main factors that promoted this growth are (1) the ending of the U.S. "Programa Bracero" in 1964, which caused the return of workers from the U.S. to Mexico, (2) the beginning of the industrialization of the Mexican side of the border in 1965 [8], and, (3) the resulting population growth demanding areas for housing and the development of infrastructure, which caused the conversion of land with native vegetation or agriculture [9].

The Tijuana River Watershed (TRW), which covers parts of both countries in this region, was particularly affected by these processes and offers an ideal setting to study socio-economic and political influences on fragmentation and connectivity of natural environments, as those factors diverge widely on both sides of the border [10]. Historically, in the TRW, the planning processes have stopped at the border of each country, and no consideration has been given to binational conservation actions and shared resources. Hence, current public and private policies, legislation at the different levels of government in both countries, and the actions and financial resources geared to the management and conservation of natural resources differ greatly in both directions from the border [11].

The presence, maintenance, loss, change, recovery and use of the natural cover and land use have been integrated in the concepts of Land Use and Land Cover [12–14]. These concepts of Land Cover and Land Use are commonly confused in most land assessments, including those derived from satellite imagery [14]. Nonetheless, these concepts have been used commonly for at least the last 25 years [13]. Some approaches, merge the two terms in a single Land Use and Vegetation Cover concept, and derived from this, the Land Use and Cover Change (LUCC) analysis. The land cover/land use couplet has been adopted by much subsequent work and has become the goal and base for many land assessments, where the differences between land cover and land use are frequently noted but not analyzed [13]. Although some authors recommend the separate use of these terms, in this work the term Land Use and Vegetation Cover (LUVC) is used, considering that land-soil classification systems in the two countries are based on the coupled land cover/land use concept. In both Mexico and the U.S., data publishers obtained land use and cover information in order to develop a land use and land cover classification system for each country, that could be used in planning land use and management activities for government and private organizations [15].

Changes in land use and cover link to the occupation of the territory by humans in the TRW, which began in 1829 when a land grant that included what is now Tijuana and parts of San Ysidro was granted. In 1876, an urbanization policy commenced in Baja California. Settlement of the region continued slowly until the 1960s and 1970s, when a boost of the business sector, investment in infrastructure in the Tijuana River and migration increased the pace of land occupation [16]. From the 1970s onward, human settlements grew and the region developed heavily. Between 1972 and 1994, the watershed lost 200 km^2 of native vegetation from 85 patches in Mexico and 110 km^2 from 87 patches in U.S. [10].

This process of change in land use from natural to urban occurs in three phases: pre-urban vegetation, construction phase, which exposes bare soil for one to three years, and mature phase with impervious surfaces and vegetation. The process has varied through time. Thus, areas recently urbanized in Tijuana (1994–2002) had a higher soil fraction (40%) and a lower impervious surface fraction (20%) than areas urbanized prior to 1938 (17% soil and 62% of impervious surface). This pattern of urbanization in Tijuana resulted in higher percentages of bare soil, and possibly high sediment production potential for decades following the later urbanization [17]. More recently, between 1992 and 2005, urban areas and grasslands grew the most in TRW, while coastal scrub and chaparral decreased in surface. The conversion of coastal scrub resulted in a landscape that was more fragmented in 2005 than in 1994 [18].

Our research addresses the change in land use and vegetation cover in the TRW between 1990 and 2011, analyzing general processes of change and addressing the differences of those changes between Mexico and the U.S. We propose that our analyses and results provide the foundation for management strategies for the conservation of ecological processes and biodiversity in both countries.

2. Study Area

The TRW is a 450,000 ha area shared between northwestern Baja California, Mexico and Southwestern California, U.S. A little over 75% is located in Mexico [10,19–21] (Figure 1). The study area included the TRW [22] plus a 10 km band around it, within the terrestrial part of Mediterranean ecoregion [23]. This buffer zone harbors most or all of the TRW's wildlife species and is representative of the vegetation types, habitats and species present inside the TRW. Such a buffer zone allows for 75% of larger area to analyze and will, hence, permit a better understanding of the phenomena under investigation. Additionally, consideration of this larger area will provide a better framework for future research on natural, ecological and socioeconomic phenomena in the region. The TRW does not have an officially recognized boundary by any of the two countries and does not consider areas for or without future urban expansion. Consideration of urban growth, at least to the year 2023, already was suggested to analyze changes in the TRW [24]. The use of this buffer area allows for it since it includes on the Mexican side, Rosarito, the Rosarito-Tijuana Corridor, part of Parque Nacional Constitucion and part of the Valle de Guadalupe-Tecate corridor, places were urban development and conservation took place in the last few years. In the United States, this band includes Chula Vista, Otay Lakes, El Descanso and part of Cleveland National Forest. The study area defined is 780,199 ha, including 445,172 ha of the TRW (other authors have given 53,200, 449,300 and 445,000 ha [10,21,25]) and 335,027 ha of the buffer zone (Figure 1). The use of this TRW buffer allows modeling of the expansion and change of land use and cover in a current context, incorporating areas of impact, growth and conservation that are influencing history and maintenance of processes of land use and vegetation cover and wildlife habitat within the TRW. The boundary is the eastern limit of the Mediterranean zone towards east from TRW.

The topography of the area is of low hills and, in some places, steep slopes. Altitude ranges from sea level to almost 2000 m above sea level [26]. The TRW occupies a region of complex geology on the western slopes of Peninsular Ranges of Southern California and Baja California, in which the dominant rock in the upper watershed consists of Plutonic rocks of Peninsular Batholith. In the lower watershed, the rocks are primarily of Oligocene, Miocene, Pliocene, and Pleistocene ages [27].

The watershed's geomorphology is a reflection of its tectonic history and morphogenetic processes [28]. The study area can be divided in four major geomorphology units: In the east, a wide area of gentle to moderately undulating terrain that extends to the central part of the watershed. In the western part of the previous, predominant flat surface change to undulating landforms, bordered by highly dissected and steeply slopes. Farther west, the landscape changes abruptly in the coastal zone, in which large mesas are comprised of loosely cemented sandstones and conglomerates. Finally, areas downstream and surrounding Tijuana River Valley are comprised of a series of terraces in the valley sides, and a large concrete channel in the valley bottom, the same conditions as in Valle de Las Palmas [28].

The climate is Mediterranean with rain occurring during the winter. TRW displays a wide range of average temperatures and in the lower elevations annual mean temperatures generally range from 16 to 19 °C. Lower temperatures, of 9 to 11 °C, occur in the extreme north and southeast portions of the TWR. Precipitation varies widely throughout the watershed, and annual mean values range from just over 200 mm in the west to nearly 1092 mm in the east [29]. The TRW is part of the California Floristic Province, one of the biodiversity hotspots of the world [1]. The dominant vegetation types are chaparral, coastal scrub, mixed conifer forest, mountain meadows and riparian forest.

Most of the land in this region is devoted to urban spaces, agriculture and the raising of livestock, with each country differing in patterns of land use and vegetation cover [10]. Rapid urban growth in San Diego County and the municipalities of Tijuana and Tecate is destroying and fragmenting natural habitats and negatively affecting the diversity of species that depend on them. Natural resources along the international border are especially vulnerable because they have not been included in regional planning processes with a binational vision [30], resulting in a high and different risk of biodiversity loss between both countries.

3. Experimental Section (Methods)

3.1. Information Compilation

Land Use and Vegetation Maps at a 1:250,000 scale by the National Institute of Statistics, Geography and Informatics (INEGI) were used for Mexico (Table 1). For our analyses, we used maps issued in 1993 (series II maps) and 2011 (series V maps). They are provided in vector format, which we converted to raster format with a resolution of 30 meters, a resolution adequate to couple them with the data for the U.S. and to maintain data resolution and accuracy according to Landsat satellite images used for their creation. For the U.S., we used information by the U.S. Geological Survey (USGS; Table 1) in raster format, with 30-meter spatial resolution, that was generated through classification of Landsat satellite images. We use Landuse and Vegetation cover data products created from 1990 and 2011 LANDSAT images.

Table 1. Source and specifications of input data for Land Use and Cover Change (LUCC) analysis.

Country	Raw Data Source	Resolution	Classification Method	# LUVC Classes
US	1990 Landsat TM	30 m	Unsupervised	21
	2011 Landsat TM	30 m	Decision tree	16
Mexico	1993 Landsat TM	30 m	Supervised	8
	2011 Landsat TM	30 m	Supervised	13

3.2. Land Use and Vegetation Cover Information Standardization

Once information on land use and vegetation cover for both countries was compiled and their georeference standardized, LUVC data was extracted using TRW limits and adjacent buffer zone. In order to analyze the patterns of change in land use and vegetation cover, we standardized the names of the categories of land use and vegetation cover between the two countries. For this, the definition for each LUVC for each country was analyzed, in terms of elements that describe communities, such as life form (trees, shrubs, herbs); function (evergreen or deciduous); and cover or density of dominant species. For urban, agricultural and livestock uses categories, subcategory names were pooled from the original classification. This was done for each country's data set and subsequently standardized between the two countries.

3.3. LUVC Change and Fragmentation Analyses

There are two ways to analyze the change of LUVC: between and within LUVC classes. Change between LUVC classes accounts for conversion from one LUVC class to a completely different class, for example from forest to urban. A change within classes is the modification of the condition of the land-cover type to other within the same category, for example from primary forest to secondary forest [31]. We analyzed the changes between classes, using the Land Change Modeler (LCM) module within IDRISI Selva software, following criteria in Pontius *et al.* [32]. We used LCM tools for the assessment of land cover change, using two land cover maps (earlier and later) with identical land-cover types legends. The change analysis panel provided a rapid assessment of quantitative change by graphing gains and losses by land cover categories. A second tool, net change, exhibited the result of adding the gains and subtracting the losses to the earlier land cover maps. The third LCM tool used was the analysis of the contribution to changes experienced by individual land cover types [33,34]. Analyses were performed at a pixel level, and focused on the transition of the different classes in the earlier land cover map to different classes in the later land cover map [33,35,36]. We created maps of observed changes per category and relationships of changes between categories, measured in hectares and interpreted according to four general processes of LUCC [37]:

1. Deforestation. Loss of primary or secondary natural vegetation cover.
2. Anthropization. Change of natural vegetation cover to any predominantly human use, such as urban or cultivated.
3. Abandonment. Change of cultivated, agricultural or livestock use to a Barren. Changes in any of these original categories to something different than barren are classified either as anthropization or as recovery, depending on the change.
4. Recovery. Change of any non-natural vegetation land use or cover to any natural vegetation cover.

We calculated basic fragmentation metrics, number of patches and mean patch area, for each category and year studied, using Fragstats 4.2.1. Software.

4. Results and Discussion

The standardization of LUCV classes produced 10 shared categories and one category only for Mexico (Table 2). Surface gains and losses were markedly different for the different categories and between the two countries (Figure 2; Table 3). LUVC categories with greatest changes in both countries were: Urban, grassland-herbaceous, shrub-scrub, cultivated, barren, and evergreen forest, which corresponded to 98% of the total area in Mexico and 91% of that in the U.S., in 2011. These categories changed in their rank of occupied surface during the study period (Table 4). In Mexico, urban went from the fifth place to the third; and cultivated from the third to the fifth. In the U.S., urban changed from fourth to third; and evergreen forest from third to fourth. Shrub-scrub remained as the LUVC's most important category in both countries, occupying >60% of total land in both countries and years analyzed (Table 3). This is the characteristic vegetation in the region and the principal habitat available for wildlife [3].

Total area persistence in its original land use and vegetation class from 1993 to 2011 for Mexico was 97% (510,316 ha) and for the United States 85% (219,495 ha). For Mexico, barren, mixed forest and riparian shrub did not persist in more than 80% of their initial surface, while for the U.S., barren, evergreen forest, shrub-scrub and cultivated did so (Table 3). Grassland-herbaceous and urban had the largest surface gain in both countries, with 5666 and 24,218 ha in Mexico; and 3434 and 20,169 ha in the U.S., respectively (Table 3). This is consistent with previous works [10,18], in which urban and grasslands gained from deforestation or vegetation degradation, and grassland-herbaceous was the initial recuperation stage after a burning, grazing, logging or clearing, activities associated with ejidos (a form of land tenure) in Mexico [18,38]. In both countries grassland-herbaceous was lost to Urban (3970 ha in Mexico, and 4927 ha in the U.S.). Gains in grassland-herbaceous were at the expense of shrub-scrub (6949 ha in Mexico and 6994 ha in the U.S.) and, to a lesser degree of cultivated areas and evergreen forest (Figure 3). The grassland-herbaceous habitat is a result of shrub-scrub deforestation or degradation, and, in the near future, it can change to urban or recover to shrub-scrub [39].

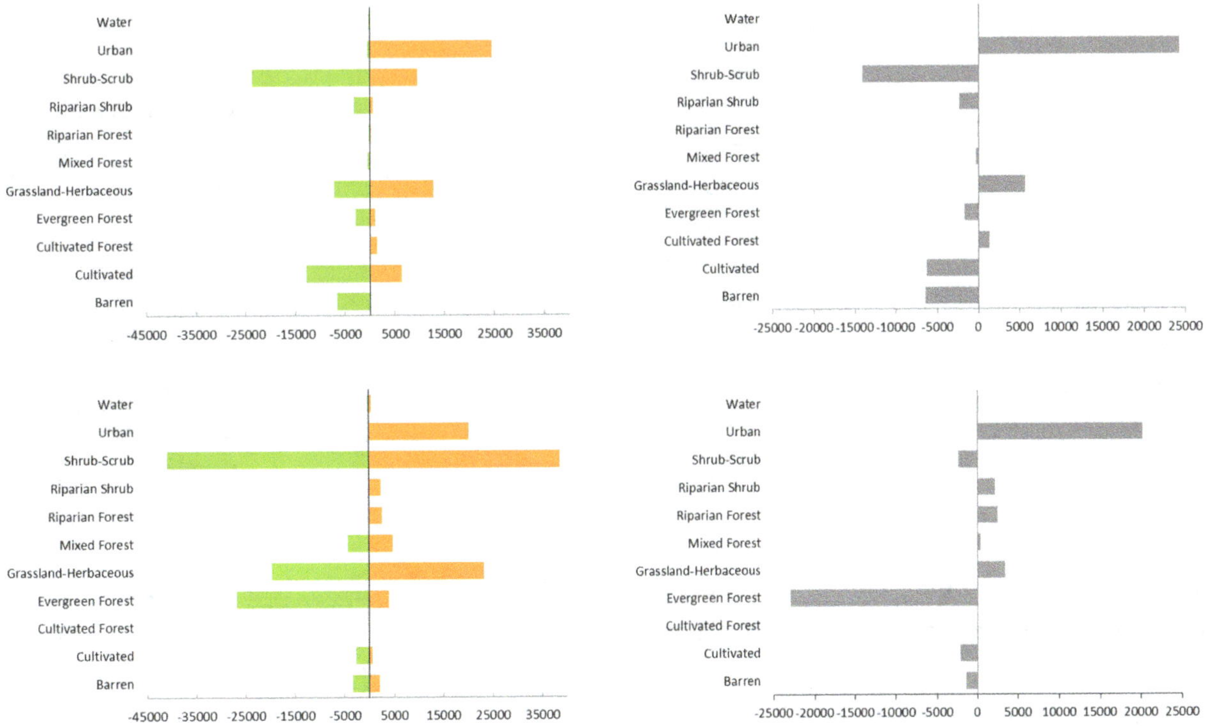

Figure 2. Gains and losses and net change per land use and vegetation cover category. Units are in hectares. (**Left**) gain and loss between 1993 and 2011 in Mexico (**top**) and between 1990 and 2011 in United States (**bottom**). (**Right**) net change in same periods and countries.

Table 2. Standardization of Land Use and Vegetation Cover (LUVC) categories names between countries.

Land Cover Classes (Mexico) as Published in Spanish	Land Cover Classes (United States)	Standardized Class Number	Standardized Class Name
Cuerpo de Agua	Open Water	1	Water
Urbano/Asentamiento Humano	Urban/Developed	2	Urban
Sin Vegetación	Barren	3	Barren
Bosque de Pino	Evergreen Forest	4	Evergreen Forest
Bosque de Pino	Evergreen Forest	4	Evergreen Forest
Bosque de Juniperus	Evergreen Forest	4	Evergreen Forest
Bosque de Pino	Evergreen Forest	4	Evergreen Forest
Bosque de Pino	Evergreen Forest	4	Evergreen Forest
Bosque de Encino	Mixed Forest	5	Mixed Forest
Bosque de Encino	Mixed Forest	5	Mixed Forest
Chaparral	Shrub/Scrub	6	Shrub-Scrub
Chaparral	Shrub/Scrub	6	Shrub-Scrub
Matorral Rosetofilo Costero	Shrub/Scrub	6	Shrub-Scrub
Matorral Desertico Microfilo	Shrub/Scrub	6	Shrub-Scrub
Matorral Desertico Rosetofilo	Shrub/Scrub	6	Shrub-Scrub
Chaparral	Shrub/Scrub	6	Shrub-Scrub
Pastizal Inducido	Grassland/Herbaceous	7	Grassland -Herbaceous
Pastizal Inducido	Grassland/Herbaceous	7	Grassland -Herbaceous
Pastizal Inducido	Grassland/Herbaceous	7	Grassland -Herbaceous
Pastizal Inducido	Grassland/Herbaceous	7	Grassland -Herbaceous
Agrícola y varios agrícola	Cultivated cropland	8	Cultivated
Agrícola y varios agrícola	Cultivated cropland	8	Cultivated
Agrícola y varios agrícola	Cultivated cropland	8	Cultivated
Bosque de Galería	Woody Wetlands	9	Riparian Forest
Bosque de Galería	Woody Wetlands	9	Riparian Forest
Vegetación de Galería	Herbaceous Wetlands	10	Riparian Shrub
Vegetación de Galería	Herbaceous Wetlands	10	Riparian Shrub
Vegetación de Galería	Herbaceous Wetlands	10	Riparian Shrub
Bosque Inducido	NA	11	Cultivated Forest

The area converted to urban was similar in both countries (Table 3). This is remarkable considering that >60% of the entire study area is located in Mexico; hence the proportion of land changed to urban was greater in the U.S. In both countries, urban areas increased at the expense of shrub-scrub, cultivated, barren, and grassland-herbaceous LUVC and, in the U.S., also from mixed forest. This result is similar to a 1992 to 2005 study [18], in which gains were dominated by urban and grassland. From 1993 to 2011 regional urban growth was similar to that for 1971–1994 [10], in that it occurred mostly around the established urban areas and along main highways, mostly in the northern and northeastern TRW.

Table 3. Initial surface and change for each LUVC category, year, and country. Superscript number in LUVC category indicates more that 80% of area persistence, 1 for Mexico and 2 for United States.

	Mexico					United States				
LUVC	1993 (ha)	% of Total	2011 (ha)	% of Total	Observed Change (ha)	1990 (ha)	% of Total	2011 (ha)	% of Total	Observed Change (ha)
Water [1,2]	694.44	0.13	705.24	0.13	10.8	11,610.00	4.51	11,754.72	4.56	144.72
Urban [1,2]	19,847.16	3.80	44,066.07	8.43	24,218.91	9035.73	3.51	29,205.63	11.34	20,169.9
Barren	6653.34	1.27	178.65	0.03	−6474.69	3613.86	1.40	2214.72	0.86	−1399.14
Evergreen Forest [1]	69,983.46	13.39	68,260.77	13.06	−1722.69	31,542.84	12.25	8563.86	3.32	−22,978.98
Mixed Forest [2]	1450.17	0.28	1092.51	0.21	−357.66	4461.39	1.73	4851.54	1.88	390.15
Shrub-Scrub [1]	352,368.45	67.43	338,304.51	64.73	−14,063.94	159,784.56	62.03	157,422.42	61.11	−2362.14
Grassland-Herbaceous [1,2]	30,573.45	5.85	36,239.49	6.93	5666.04	34,361.91	13.34	37,796.4	14.67	3434.49
Cultivated[1]	36,333	6.95	30,023.37	5.74	−6309.63	2941.74	1.14	819.27	0.32	−2122.47
Riparian Forest [1,2]	516.15	0.10	507.24	0.10	−8.91	42.66	0.02	2553.48	0.99	2510.82
Riparian Shrub [2]	4185.81	0.80	1782.99	0.34	−2402.82	199.26	0.08	2411.91	0.94	2212.65
Cultivated Forest	0	0.00	1444.59	0.28	1444.59	0.00	0.00	0	0.00	0
TOTAL	522,605.43	100	522,605.43	100		257,593.95	100	257,593.95	100	

Table 4. Rank by occupied surface of LUVC categories with greatest changes. Shaded cells indicate LUVC categories with greatest surface changes.

LUVC Categories	Rank by Occupied Area in Mexico		Rank by Occupied Area in the U.S.	
	1993	2011	1990	2011
Shrub-Scrub	1	1	1	1
Evergreen Forest	2	2	3	4
Cultivated	3	5	6	6
Grassland-Herbaceous	4	4	2	2
Urban	5	3	4	3
Barren	6	6	5	5

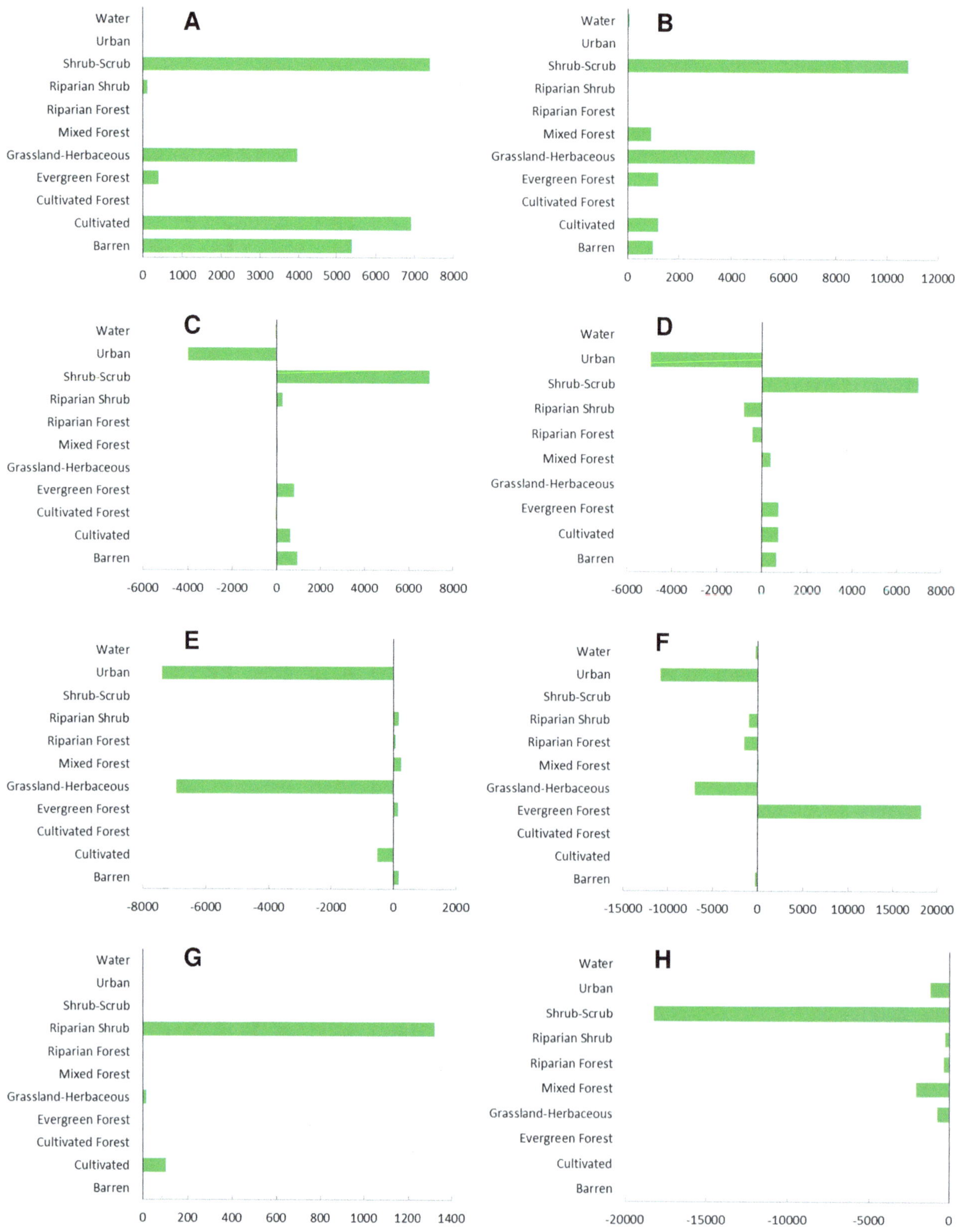

Figure 3. Contributors to net change for five land use and vegetation cover categories with greatest change. Units are in hectares. Mexico (**left**); United States (**right**). (**A**) and (**B**) are for Urban; (**C**) and (**D**), Grassland-Herbaceous; (**E**) and (**F**), Shrub-Scrub; (**G**), Cultivated Forest (only in Mexico); and (**H**), Evergreen Forest (only in the United States).

Shrub-scrub covers >60% of the study area's total surface and is the dominant category and exhibited the greatest change (surface loss and gains) in both countries. Nevertheless, Mexico had the greatest net surface loss in shrub-scrub, with 14,063 ha lost (6949 to grassland-herbaceous and 7396 ha to urban), compared with 2363 ha of shrub-scrub lost in the U.S. (Figure 2; Table 3). Our results were similar to those of other studies, which reported that this was the category that changed most [10,18].

In Mexico, the next two categories with the greatest loss were cultivated and barren (6309 and 6474 ha), but in the U.S., these categories lost less surface (2122 and 1399 ha). In other time periods, these categories exhibited a similar tendency in Mexico [10,18], in the first case due to a lack of irrigation water and in the second, due to its conversion to urban areas.

Lastly, evergreen forests exhibited substantial changes only in the U.S., where it lost 22,978 ha, mostly through conversion to shrub-scrub (18,239 ha), with lower conversion into mixed forest (2080) and urban (1202). In Mexico, only 1722 ha of this category were lost during the period (Figure 3). A new cover category emerged during this period in Mexico, but not in the U.S.: Cultivated forest, which gained surface from riparian shrub, shrub-scrub, cultivated and grassland-herbaceous. Thus, overall, shrub-scrub and evergreen forest exhibited the greatest decrease flows (loss of surface to other categories), while urban and grassland-herbaceous are the LUVC categories gained most surface from other categories (Figure 4).

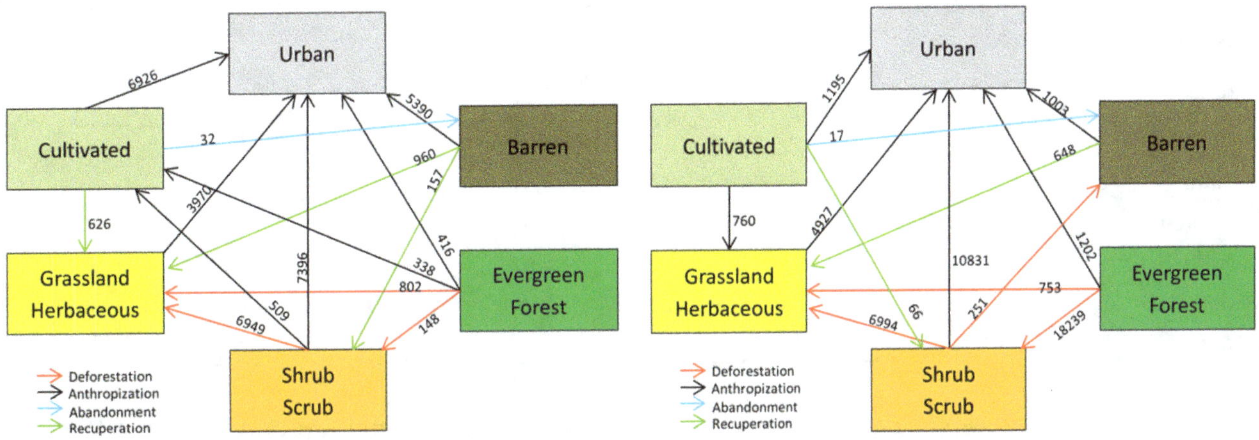

Figure 4. Land use and vegetation cover category transitions and transition processes. Numbers indicate change in hectares. Mexico (**left**); United States (**right**).

The four general change processes described (deforestation, anthropization, abandonment and recuperation) were present in the study area (Figure 4). Anthropization and deforestation were the principal LUCC processes in both countries, considering the number of flows of surface gain and loss. The areas with natural land cover (shrub-scrub, evergreen forest and grassland-herbaceous) lost 29,840 ha compared to the 25,300 in 1970–1994 [10] but differ in the proportion of surface changed in each country: 42% occurred in Mexico and 58% in the U.S., which is strikingly different from the 80% and 20%, respectively, for 1970–1994 [10]. Recuperation in our study was similar to that in 1970–1994 [10], with 61% of habitat recuperation in Mexico and 39% in the U.S.

Between 1993 and 2011, there was a decrease in the number of patches and in mean patch area (Table 5) in Mexico. The first of these evidences defragmentation [40]. Riparian shrub and shrub-scrub had the largest decrease in number of patches and were the LUVC categories that contributed to the emergence of the previously not present cultivated forest (Table 6; Figure 3).

Table 5. General fragmentation metrics.

	Number of Patches		Mean Patch Area (ha)	
Country/Year	**1993**	**2011**	**1990**	**2011**
Mexico	373	359	1422	1029.76
U.S.	124364	14635	8.5	16.27

Table 6. Fragmentation metrics by LUVC and country.

	Mexico				United States			
	1993		2011		1990		2011	
	Number of Patches	Mean Patch Area (ha)	Number of Patches	Mean Patch Area (ha)	Number of Patches	Mean Patch Area (ha)	Number of Patches	Mean Patch Area (ha)
Water	3	231.48	4	176.31	154	75.39	274	42.90
Urban	18	1102.62	38	1159.63	3037	2.98	923	31.64
Barren	1	6653.34	2	89.33	12755	0.28	721	3.07
Evergreen Forest	65	1076.67	55	1241.10	37191	0.85	1330	6.44
Mixed Forest	22	65.92	10	109.25	23407	0.19	1733	2.80
Shrub-Scrub	66	5338.92	56	6041.15	14367	11.12	2528	62.27
Grassland-Herbaceous	102	299.74	104	348.46	29468	1.17	5730	6.60
Cultivated	71	511.73	73	411.28	3337	0.88	52	15.76
Riparian Forest	3	172.05	3	169.08	262	0.16	772	3.31
Riparian Shrub	22	190.26	13	137.15	386	0.52	572	4.22
Cultivated Forest	0	0.00	1	1444.59	0	0.00	0	0.00
TOTAL	373		359		124364		14635	

There was greater landscape fragmentation in the U.S. than in Mexico and more in the early 1990s than in 2011. Mean patch area increased and patch number decreased between these dates. These results indicate land defragmentation, cover recuperation in some sites and LUVC compaction [39,40]. Recuperation involved transformation into the same land use and cover class as in adjacent cells in the 1990s (Figure 5). Defragmentation, despite the loss of natural vegetation, is favorable for wildlife habitat as it enlarges some habitats, compacts the area occupied by a category of LUVC and densifies the available resources for wildlife but probably also causes the loss of some habitat [39]. An example of defragmentation and compaction is the observed urban growth in sites adjacent to those where urban occupation persisted between the early 1990s and 2011 (Figure 5), mainly around Tijuana and Tecate and in Southern California near the border. Additionally, this urban expansion within Northern Tijuana resulted in the loss of natural areas within the city and in the suburbia. In the last ten years urban, agriculture and rural urban land use increase in both countries [41] expanding the previous surface of these land uses, despite similarity in patterns of defragmentation and compaction in some LUVC classes, there are great differences in socio-economic and legal systems, financial resources for conservation and number and surface of protected areas between Mexico and the U.S. [41]. For example, in the U.S., several federal, state, and county conservation programs have existed since the 1990s, with additional programs coming into operation in the last 10 years. In contrast, in Mexico there is a scarcity of new programs, with only those established since or before 1990s still in existence [41,42]. The combinations of these factors have resulted in different fragmentation patterns in the two countries, indicating a different dynamics of change resulting for different conservation policies and planning efforts.

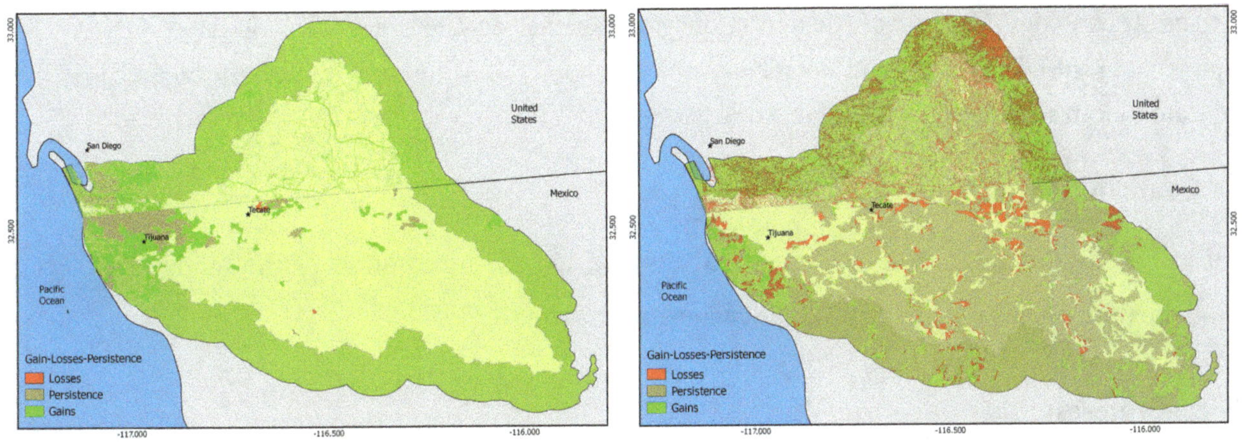

Figure 5. Surface gains, persistence and loss. **(Left)** shows urban land use and vegetation cover category, one which greatest surface gained and **(right)** shows shrub-scrub land use and vegetation cover category, the one with greatest surface lost.

5. Conclusions

(1) Patterns of land use change, change trends and transition processes by LUVC category in the Tijuana River Watershed from the early 1990s to 2011 differ between Mexico and the United States, largely resulting from different conservation policies and planning efforts between two countries.

(2) Changes were toward the anthropization, mainly by the growth of urban areas. In both countries, urban gained surface from shrub-scrub, cultivated, barren, and grassland-herbaceous LUVC categories.

(3) Three categories shared between the two countries had the main gains and losses: Grassland-herbaceous and urban (gain) and shrub-scrub (loss).

(4) The evergreen forest had significant surface changes only in United States, despite having a wide cover in the Mexican part of the TRW.

(5) Shrub-scrub was the dominant wildlife habitat in the area, which makes it the most important LUVC for conservation. It is necessary to analyze the LUVC as wildlife habitats for change, fragmentation and connectivity in order to know the conservation status in the region and contribute to management and conservation efforts.

(6) In Mexico, cultivated forest emerged in 2011 as a new category gaining surface from riparian shrub, shrub-scrub, cultivated, and grassland-herbaceous.

(7) Ten land-use and cover types provide a suitable standardization of LUVC names that can be applied to studies for the entire border between Mexico and the United States.

(8) A 10 km buffer zone surrounding the core of the study area proved helpful, as it comprised larger political, socioeconomical and ecological framework but still reflects the processes occurring within the core area.

(9) There was more fragmentation in U.S. than in Mexico, but, in both countries, we found a trend toward defragmentation.

Acknowledgments

This work is part of the Ph.D. dissertation of Ricardo Eaton-Gonzalez, who was supported by National Science and Technology Council in Mexico (Consejo Nacional de Ciencia y Tecnologia—CONACyT; grant number 111168). We thank John Rotenberry and University of Minnesota for a short research stay by Ricardo Eaton-Gonzalez, during which this work started.

Author Contributions

Ricardo Eaton-Gonzalez conducted data analysis and led the preparation of the manuscript. Eric Mellink helped in structuring and editing the manuscript.

Conflicts of Interest

The authors declare no conflict of interest.

References

1. Myers, N.; Mittermeier, R.A.; Mittermeier, C.G.; da Fonseca, G.A.; Kent, J. Biodiversity hotspots for conservation priorities. *Nature* **2000**, *403*, 853–858.
2. Riemann, H.; Exequiel, E. Endemic regions of the vascular flora of the Peninsula of Baja California, Mexico. *J. Veg. Sci.* **2007**, *18*, 327–336.
3. Conservation International. Biodiversity Hotspots: California Floristic Province. Available online: http://www.cepf.net/resources/hotspots/North-and-Central-America/Pages/California-Floristic-Province.aspx (accessed on 18 March 2015).
4. Lesschen, J.P.; Verburg, P.H.; Staal, S.J. *Statistical Methods for Analysing the Spatial Dimension of Changes in Land Use and Farming Systems*; International Livestock Research Institute: Nairobi, Kenya; Wageningen, The Netherlands, 2005.
5. Bürgi, M.; Hersperger, A.M.; Schneeberger, N. Driving forces of landscape change—Current and new directions. *Landsc. Ecol.* **2004**, *19*, 857–868.
6. Hannah, L.; Carr, J.L.; Lankerani, A. Human disturbance and natural habitat: A biome level analysis of a global data set. *Biodivers. Conserv.* **1995**, *4*, 128–155.
7. Ganster, P.; Sweedler, A.; Clement, N. Development, growth, and the future of the border environment. In *The Us-Mexican Border Environment: A Road Map to a Sustainable 2020*; Ganster, P., Ed.; San Diego State University Press: San Diego, CA, USA, 2000; pp. 73–103.
8. Ganster, P.; Pijawka, D.; Rasmussen, P.W.; Van Schoik, R. Overview. In *The Us-Mexican Border Environment: A Road Map to a Sustainable 2020*; Ganster, P., Ed.; San Diego State University Press: San Diego, CA, USA, 2000.
9. Spalding, M. Addressing border environmental problems now and in the future: Border XXI and related efforts. In *The Us-Mexican Border Environment: A Road Map to a Sustainable 2020*; San Diego State University Press: San Diego, CA, USA, 2000; pp. 105–137.

10. Ojeda-Revah, L.; Bocco, G.; Ezcurra, E.; Espejel, I. Land-cover/use transitions in the binational Tijuana River Watershed during a period of rapid industrialization. *Appl. Veg. Sci.* **2008**, *11*, 107–116.
11. Atkinson, E.E. *Linking Land Use and Policy in the Tijuana River Watershed*; San Diego State University: San Diego, CA, USA, 2009.
12. Comber, A.J.; Fisher, P.F.; Wadsworth, R.A. What is land cover? *Environ. Plan. B Plan. Des.* **2005**, *32*, 199–209.
13. Fisher, P. *Re-Presenting GIS*; John Wiley & Sons: Hoboken, NJ, USA, 2005.
14. Comber, A.J. The separation of land cover from land use using data primitives. *J. Land Use Sci.* **2008**, *3*, 215–229.
15. Anderson, J.R.; Hardy, E.E.; Roach, J.T.; Timer, R.E. *A Land Use and Land Cover Classification System for Use with Remote Sensor Data*; Geological Survey Professional Paper 964; U.S. Government Printing Office; Washington, DC, USA, 1976.
16. Ojeda-Revah, L.; Espejel-Carbajal, M.I. La cuenca binacional del río tijuana: Un enfoque biohistórico. *Econ. Soc. Territ.* **2008**, *8*, 517–548.
17. Biggs, T.W.; Atkinson E.; Powell, R.; Ojeda-Revah, L. Land cover following rapid urbanization on the U.S.–Mexico border: Implications for conceptual models of urban watershed processes. *Landsc. Urban Plan* **2010**, *96*, 78–87.
18. Farley, K.A.; Ojeda-Revah, L.; Atkinson, E.E.; Eaton-González, B.R. Changes in land use, land tenure, and landscape fragmentation in the Tijuana River Watershed following reform of the ejido sector. *Land Use Policy* **2012**, *29*, 187–197.
19. Bocco, G. The Tijuana River Watershed Geographic Information System: A Tool for Shared Management. In *Identifying Priorities for a Geographic Information System (GIS) for the Tijuana River Watershed: Applications for Land Use, Planning and Education*; Wright, R., Ries, K., Winckell, A., Eds.; Institute for Regional Studies of the Californias, San Diego State University: San Diego, CA, USA, 1995; pp. 15–18.
20. Zedler, J.B.; Nordby, C.S.; Kus, B.E. *The Ecology of Tijuana Estuary, California: A National Estuarine Research Reserve*; NOAA Office of Coastal Resource Management, Santuaries and Reserves Division: Washington, DC, USA, 1992.
21. Ganster, P. Introduction. In *Tijuana River Watershed Atlas/Atlas de la Cuenca del río Tijuana*; Wright, R., Vela, P., Ganster, P., Eds.; San Diego State University Press: San Diego, CA, USA, 2005; p. 1.
22. Center for Earth System Analysis Research (CESAR). Tijuana River Watershed Digital Boundary File. Vector digital file from Tijuana River Watershed GIS Database. Available online: http://geography.sdsu.edu/Resources/Data/Clearinghouse/trw.php (accessed on 17 April 2012).
23. Commission for Environmental Cooperation (CEC). Terrestrial Ecoregions. Vector Digital Data. Available online: http://www.cec.org/naatlas/ (accessed on 17 April 2012).
24. Institute for Regional Studies of the Californias and the Department of Geography at San Diego State University. A Binational Vision for the Tijuana River Watershed. 2005. Available online: http://irsc.sdsu.edu/docs/pubs/Tijuana_River_Watershed_Binational_Vision.pdf (accessed on 7 August 2015).

25. Wright, R. Introduction. In *Identifying Priorities for a Geographic Information System (GIS) for the Tijuana River Watershed: Applications for Land Use, Planning and Education*; Wright, R., Ries, K., Winckell, A., Eds.; Institute for Regional Studies of the Californias, San Diego State University: San Diego, CA, USA, 1995; pp. 1–2.
26. Wright, R. Topography. In *Tijuana River Watershed Atlas/Atlas de la Cuenca del río Tijuana*; Wright, R., Vela, P., Ganster, P., Eds.; San Diego State University Press: San Diego, CA, USA, 2005; p. 5.
27. Deméré, T.A. Geology: Generalized rock types and faults. In *Tijuana River Watershed Atlas/Atlas de la Cuenca del río Tijuana*; Wright, R., Vela, R., Ganster, P., Eds.; San Diego State University Press: San Diego, CA, USA, 2005; p. 6.
28. Winckel, A. Geomorphology. In *Tijuana River Watershed Atlas/Atlas de la Cuenca del río Tijuana*; Wright, R., Vela, R., Ganster, P., Eds.; San Diego State University Press: San Diego, CA, USA, 2005; p. 7.
29. Aguado, E. Temperature and precipitacion. In *Tijuana River Watershed Atlas/Atlas de la Cuenca del río Tijuana*; Wright, R., Vela, R., Ganster, P., Eds.; San Diego State University Press: San Diego, CA, USA, 2005; pp. 10–11.
30. Conservation Biology Institute. La Posta Linkage Portfolio, San Diego County, California. 2003. Available online: http://www.sandiegocounty.gov/content/dam/sdc/pds/ceqa/Soitec-Documents/Final-EIR-Files/references/rtcref/ch9.0/rtcrefaletters/O14%202014-12-19_CBI2003.pdf (accessed on 14 August 2013).
31. Lu, D.; Moran, E.; Hetrick, S.; Li, G. Land-use and land-cover change detection. In *Advances in Environmental Remote Sensing: Sensors, Algorithms, and Applications*; CRC Press: New York, NY, USA, 2011; pp. 273–290.
32. Pontius, R.G.; Shusas, E.; McEachern, M. Detecting important categorical land changes while accounting for persistence. *Agric. Ecosyst. Environ.* **2004**, *101*, 251–268.
33. Eastman, J.R. *IDRISI Selva Manual, Manual Version 17*; Clark University: Worcester, MA, USA, 2012.
34. Mishra, V.N.; Rai, P.K.; Mohan, K. Prediction of land use changes based on land change modeler (LCM) using remote sensing: A case study of muzaffarpur (Bihar), India. *J. Geogr. Inst. Jovan Cvijic SASA* **2014**, *64*, 111–127.
35. Pontius, R.G., Jr.; Chen, H. *Land Change Modeling with GEOMOD*; Clark University: Worcester, MA, USA, 2006.
36. De Alba, H.; Barros, J. Deforestation in the kayabi indigenous territory: Simulating and predicting land use and land cover change in the brazilian amazon. In Proceedings of the III Coloquio Internacional en Geografia, Michoacán, México, 21–24 March 2012.
37. Garibay, C.; Bocco, G. *Cambios de uso del Suelo en la Meseta Purépecha (1976–2005)*; INE-SEMARNAT & CIGA-UNAM: Mexico, D.F., Mexico, 2011.
38. Bonilla-Moheno, M.; Redo, D.J.; Aide, T.M.; Clark, M.L.; Grau, H.R. Vegetation change and land tenure in Mexico: A country-wide analysis. *Land Use Policy* **2013**, *30*, 355–364.
39. Lambin, E.F.; Turner, B.L.; Geist, H.J.; Agbola, S.B.; Angelsen, A.; Bruce, J.W.; Xu, J. The causes of land-use and land-cover change: Moving beyond the myths. *Glob. Environ. Chang.* **2001**, *11*, 261–269.

40. Fahrig, L. Effects of habitat fragmentation on biodiversity. *Annu. Rev. Ecol. Evol. Syst.* **2003**, *34*, 487–515.
41. Stallcup, J.A.; Randall, J.M.; Smith, T.; Cohen, B.S.; Guerrero Avila, C.; Vargas, M.A.; Morrison, S.A. *Las Californias Binational Conservation Initiative 2015: A Decadal Review of Conservation Status of the California—Baja California Border Region*; The Nature Conservancy: Arlington County, VA, USA, 2015.
42. Castro-Ruiz, J.L.; Ganster, P.; Wright, R.; Comer, K.; Silvan, L.; Saxod, E.; Zúñiga, W. Beyond the generation of data for watershed management: The experience of the Tijuana River Bi-National Watershed Advisory Council. Presented at the Border Institute VIII Cooperation on Planning and Operation of Trans-boundary Watersheds: Sharing Data across Borders, Rio Rico, AZ, USA, 22–24 May 2006. Available online: http://scerpfiles.org/bi/bi-viii/papers.htm (accessed on 7 August 2015).

7

Why Don't More Farmers Go Organic? Using A Stakeholder-Informed Exploratory Agent-Based Model to Represent the Dynamics of Farming Practices in the Philippines

Laura Schmitt Olabisi [1,*], Ryan Qi Wang [2] and Arika Ligmann-Zielinska [3]

[1] Department of Community Sustainability and Environmental Science & Policy Program, Michigan State University, 151 Natural Resources Building, 480 Wilson Road, East Lansing, MI 48824, USA

[2] Department of Civil and Environmental Engineering, Virginia Tech, 113 Patton Hall (0105), Blacksburg, VA 24061, USA; E-Mail: wangqi@vt.edu

[3] Department of Geography and Environmental Science & Policy Program, Michigan State University, Room 121 Geography Building, 673 Auditorium Rd., East Lansing, MI 48824, USA; E-Mail: ligmannz@msu.edu

* Author to whom correspondence should be addressed; E-Mail: schmi420@msu.edu

Academic Editors: James Millington and John Wainwright

Abstract: In spite of a growing interest in organic agriculture; there has been relatively little research on why farmers might choose to adopt organic methods, particularly in the developing world. To address this shortcoming, we developed an exploratory agent-based model depicting Philippine smallholder farmer decisions to implement organic techniques in rice paddy systems. Our modeling exercise was novel in its combination of three characteristics: first, agent rules were based on focus group data collected in the system of study. Second, a social network structure was built into the model. Third, we utilized variance-based sensitivity analysis to quantify model outcome variability, identify influential drivers, and suggest ways in which further modeling efforts could be focused and simplified. The model results indicated an upper limit on the number of farmers adopting organic methods. The speed of information spread through the social network; crop yields; and the size of a farmer's plot were highly influential in determining agents' adoption rates. The results of this stylized model indicate that rates of organic farming adoption are highly sensitive to the yield drop after switchover to organic techniques, and to the speed of information spread through existing social networks. Further research and model development should focus on these system characteristics.

Keywords: organic agriculture; agent-based modeling; sensitivity analysis; social networks; Philippines

1. Introduction

Acreage under organic farming methods is increasing globally, as government support for organic farmers and market demand for organic products grows [1]. A major driving force behind both of these trends is the recognition of the negative impacts of chemically-intensive farming methods on the environment and human health [2,3]. Approximately one-third of global land under organic production is located in the developing world [1]. In many developing nations, agricultural production is dominated by smallholder farmers, who often lack access to crop insurance or inexpensive credit. Some have argued that organic agriculture can benefit smallholder farmers by eliminating their reliance on expensive, fossil fuel-derived chemical inputs [4]. Organic farming therefore could make smallholder farmers more resilient to input price shocks, which are a significant source of insecurity for them [5]. This argument is bolstered by research from the United States, which indicates that organic farmers spend less on inputs than conventional farmers [6]. Policy makers, extension agents, and non-governmental organizations that serve developing-world farmers need a clear understanding of farmers' motivations and challenges in converting to organic agriculture, so that efforts to promote organic farming will be maximally effective.

Few organic adoption studies to date have portrayed a dynamic and complex decision environment, rather than a static snapshot of farmer decision-making [7]. Farmers' demographic characteristics, their economic motivations, their concern for the environment and for their families' health, all interact in complex and heterogeneous ways as they consider their options and choose their production method. Moreover, farmers' choices may be different year to year, depending on the decision context. They may choose to convert to organic, and then return to conventional methods when their situation changes. Most previous studies have not taken advantage of modeling tools that could portray this type of decision-making in dynamic contexts, such as agent-based modeling (ABM).

ABM has been used extensively in studying land-use conversion decisions, and has provided insights into how decision/environment feedback loops operate to produce non-intuitive outcomes [8–15]. Another advantage of ABM is its ability to represent the dynamics of social influences and information propagation through social networks via peer-to-peer, or "word-of-mouth", communications [16,17]. Previous studies have found that a farmer's source of information about farming and organic techniques influences his or her decision to adopt organic methods [18]. Peer-to-peer sharing through social networks can be an important source of information for farmers [19]. Modeling the spread of information about organic agricultural techniques through social networks could therefore provide insight into how best to promote these technologies and support farmer adoption.

ABM has been successfully applied to agricultural systems all over the world [20–22], to study the complex interactions between farmers, global and regional crop markets, and biophysical (especially hydrological and crop-soil) systems [8,13,20,23,24]. Agricultural applications vary from new practice adoption [8,14,20,24–27], changes in agricultural production and its viability [14,25,28], the impact of different decision making practices on agricultural land use/land cover change [14,23,25,28],evaluation of landscape structure [13,28], farmers' imitative behavior [8,29], and explicit analysis of agricultural policies [8,20,30]. For example, Berger [8] uses the concept of innovation diffusion to study various farming production alternatives and their influence on local hydrology in Chile. Happe *et al.* [30] present an ABM for agricultural policy analysis of different farm structures in Germany, Schreinemachers *et al.* [31] build an ABM to analyze the diffusion of greenhouse agriculture in Thailand, and Evans *et al.* [25] develop an empirically-rich ABM to explore the transition from shifting cultivation to rubber production in Laos. Many researches stress the usefulness of ABM to represent the diversity of farmer decision-making [13,20,27,30].

Of particular interest to our study are ABM applications that simulate organic farming adoption [26,27]. For example, Kaufmann *et al.* [27] develop an agent-based model of agricultural decision making that utilizes the theories of planned behavior and innovation diffusion, coupled with a survey-informed social network sub-model, to evaluate the diffusion of organic farming in Latvia and Estonia. The goal of their modeling exercise is to test how the economic changes are intertwined with agent-agent interactions in the formation of beliefs concerning the transition to organic farming. They conclude that mere social influence is not sufficient when modeling conversion from traditional to organic farming. Without exogenous economic factors, in the form of subsidies, "*organic farmers remain organic, and conventional farmers remain conventional*" (p. 2589). In a comparable study, Deffuant *et al.* [26] develop an ABM to model organic farming in a selected region of France. The novelty of their approach lies in introducing an auxiliary *institution* agent that evaluates the farm potential and assists the farmer in the decision to go organic. Similarly to Kaufmann *et al.* [27], they conclude that organic farming adoption results from complex interactions between economic and social processes.

In this paper, we define ABM as a simulation environment composed of heterogeneous computational entities (called agents) that represent Philippine smallholder farmers, for the purpose of exploring group dynamics around organic agriculture adoption. The agents are situated in a common agricultural environment of an upland paddy rice system, where organic farming practices are being actively promoted by a local non-governmental organization (NGO). Agents' decision making to adopt and maintain organic agriculture is constrained by limited access to information about the optimal farming strategies and the economic resources available. The agents are driven by their individual goals and social behaviors, and they constantly adapt to changing agro-economic and ecological conditions [8,12,32]. Previous agricultural ABM research suggests that, to account for a fuller complexity of farmer decision making and provide room for experimentation aimed at sustainable resource use, agent behavior should be informed by both normative science (economics) and social science that more realistically represents the actual resource-use decision making [33]. Thus, in our model, the socioeconomic micro-decisions of agents, strengthened by the interactions among them, generate macro-structures of the system including the level of organic farming adoption. The goal of this structure is to determine how landscape-scale patterns of adoption emerge from individual farmer decision-making.

2. Study Area and Research Questions

The Philippines, like many countries in the developing world, has passed legislation supporting organic agriculture, citing its benefits to human health and the environment [34]. Acreage under organic agriculture has risen steeply in recent years, from approximately 3500 hectares in 2004 to 52,400 hectares in 2009 [35]. This represents only 0.6% of the country's 9.2 million hectares of cultivated land area. The Philippine commitment to organic agriculture is nonetheless significant, given the country's history of promoting chemically intensive technologies to raise yields of rice, a staple crop [36]. Achieving self-sufficiency in rice production is another important goal of the Philippine government (the country consistently ranks as one of the top five rice importers) [37]. This necessitates the country increasing rice yields to the levels of some of its more productive Asian neighbors [38]. Achieving both higher rice yields and greater amounts of land under organic production will require investments in programs that target smallholder farmers, who constitute the majority of rice producers in the Philippines.

We chose Negros Island, in the central Philippines, as a study site (Figure 1). The island is located in the country's agricultural belt; major crops include sugarcane, maize, rice and coconut. While the rice-growing areas just north of Manila are known as the country's rice bowl, rice lands on Negros are also highly productive, according to the Philippine Bureau of Agricultural Statistics (www.bas.gov.ph); three harvests per year are standard.

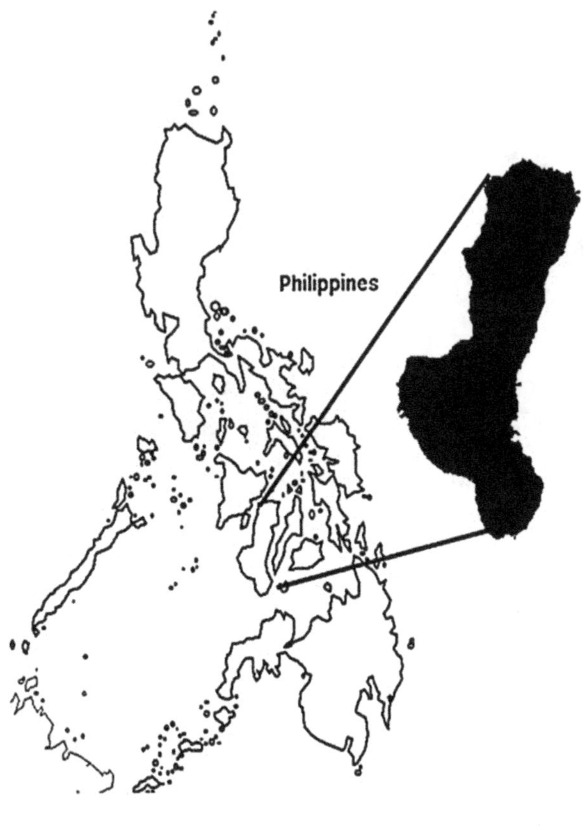

Figure 1. Map of the Philippines showing location of Negros Island where the farmer focus groups were conducted in an upland rice-growing area.

The Negros Institute for Rural Development (NIRD) is an internationally funded and locally governed NGO, which has been promoting organic agricultural techniques in an upland rice producing area on Negros since 1999. The municipality of Canlaon, where NIRD is located, is therefore an ideal location for studying the spread of organic technology adoption. According to NIRD staff, some farmers in the area have adopted organic techniques and used them consistently, while a larger number of farmers converted to organic methods but then went back to conventional methods after some time. This presents an intriguing research question which we addressed in our model: Why do some farmers on Negros go organic, while others do not? Put another way, what are the driving factors determining the rate of organic adoption over time which Negros has experienced?

3. Data Collection

Focus Group Discussions and Model Parameterization

We collected both qualitative and quantitative data from the study site to parameterize the model. We organized focus groups in November, 2010, to assess the state of organic agriculture adoption in the Canlaon region, and to identify barriers to adoption. Because NIRD introduced organic methods to this region and conducts periodic workshops and trainings in organic techniques, they are in contact with most of the organic farmers in the area. Ten farmers who had practiced organic methods for at least the past three years were able to attend the focus group sessions, and we constructed groups of similar size to represent conventional farmers (eight participants) and farmers who had tried organic farming but had returned to conventional methods (ten participants). Organic farmers were over-represented in this sample compared to their presence on the landscape, because the purpose of the focus group exercise was to examine the motivations and decision processes of farmers who choose to go organic.

The groups were asked to discuss the questions: (1) Why do you farm using your chosen method (organic/conventional)? (2) What are the benefits of using your chosen farming method? and (3) What are the challenges of using your chosen farming method? A note-taker was present in each group to record key concepts as the discussions took place. Afterward, the three groups were brought together in a common forum to share their observations and ask one another questions. These sessions were recorded and coded using three categories (Farm Characteristics, Organic Adoption, and Challenges) and fifteen sub-categories (farm size; family size; years farming; general history (any information about the farm not covered under other categories); markets (where farm products are sold); off-farm income; livestock; fertilizer use; history using organic; perceived benefits of organic; characteristics of organic (aspects of organic technologies that farmers discussed, other than benefits); pest management; yield; knowledge of organic; labor). All conversations took place in Cebuano-Visayan, the local language. The model decision structure represented in Figure 2 was developed to reflect, as closely as possible, the decision-making process around organic conversion described to us by farmers. Both organic and conventional farmers agreed on the importance of an "experimentation" period for organic conversion, in which they would grow organically on one portion of land to test the efficacy of this method. In addition, all farmers agreed that, in order to farm organically, one has to know how to use organic methods and be able to bear the initial cost of changeover from conventional to organic methods.

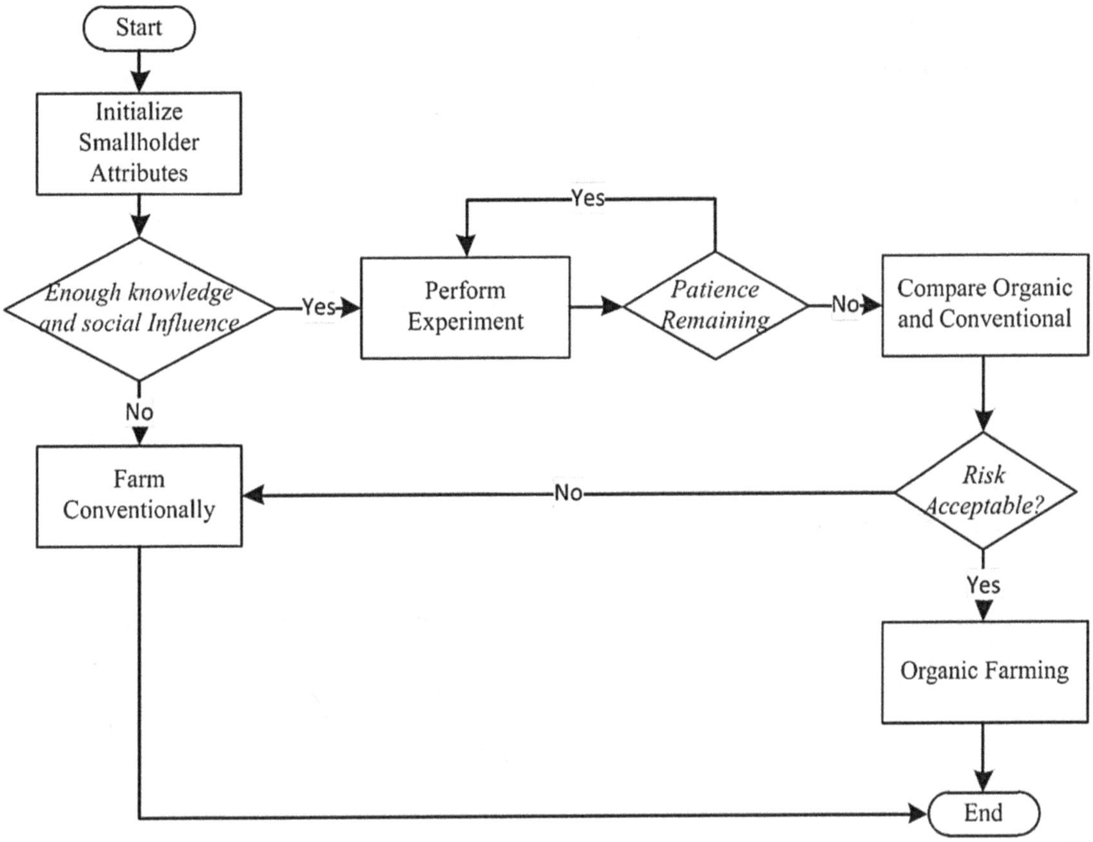

Figure 2. The decision-making process used by farmer agents in the agent-based modeling (ABM). In order to experiment with organic agriculture (plant half of their land using organic methods), farmers must have knowledge of organic techniques. Once they begin an experiment with organic techniques, they wait a certain number of rice seasons to compare their yields under organic production with their yields under conventional production. The number of seasons they are willing to wait is designated by "patience", a characteristic of each farmer agent that describes his/her dedication to organic production. After the finished experiment, they evaluate both methods while considering the risk imposed by external climatic and economic conditions. Then they chose which method to adopt.

From June 2010 through December 2010 (the length of one cropping season), NIRD field staff assisted the authors in collecting information on crop yields and farm size from the farmers who participated in the focus groups. A total of twenty-eight farmers participated in this data collection (ten organic farmers, eight conventional farmers, and ten mixed-method farmers). "Mixed-method" refers to farmers who at some point have tried organic farming, but do not do so consistently and may have reverted to conventional farming at the time of data collection. This sample was not intended to be statistically representative for the larger population of rice farmers; rather, the information was to be used to develop the farm characteristics represented in the model.

The average farm size of the participants was 0.9 hectares (0.63 ha for the organic group; 0.85 ha for the conventional group, and 1.1 ha for the mixed-methods group). This is smaller than the 2.8 ha average farm size for the island [39]. Given the highly unequal distribution of farm acreage between large-scale sugarcane plantations and smallholder rice and vegetable farms on Negros Island, the average farm size of the study participants represents the landholdings of small-scale farmers [40]. The average yield for one cropping season reported by the organic group was 3.0 MT/ha; for the conventional farmers it was 5.4 MT/ha; and for the mixed-methods farmers it was 4.5 MT/ha. We believe the organic yields were lower for two reasons: first, some of the organic farmers reported that access to organic fertilizers was a problem, and they may not have been adding adequate levels of nutrients to ensure yields. Secondly, some of the farmers who converted to organic methods in the recent past may have been suffering from the initial yield loss at changeover (see below for more information). However, we would caution against using one cropping season's data from a limited sample of farmers to depict an overall pattern. For this reason, we used a wider yield distribution drawn from regional datasets to parameterize our model simulation.

Among organic farmers who participated in the focus group, concern for the environment and for their family's health were agreed to be key reasons for choosing organic methods. Participants also noted that organic methods reduce input costs and therefore improve net profit. In addition, ducks and fish can be integrated into paddy rice production when organic methods are used, while chemical herbicides and pesticides are harmful to these animals. The incorporation of animal agriculture/aquaculture with rice production was clearly seen as desirable by organic farmers, who mentioned the enhanced productivity that this integration affords, and the weed and pest-controlling characteristics of ducks. These observations are consistent with literature studies of paddy rice agriculture elsewhere in Asia [41]. Price premium for organic rice was not a determinant in the choice of organic methods for focus group participants, because the farmers do not have access to customers who might be willing to pay a premium.

Farmers who used conventional techniques, or who had switched between organic and conventional techniques, were not as strongly devoted to their method of production as the organic farmers. In fact, most of the farmers in these groups stated that they try to incorporate some organic techniques, such as mixing animal manure with chemical fertilizer, or using organic pest control methods. The most significant barriers to organic farming mentioned were labor requirements (mainly for composting and weeding), lack of nutrient inputs (particularly for farmers without animals), and lack of information about organic techniques. Some conventional farmers, but not all, believed that high-yielding varieties of rice (HYVs) were preferable to traditional varieties for their consistent high yields, and these HYVs require chemical pest control. Farmers also stated that they are sometimes forced to use chemical inputs by creditors, who view these inputs as necessary to guarantee their return on investment. Although we didn't explicitly model this requirement as part of the simulation, we noted that without access to credit, most farmers would not be able to produce. This statement is therefore well worth further examination in the context of promoting organic agriculture.

4. Methods

4.1. Agent-Based Model

The ABM simulates a virtual farming area containing 2500 farmer agents who must choose between organic and conventional farming methods over the course of 100 cropping seasons (approximately 33 years). We chose the number 2500 because this is a reasonable approximation of the number of rice farmers that live in the Canlaon region; this number therefore represents the "upper limit" of organic diffusion, if all farmers were to adopt organic technologies [39]. Each farmer agent represents a farm household, and all agents operate in a virtual "social space" (a concept which is relevant for the social network learning described below), but the model is not spatially explicit. We built the model using Repast Simphony software and programmed it using Java language. The farmer agents all have the same decision-making rule (see Figure 2), but since their attributes are different, their decision outcomes are also different. This decision-making framework was developed based on the focus group discussions, and reflects the considerations of the farmers as described in the groups.

4.1.1. Farmer Agent Specification

The literature on organic farming adoption indicates that economic benefits can be a significant driver in farmers' decisions to convert to organic methods [42]. Several studies have also pointed to farmer characteristics that influence whether farmers choose to convert to organic methods. These include age, gender and education of the producer [18,43]; farmers' social and ecological values [44,45]; sources of information about farming [18]; and, possibly, the desire to become more resilient to input price shocks or climatic shocks [46]. All of these empirical observations were used during agent conceptualization of our ABM of organic farming, together with the focus group results.

The agents were assigned values for land area, animal ownership, and number of people in the household (corresponding to household food demand and labor availability), taken from the range of observed values for these variables in the field and in Philippine agricultural statistics, and assuming a uniform distribution across the range [47] (See Table 1 for range of values assigned to agent characteristics; more details about the model and its parameterization can be found in Supplementary Section 1). We used uniform distributions for all our input variables because the small sample of participants that served as the source of input to our agent attributes was insufficient to generate more complex empirical probability density functions. The rationale behind this distribution is that it requires at most two parameters (*i.e.*, upper/lower bounds) identified from data. Initial capital assets were also randomly distributed across a uniform range, but after the start of model simulation capital assets were calculated for each time step based on yield.

We constructed a stylized social network for the farmer agents to represent the ways in which farmers influence their social connections and share information about organic techniques. The structure of the network was informed by qualitative data (*i.e.*, how farmers described receiving information). The farmer agents were assigned values for "social reach", the unitless maximum distance between nodes which, in this case, were connected farmers (*i.e.*, the extent of influence). Any two farmers within each other's social reach were connected by a link. Each link creates a dyad and each farmer in the dyad is a social connection to the other. These links were the channels through which the model passed information about organic methods; farmers could not experiment with organic techniques unless they were connected to a farmer who already used them. The links represent the farmer's social influence on his/her networks. The values of social reach were initially set at 5, and then drawn from a uniform distribution during our computational experiment of sensitivity analysis, because we did not collect information on the structure of the social network at the field site [48].

Table 1. Factors (parameters) used in the model with their respective probability distributions (U—continuous uniform, D—discrete uniform).

Input Factor	Definition and Units	Distribution
Area of land (a)	Hectares	U = (0.25, 10)
Labor availability (la)	People in household	D = {1, 2, 3, …, 8}
Social reach (sr)	Distance in agent's social space (unitless)	U = (5, 25)
Social influence (si)	Density of social connections among agents	D = {1, 2, 3, …, 30}
Animal ownership (ao)	Proportion of farmers owning animals (unitless)	U = (0, 1)
Patience (p)	Number of farming seasons (3 seasons/year)	D = {1, 2, 3, …, 15}
Risk taken (r)	Attitude towards risk related with the adoption of organic farming (unitless)	U= (0, 1)
Cost of conventional fertilizer (ccf)	Philippine Pesos per kilo	U = (40, 400)
Cost of organic fertilizer (cof)	Philippine pesos per kilo	U = (40, 400)
Average cost of other conventional input (coci)	Philippine pesos per hectare	U = (3000, 7500)
Average cost of other organic input (cooi)	Philippine pesos per hectare	U = (1100, 2900)
Average cost of labor (cl)	Philippine pesos	U = (6000, 13,000)
Average price of rice (p)	Philippine pesos per kilo	U = (7.44, 17.86)
Average yield (y)	Kilos per hectare	U = (1200, 6000)
Organic fertilizer threshold (oft)	Hectares	U = (0,10)
Land area for one labor (lal)	Hectares/person	U = (0.1,10)

Farmer agents were initially randomly distributed in unitless social space. To initialize the model, we set up the network structure by applying Hamil and Gilbert's methods [48]. A virtual grid space is created with the dimension of 100 × 100. Therefore, the space is divided into 10,000 grid cells. Then 2500 agents are created, and each of them occupies a cell. Each cell can only contain one agent. The values of the social attributes are initialized following the distributions mentioned above. After the attributes are set, agents are connected based on the values of social reach. Any two agents ***a*** and ***b*** are connected if Equation (1) is satisfied.

$$distance(\boldsymbol{a},\boldsymbol{b}) = \sqrt{(x_a - x_b)^2 + (y_a - y_b)^2} \leq \min(social\ reach(\boldsymbol{a}), social\ reach(\boldsymbol{b})) \quad (1)$$

The model was "seeded" with between 5 and 10 organic farmer agents. This is representative of the pioneer smallholder farmers in the Canlaon region who learned organic farming through NIRD when the NGO first was established there. Figure 3 depicts the social network for one model run.

Farmer agents were also assigned a value for the attribute "patience", and a value for "risk attitude". The patience attribute represented the number of rice cropping seasons a farmer was willing to experiment with organic methods before comparing his or her net profit under organic farming with net profit under conventional methods. "Patience" therefore represented all of the values a farmer might hold that might prompt him or her to consider organic farming, even if it was not initially as productive as conventional farming.

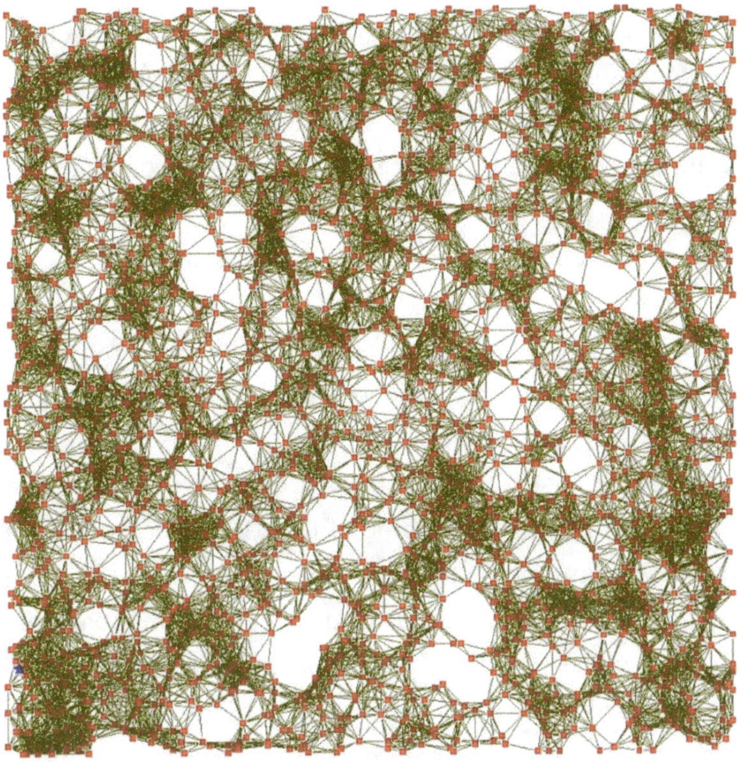

Figure 3. Social network depicted at the end of a model run with the "social reach" variable set at 20. The agents are distributed randomly in social space; circles represent farmer agents, and lines represent social connections between the farmer agents.

"Risk attitude" was an index assigned to each farmer agent reflecting their willingness to try a new farming method after their field experiment (in this case, organic farming). This index varied between 0 and 1 for each agent, and was compared to a randomly generated "risk index" between 0 and 1 for each farming season, representing the time-dependent environmental riskiness of trying a new technology (encompassing, for example, weather and economic conditions, political instability, *etc.*). Notice that "risk attitude" was only used to determine which farming method to choose when a given farmer agent expected profit from both organic farming and conventional farming. If a farmer had a relatively low risk attitude, indicating he/she was risk taking, he/she would be willing to try new technologies when expected profits were both positive. Conversely, if a farmer had a relatively high risk attitude, indicating risk aversion, he/she would almost never be willing to try organic farming. The qualitative and quantitative evidence from the field support this depiction of conversion to organic methods as a significant risk. Studies suggest an initial yield loss associated with organic conversion, which represents a risk to a farmer's income stream and household food supply [4].

4.1.2. Decision Algorithm

Before the start of each time step, a farmer agent must decide whether to experiment with organic farming (Figure 2). This can only take place if the farmer is connected to other farmer agents with adequate knowledge of organic techniques. The model assumes that this knowledge is disseminated through the social network, so if a farmer agent is connected to a certain number of organic farmers or farmers experimenting with organic techniques, he/she is able to begin an experiment. We assumed the number, which is called "influence threshold", could be drawn from a discrete uniform distribution between 1 and 8, representing a minimum and maximum number of people a farmer might reasonably turn to for farming advice. Once the condition was satisfied, the farmer agent would begin his/her experiment. This means that he/she plants half of his/her land using the organic techniques and half using the conventional techniques.

Input costs were calculated for each farmer agent using a given farming technique at each model time step. These costs included the cost of fertilizer, the cost of other inputs, and the cost of labor (if labor provided by the household was not sufficient). To simulate the fluctuation of input costs across the model run time, the amount in Philippine pesos assigned to labor, fertilizer and input costs for a given time step was drawn randomly from a uniform distribution based on data collected in the field and from the Philippine Bureau of Agricultural Statistics (see Table 1). Conventional farmers must purchase all of their fertilizer, but organic farmers may derive some of their fertilizer from composting agricultural waste (such as manure and rice straw), and purchase organic fertilizer for their remaining needs. We assumed that farmers with at least one hectare of land and one animal were able to produce all of their fertilizer on-farm, up to 10 hectares (this threshold was based on empirical observations, and tested with the sensitivity analysis described below). The amount of fertilizer applied was assumed to be up to 1.5 times the average nitrogen application for the region recorded in 2010 agricultural statistics (organic and conventional fertilizer types are adjusted by their respective nitrogen contents), based on what the farmer-agent could afford.

Yield was calculated as a function of the nitrogen application for either conventional or organic methods, using a Michaelis-Menten equation applied to rice yields and fertilizer application rates from provincial statistics:

$$yield = (land\ area \times 1.0045^{seasoncount} \times 4800 \times fertilizer\ amount)/(fertilizer + 6.1) \tag{2}$$

where *fertilizer amount* is the maximum nitrogen application rate a farmer can afford, and *seasoncount* is the model time step [49]. Data from the region indicate that conventionally grown rice yields increase at a rate of 0.45% per season, and this annual increase was simulated in the model. When a farmer agent switched their land to organic methods (or planted half of their land in organic methods as an experiment), the modeled yield initially dropped by 50% before recovering at a rate of 5% per season to the yields described by the Michaelis-Menten equation above. Therefore, after approximately 10 seasons (or three years), yields for a farmer who had converted to organic methods would be no different than if he/she had continued to farm conventionally. This initial yield drop represents a high estimate of initial yield losses reported in field experiments by organic farmers, due to the need for soil organic matter and soil chemistry to recover post-chemical fertilizer application [4].

A farm household consumes part of the rice (enough to feed the household members during the cropping season) and sells the remainder, if any. The price of rice was also drawn randomly from a uniform distribution based on recent farmgate prices reported by the Bureau of Agricultural Statistics (see Table 1). The sale of the rice provides the capital input necessary for the rice season that follows. It is important to note that organic rice is sold at the same price as conventionally grown rice in our modeled environment, because farmers in the Canlaon region do not have access to markets at which organic rice might fetch a premium due to transportation constraints.

As mentioned above, each farmer agent was randomly assigned a value for the variable "patience", which represents the length of the experiment he/she conducts before comparing the net profit from conventional farming methods to the net profit from organic farming methods and choosing the method that is more profitable. In the model, this number ranged from one to 15 farming seasons (based on the farmer focus groups, we thought five years would be the maximum amount of time a farmer would experiment with an organic technology). The more "patience" an agent has, the more likely he/she is to choose organic farming, because once the yields have recovered from the initial loss caused by switching to organic methods, input costs tend to be lower for organic farmers, who can produce inputs on-farm rather than buying them.

Based on the presented specification, our model can be placed somewhere on the continuum between the highly abstract agricultural ABM that result in stylized simulations [29] and empirically-rich ABM equipped with microeconomic mechanisms and complex agricultural markets [20,30]. This mixed approach, combining observational and simulated data in computational experimentation, has been successfully applied in other agricultural ABM [14,23].

4.2. Exploring Model Outcome Variability with Sensitivity Analysis

Sensitivity analysis of agricultural ABM is rarely undertaken. Notable examples include Happe *et al.* [30] and Schouten *et al.* [50] where regression-based metamodels are developed to evaluate the influence of the uncertain model inputs on outputs. In their study, Schouten *et al.* [50] also employ a simple one-parameter-at-a-time (OAT) sensitivity analysis and compare it with the regression metamodel.

The ABM described above is an example of a dynamic model that emulates a complex agricultural system. It is, therefore, imperative to apply a simulation procedure appropriate to complex nonlinear models, which would account for input and output variability as well as the potential input interactions. Neither regression nor OAT meet these requirements. Crosetto *et al.* [51], Gomez-Delgado and Tarantola [52], Chu-Agor *et al.* [53], and Ligmann-Zielinska *et al.* [11] among others, postulate the use of Monte Carlo simulation that incorporates global sensitivity analysis (GSA) as a part of computational experiments. GSA is a method of experimentation in which the variability of model results is quantified based on simultaneous sampling of the whole set of input variables, which are then examined individually and in combinations [54].

The most comprehensive method of GSA is based on model-independent output variance decomposition, in which model outcome variability (represented using variance V) is apportioned to various model inputs [54,55], so that the underlying causes of variable outcomes can be explicitly identified. The procedure starts from generating M samples of input variable values using a selected experimental design. The model is then executed for each sample $m \in M$ and the result is recorded. These output values form a distribution which can be summarized using descriptive statistics like mean and variance (V). GSA then uses output variance decomposition which partitions V based on the contribution of each input variable to V. This partitioning (aka decomposition) is accomplished by estimating the conditional variances of every input variable k. By calculating the ratio of conditional variance due to k to the total V, we obtain a first order sensitivity index (Sk). If a given variable k has a relatively high value of Sk, its single influence on output variability is substantial. To express the interactions among variables, which are ubiquitous in complex system models, we also compute a total effect index for every k (STk), which accounts for all higher-order effects of inputs [56]. Consequently, variables with relatively low values of the (Sk,STk) pairs are deemed unimportant in shaping output variability and, in the consecutive experiments, can be set to constant values contributing to model simplification. In addition, the (Sk,STk) pairs provide valuable information on the mechanisms that affect the dynamics of the model and, consequently, can serve as quantitative indicators explaining model drivers. For details on (Sk,STk) calculation the reader is referred to Saltelli *et al.* [54].

To calculate the (Sk,STk) pairs we employed the quasi-random Sobol experimental design described in Saltelli (2002), which proved to be the most effective in approximating the values of sensitivity indices [57]. To compute the indices, we used SimLab open source software (http://ipsc.jrc.ec.europa.eu/?id=756).

4.3. Computational Experiments

Following the quasi-random experimental design mentioned above, we employed Monte Carlo simulation and executed the model M times, where M was set to 17,408 runs based on the procedure described in Ligmann-Zielinska and Sun [58]. At the end of each simulation, we recorded the number of agents who adopted organic farming (N) and compared that to the total number of farmers (T) in the model using a simple ratio N/T, referred to as RATIO in the following sections. Given that the focus of our experimentation is organic farming adoption, we set the number of farmers to 2500 (*i.e.*, T = 2500), with a time step of one cropping season, and ran the model for 100 seasons. The results were analyzed in two ways. First, we compiled the RATIO values into a probability distribution and summarized it using descriptive statistics (including average and variance). Second, we performed variance decomposition of the RATIO distribution, and calculated Sk and STk for each of the 16 input variables (a total of 32 indices plus an interaction index) to investigate the overall model behavior and determine its critical drivers. The interpretation of the pairs of indices (Sk, STk) followed the procedure described in Ligmann-Zielinska and Sun [58].

5. Results

Agent-Based Model

By the end of the model simulation, all farmer agents had the opportunity to experiment with organic techniques, as the knowledge of organic agriculture spread throughout the social network. Experimentation concluded by rice season 23, at which point all farmer agents selected a farming type. The number of organic adopters in the model leveled out after experimentation in the model run using default parameter values (Table 1, Supplementary Section 1) (Figure 4). This ratio of organic farmers to total farmers was robust to the seasonal variation in labor, fertilizer, and input prices, as well as model-simulated seasonal variation in yields, as indicated by the fact that after the initial decision period, the ratio of organic to total farmers did not change significantly. This result is qualitatively similar to the patterns of adoption reported to us by the NGO operating in the Canlaon region, and by the farmers who participated in the focus group. Both NGO workers and farmers confirmed that, after an initial period when many farmers seemed to be adopting organic methods, the adoption rate leveled off and most farmers did not switch to organic despite learning about organic methods. This observation is similar to the conclusion derived from the study by Deffuant *et al.* [26] which suggests that, while sympathetic to organic agriculture, farmers did not convert to organic even though they were exposed to positive messages from the public.

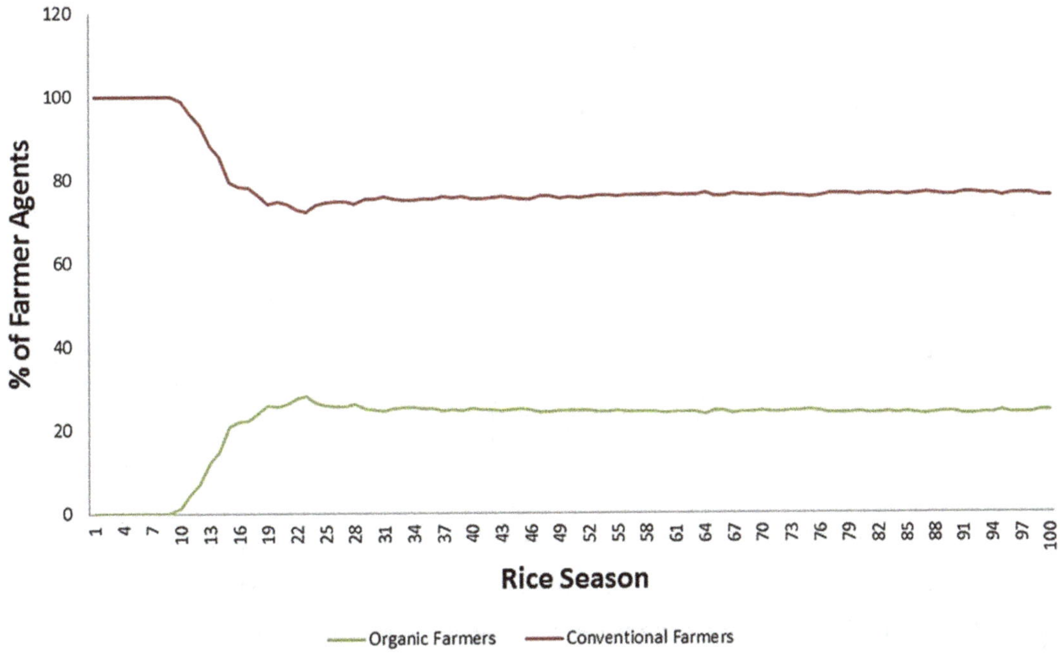

Figure 4. Model output depicted number of organic farmers and number of conventional farmers over the 100 cropping seasons simulated in the model, at default parameter values (see supplementary for description).

The box plot in Figure 5 summarizes the distribution of RATIO. The Monte Carlo simulations result in RATIO ranging from 0% adoption of organic farming to the maximum of 60% organic farming adoption. The mean adoption of organic farming equals 31% of the total agent population, with a substantial level of variability (std = 33%), which we analyze in the following section.

Sensitivity Analysis: Variance Decomposition of Model Results

The results of variance decomposition of the ratio of organic farmers to total farmer agents (RATIO) are depicted in Figure 6. As mentioned above, model sensitivity to different inputs is analyzed separately for their individual influence (Sk) and their total (STk) influence. With the first order sensitivity index (Sk) we look for important input variables that, if fixed independently, would reduce the variance of RATIO the most. In other words, the variables with relatively high S values have the most impact on the variability of the adoption of organic farming. In our experimentation, the social reach (defined in Section 4.1.1) scores highest (Ssr = 20% of total variance), followed by yield (Sy = 15%), and area of land under cultivation (Sa = 12%).

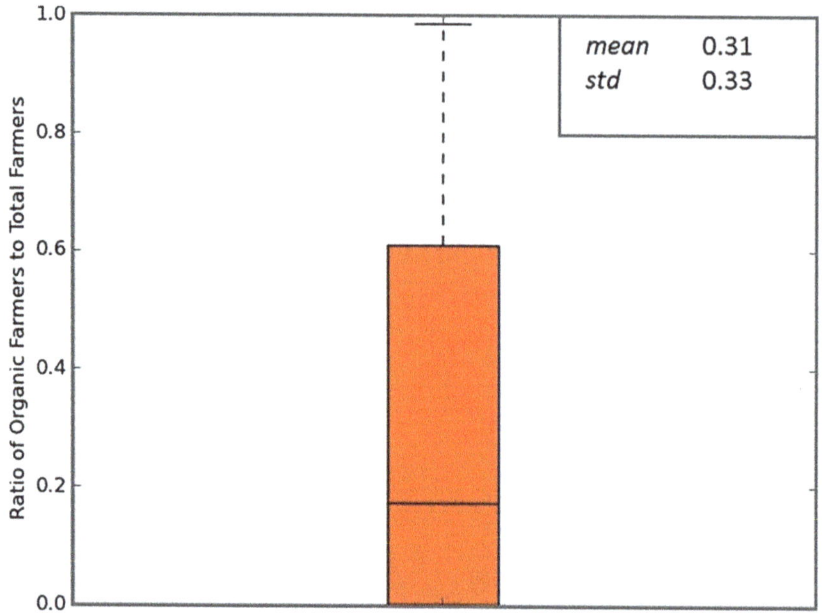

Figure 5. Distribution of the ratio of organic farmer agents to total farmer agents at the end of the model run (100 cropping seasons), over all Monte Carlo sensitivity simulations. This ratio ranges from 0% of farmer agents adopting organic methods to a maximum of 60% of farmer agents adopting organic methods, with a mean adoption rate of 31%.

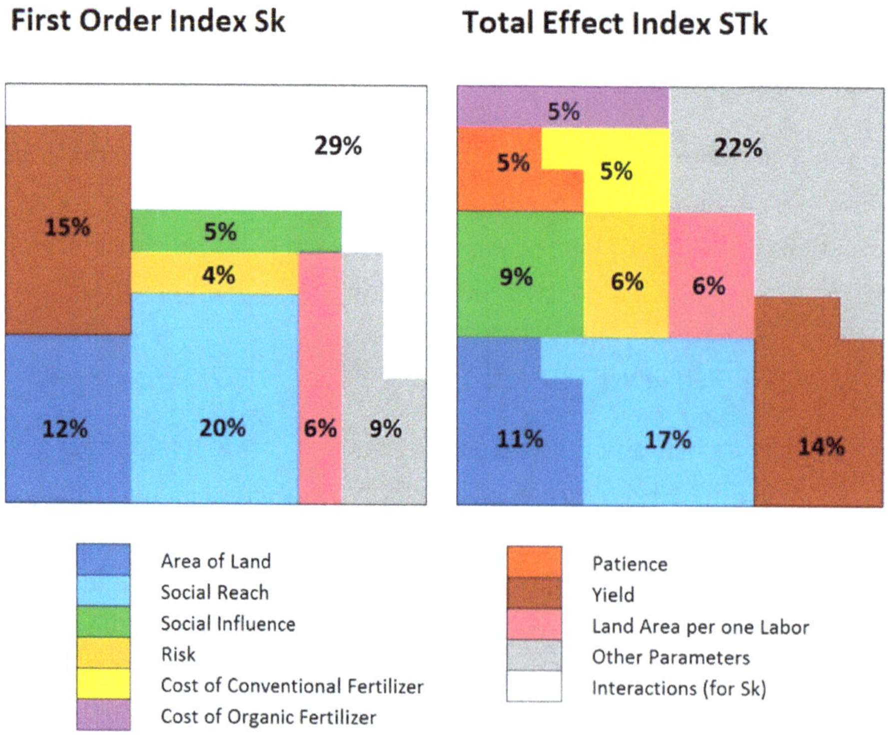

Figure 6. Variance decomposition of variable effects on model output (ratio of organic-adopting farmer agents to total farmer agents). Model sensitivity to different input variables is analyzed separately for their first order influence (Sk) and their total effects (STk) influence, in which variable interactions are explicitly quantified. The variables with relatively high S values have the most impact on the variability of the proportion of farmer agents who adopt organic methods.

The model's nonlinear behavior is expressed by the relatively high level of variable interactivity, which is derived from the percentage sum of all first order indices. In our simulation, this sum amounts to 71% of RATIO variability (Figure 6). Therefore, only 71% of the variance in organic farming adoption rates can be explained by analyzing the variables in isolation. The remaining 29% of V is attributed to variable interdependence, which is expressed using the total effect index (STk).

Over half of the variables (9 out of 16) contribute substantially to the complex behavior of the model. The highest interaction effect can be observed for social reach (STsr = 17%), yield (STy = 14%), area of farmland (STa = 11%), and influence threshold *i.e.*, the density of social connections of any given agent (STsi = 9%). Consequently, not only do these four variables individually contribute to the variability of the RATIO, but they also have a strong influence when analyzed as a group.

Other variables that somewhat contribute to RATIO variability include the cost of fertilizer (both conventional, where STccf = 5%, and organic, where STcof = 5%), labor availability (STlal = 6%), attitude to risk taking (STr = 6%), and patience (STp = 6%).

6. Discussion

There appears to be an upper limit on the number of farmers choosing to go organic that is fairly robust to fluctuations in prices and the external risk environment. The results of the sensitivity analysis, which reveal how influential the various model variables are in determining the ratio of organic adopting farmers to total farmers, gives us insight into why this limit exists. The sensitivity analysis suggests that the decision to shift from traditional to organic farming is mainly driven by yield, and by imitative behavior, that is, by the number of farmer agents that have already adopted organic farming, as indicated by the strong effects of social reach and influence threshold on the organic adoption ratio (Figure 6). Consequently, our ABM is a classic example of simulating innovation diffusion, where the decision made by agent (A) is influenced by the decisions of proximal agents in A's social space [59–61]. The fact that social reach and influence threshold account for 26% of output variance corroborates our postulate about the role of diffusion of innovation in explaining organic farming adoption in our ABM.

Particularly, social reach and influence threshold affect the rate at which farmers "learn" about organic farming, a prerequisite to their experimentation with organic technologies. This is important in the model context because of the assumption that conventional yields rise over time. The more rice cropping seasons that go by before a farmer agent learns about organic technologies and is able to experiment with them, the higher the baseline conventional yields to which he/she compares organic yields—therefore, the less likely it is that he/she will choose to switch to organic. When yields are higher, the loss of 50% of yield during switchover is a greater absolute yield loss compared to when yields are lower. For example, if conventional yields are one ton/ha, upon switchover organic yields would be 0.5 ton/ha. Given the lower costs of organic farming, the farmer might be able to make up this 0.5 ton/ha difference between organic and conventional yields in net profit by going organic. However, if yields are 5 tons/ha, the yield loss upon switchover would be 2.5 tons/ha—a greater profit gap to make up.

We can also conclude that the adoption decision is dependent on the land area an agent farms. Land area constraints on organic farming adoption were discussed in the farmer focus groups, and the model's sensitivity to this variable confirms that farmers with limited land may not be able to generate the nutrients they need to use as inputs for organic production. However, the model was not sensitive to the "organic threshold", or the amount of land required to produce on-farm organic fertilizer. This provides evidence for rejecting our hypothesis that only farmers who can produce their own fertilizer on-farm will be able to use organic methods. Interestingly, none of the variables that would seem to have a direct influence on the variability of organic farming (like cost of organic fertilizer and cost of other organic inputs) drive the switch from conventional to organic farming, suggesting that cost comparisons between farming methods may not comprise a farmer's main motivation for adopting organic.

Patience and fertilizer costs play particularly intriguing roles in determining organic adoption rates. When treated independently, these variables have negligible influence on the adoption of organic farming (their Sk values equal zero). However, their influence becomes much more pronounced when analyzed in interactions with other variables, suggesting that they play a complex intertwined role in the decision to switch cultivation to organic farming. This suggests that value orientations towards promoting family health and health of the environment, which were prominent topics of discussion in the focus groups, do not by themselves determine whether a given farmer will choose to adopt organic methods. Rather, knowledge of organic methods and the ability to produce organic fertilizer may be pre-requisites for a farmer putting his or her pro-health or pro-environment values into action. Farmers who have not been able to practice organic methods may nonetheless be interested in integrated pest management techniques, and this was a major topic of discussion in the focus group sessions [62].

Based on the sensitivity analysis we can also suggest the direction of ABM simplification. In particular, animal ownership, labor availability, crop price, labor costs, the amount of land needed to produce organic fertilizer on-farm (organic threshold) and other organic and conventional costs proved to be unimportant in shaping the variability of the RATIO statistics. Consequently, an equivalent simplified model would have these variables set to constant average values, reducing the dimensionality of the model from 16 to 9 uncertain inputs. This simplified model would be equivalent to its predecessor in that it would maintain the variability of RATIO (*i.e.*, with V unchanged) [11,54]. This property can be of value in policy analysis where a certain level of result variability is required to account for any unanticipated events (surprises) in the target system. At the same time, a model with a smaller number of variables can be easier for stakeholders and decision-makers to understand. The drawback of our model simplification is that it is dependent on the type and form of the output variable. Every time we substitute RATIO with a different output of interest (e.g., we introduce different categories of organic farmers based on their land management practices or household income), we should revisit the initial ABM formulation and perform a separate variance-based model simplification.

Implications and Limitations

Quantitative data on organic adoption over time, which would allow for more extensive validation of the model, is not yet available from this region. However, these model results are consistent with anecdotal observations collected by NIRD and expert opinion of in-region farmers—for example, the NGO has observed that, despite a long-term presence in the region, the number of farmers who have decided to convert to organic methods remains limited. Similarly, the model depicted nearly all farmers experimenting with organic methods after encountering these methods through their social network, but only a minority of the farmer agents ended the model run by adopting organic. This confirms the model's usefulness as a scoping exercise that can shed light on where future research and farmer support activities should be focused.

We can suggest that further research would be warranted into the social network among farmers in this region, and farmers' means of sharing information, since social reach and influence threshold were both found to be significant in determining rates of organic adoption. Our model simulation used a uniform, random social network to represent information sharing, which obviously does not reflect the network topology in a typical community [27]. Moreover, there is no feedback effect from farmers observing organic experiments which have been abandoned—it is possible that if a significant number of these occurred in their social network, they might be more reluctant to try organic agriculture in the first place. However, it is worth noting that this dynamic was not mentioned in the focus groups as a factor in determining organic experimentation.

An interesting implication of the model output seems to be that, as conventional yields rise due to improved crop varieties and other variables, at some point it may be "too late" for farmers to consider going organic, because conventional yields would be high enough that the 50% loss occasioned by organic switchover would represent too large of a yield gap for the farmer. The significance of the yield variable in determining organic adoption rates confirms this interpretation of the model. Therefore, if organic adoption is desirable in a given area, informing interested farmers about organic techniques relatively quickly (possibly through strategic use of social networks), while minimizing the initial yield losses from switching to organic methods and boosting organic yields as quickly as possible to compete with conventional yields (for example, through intensive soil fertility amendments), would seem to be the best set of strategies. To return to the original research question proposed by our NGO partner NIRD, we seem to have identified two major drivers of organic adoption: yield and information sharing through social networks. Additional research on how these variables change over time should help to confirm this determination.

We did not collect detailed demographic data on the farmers who participated in the focus groups, which limited the degree to which we were able to assess how organic adoption rates may differ by gender, age, years spent farming, *etc.* It is likely that these demographic variables also affect the nature of the social networks in which farmers are embedded. The topology of the social networks represented in the model was therefore totally under-developed, and warrants further exploration in a more sophisticated modeling framework. In addition, our data collection was limited to one rice cropping season by the time and budget available for this study. Given that we are simulating a dynamic system, observing empirical trends over multiple seasons would give us more insight into whether the behavior simulated in the model is similar to the behavior of the real-world system. One of the reasons organic adoption is being promoted worldwide is for resilience to climate change-induced drought and other extremes in precipitation and temperature patterns [63]. An improved version of this model would include such shocks, in order to observe their effects on organic adoption and yields compared to conventional yields. In addition, economic shocks such as restriction to credit or large spikes in prices (larger than we modeled here) could shed light on the relative resilience of organic techniques compared to conventional techniques.

Another shortcoming of our model is its lack of spatial information. Clearly, many aspects of farm production are highly dependent on that farm's location in the landscape, and one would expect organic adoption to be affected by spatial characteristics such as farm slope, proximity to roads and water sources, soil characteristics, *etc.* We hope to include spatial aspects in a future version of this model.

Importantly, we were using the ABM as an exploratory, rather than a predictive, model [64,65]. Given the complex and uncertain nature of the simulated upland farming system, and the paucity of data from the region, we believe a traditional quantitative model that seeks to make predictions about systemic outcomes is inappropriate [66]. Rather, our goal was to shed light on the dynamic aspects of the system described in questions 1 and 2 in the introduction, so that we might target future research and modeling efforts. This approach is philosophically different from the traditional use of statistical or optimization techniques to design a "best" model which produces a "right" answer [66], but it is no less powerful. Systems modelers have argued for decades that an optimization approach to a complex system with high levels of uncertainty can lead to incomplete consideration of system drivers at best [67], and misleading or erroneous conclusions at worst [68]. A dynamic simulation of a complex system, as undertaken here, can be used for theory development and hypothesis testing [69]. With more empirical data with which to calibrate and validate the model, however, this model could be used to make more targeted projections to inform policy in a given location.

7. Conclusions

We built a stylized agent-based model to explore the reasons behind low adoption rates of organic agriculture in a productive rice-growing region in the upland Philippines, where organic agriculture has been promoted for fourteen years. The model was run for multiple input variable sets that denote different socioeconomic, behavioral, and ecological characteristics of the farmers and the agricultural system. Given the uncertainty of the system, we explored the variability of the results using a variance-based sensitivity analysis framework to identify the core drivers of the decision to adopt organic farming. We found that the speed of information spread through the social network was highly influential in agents' decisions to adopt organic agriculture, because the longer it takes for them to hear about organic techniques and experiment with them, the more progress conventional varieties and techniques achieve in boosting yields, leading to opportunity costs for farmers who convert to organic. Land area is a constraint to organic adoption, because farmers with small fields are not able to generate on-farm organic inputs that are sufficient to maintain yields. The model revealed a high degree of complexity in farmers' decision-making, with interactive effects between decision variables explaining one-third of the variation in organic adoption rates. Yield improvements and information spread through social networks appear to be the major drivers of organic adoption, warranting further research and modeling attention, and perhaps programmatic targeting by our partner NGO. In addition, more research on farmers' decision-making processes is warranted, as these processes are complex and driven by interactions between economic, social and ecological factors. Continued interaction with farmers in the region will allow us to gain more insight into their decision-making environment and constraints.

Acknowledgments

We thank the Center for Advanced Studies in International Development at Michigan State University, which funded this research, and the Negros Oriental Institute for Rural Development.

Author Contributions

Laura Schmitt Olabisi oversaw the overall development of this project and is the lead writer of this paper. Laura Schmitt Olabisi performed the field interviews with stakeholders and processed the data. Ryan Qi Wang was the primary developer of the model and performed its calibration. Ryan Qi Wang is also the lead author of the Methods section and the Supplementary. Laura Schmitt Olabisi and Arika Ligmann-Zielinska conceived and designed the experiments. Ryan Qi Wang performed the experiments. Arika Ligmann-Zielinska performed and wrote the sensitivity analysis sections.

Conflicts of Interest

The authors declare no conflict of interest.

References

1. Willer, H., Kilcher, L., Eds. *The World of Organic Agriculture: Statistics & Emerging Trends 2009*; International Federation of Organic Agriculture Movements (IFOAM): Bonn, Germany, 2009.
2. Horrigan, L.; Lawrence, R.S.; Walker, P. How sustainable agriculture can address the environmental and human health harms of industrial agriculture. *Environ. Health Perspect.* **2002**, *110*, 445–456.
3. Pimentel, D. Green revolution agriculture and chemical hazards. *Sci. Total Environ.* **1996**, *188*(S1), S86–S98.
4. Mendoza, T.C. Evaluating the benefits of organic farming in rice agroecosystems in the Philippines. *J. Sustain. Agric.* **2004**, *24*, 93–115.
5. Eakin, H. Smallholder maize production and climatic risk: A case study from Mexico. *Clim. Change* **2000**, *45*, 19–36.
6. Uematsu, H.; Mishra, A.K. Organic farmers or conventional farmers: Where's the money? *Ecol. Econ.* **2012**, *78*, 55–62.
7. Burton, M.; Rigby, D.; Young, T. Modelling the adoption of organic horticultural technology in the UK using duration analysis. *Aust. J. Agric. Resour. Econ.* **2003**, *47*, 29–54.
8. Berger, T. Agent-based spatial models applied to agriculture: A simulation tool for technology diffusion, resource use changes and policy analysis. *Agric. Econ.* **2001**, *25*, 245–260.
9. Parker, D.C.; Manson, S.M.; Janssen, M.A.; Hoffman, M.J.; Deadman, P. Multi-agent systems for the simulation of land-use and land-cover change: A review. *Ann. Assoc. Am. Geogr.* **2003**, *93*, 314–337.
10. An, L.; Linderman, M.; Qi, J.; Shortridge, A.; Liu, J. Exploring complexity in a human-environment system: An agent-based spatial model for multidisciplinary and multiscale integration. *Ann. Assoc. Am. Geogr.* **2005**, *95*, 54–79.
11. Ligmann-Zielinska, A.; Kramer, D.B.; Cheruvelil, K.S.; Soranno, P.A. Using uncertainty and sensitivity analyses in socioecological agent-based models to improve their analytical performance and policy relevance. *PLoS ONE* **2014**, *9*, e109779.

12. Acosta-Michlik, L.; Espaldon, V. Assessing vulnerability of selected farming communities in the Philippines based on a behavioural model of agent's adaptation to global environmental change. *Glob. Environ. Change* **2008**, *18*, 554–563.
13. Valbuena, D.; Verburg, P.; Veldkamp, A; Bregt, A.K.; Ligtenberg, A. Effects of farmers' decisions on the landscape structure of a Dutch rural region: An agent-based approach. *Landsc. Urban Plan.* **2010**, *97*, 98–110.
14. Bert, F.E.; Podesta, G.P.; Rovere, S.L.; Menendez, A.N.; North, M.; Tatara, E.; Laciana, C.E.; Weber, E.; Toranzo, F.R. An agent based model to simulate structural and land use changes in agricultural systems of the argentine pampas. *Ecol. Model.* **2011**, *222*, 3486–3499.
15. Huang, Q.; Parker, D.C.; Sun, S.P.; Filatova, T. Effects of agent heterogeneity in the presence of a land-market: A systematic test in an agent-based laboratory. *Comput. Environ. Urban Syst.* **2013**, *41*, 188–203.
16. Goldenberg, J.; Libai, B.; Muller, E. Talk of the network: A complex systems look at the underlying process of word-of-mouth. *Mark. Lett.* **2001**, *12*, 211–223.
17. Allsop, D.T.; Bassett, B.R.; Hoskins, J.A. Word-of-Mouth research: Principles and applications. *J. Advert. Res.* **2007**, 47, 398–411.
18. Burton, M.; Rigby, D.; Young, T. Analysis of the determinants of adoption of organic horticultural techniques in the UK. *J. Agric. Econ.* **2008**, *50*, 47–63.
19. Conley, T.; Udry, C. Social learning through networks: The adoption of new agricultural technologies in Ghana. *Am. J. Agric. Econ.* **2001**, *83*, 668–673.
20. Schreinemachers, P.; Berger, T. An agent-based simulation model of human-environment interactions in agricultural systems. *Environ. Model. Softw.* **2011**, *26*, 845–859.
21. Matthews, R.B.; Gilbert, N.G.; Roach, A.; Polhill, J.G.; Gotts, N.M. Agent-based land-use models: A review of applications. *Landsc. Ecol.* **2007**, *22*, 1447–1459.
22. Kaye-Blake, W.; Li, F.Y.; Martin, A.M.; McDermott, A.; Rains, S.; Sinclair, S.; Kira, A. *Multi-Agent Simulation Models in Agriculture: A Review of Their Construction and Uses*; Lincoln University: Christchurch, New Zealand, 2010.
23. Bithell, M.; Brasington, J. Coupling agent-based models of subsistence farming with individual-based forest models and dynamic models of water distribution. *Environ. Model. Softw.* **2009**, *24*, 173–190.
24. Matthews, R. The People and Landscape Model (PALM): Towards full integration of human decision-making and biophysical simulation models. *Ecol. Model.* **2006**, *194*, 329–343.
25. Evans, T.P.; Phanvilay, K.; Fox, J.; Vogler, J. An agent-based model of agricultural innovation, land-cover change and household inequality: The transition from swidden cultivation to rubber plantations in Laos PDR. *J. Land Use Sci.* **2011**, *6*, 151–173.
26. Deffuant, G.; Huet, S.; Bousset, J.P.; Henriot, J.; Amon, G.; Weisbuch, G. Agent based simulation of organic farming conversion in Allier dèpartement. In *Complexity and Ecosystem Management: The Theory and Practice of Multi-Agent Systems*; Janssen, M.A., Ed.; Edward Elgar Publishers: Chelten-Ham, UK, 2002; pp. 158–187.
27. Kaufmann, P.; Stagl, S.; Franks, D.W. Simulating the diffusion of organic farming practices in two New EU Member States. *Ecol. Econ.* **2009**, *68*, 2580–2593.

28. Bakker, M.M.; van Doorn, A.M. Farmer-specific relationships between land use change and landscape factors: Introducing agents in empirical land use modelling. *Land Use Policy* **2009**, *26*, 809–817.
29. Gotts, N.M.; Polhill, J.G. When and how to imitate your neighbours: Lessons from and for FEARLUS. *J. Artif. Soc. Soc. Simul.* **2009**, *12*, 2.
30. Happe, K.; Kellermann, K.; Balmann, A. Agent-based analysis of agricultural policies: An illustration of the agricultural policy simulator AgriPoliS, its adaptation, and behavior. *Ecol. Soc.* **2006**, *11*, 49.
31. Schreinemachers, P.; Berger, T.; Sirijinda, A.; Praneetvatakul, S. The diffusion of greenhouse agriculture in northern Thailand: Combining econometrics and agent-based modeling. *Can. J. Agric. Econ.* **2009**, *57*, 513–536.
32. Valbuena, D.; Verburg, P.; Veldkamp, A.; Bregt, A.K.; Ligtenberg, A. An agent-based approach to explore the effect of voluntary mechanisms on land use change: A case in rural Queensland, Australia. *J. Environ. Manag.* **2010**, *91*, 2615–2625.
33. Jager, W.; Janssen, M.A.; De Vries, H.J.M.; De Greef, J.; Vlek, C.A.J. Behaviour in commons dilemmas: Homo economicus and Homo psychologicus in an ecological-economic model. *Ecol. Econ.* **2000**, *35*, 357–379.
34. Congress of the Philippines. An Act Providing for the Development and Promotion of Organic Agriculture in the Philippines and for Other Purpose, in Republic Act 10068. Fourteenth Congress: Manila, Philippines, 2009.
35. Lesaca, P.R.A. Organic Agriculture in the Philippines: Going Back to Basic. BAR Digest 2012 13 February 2014. Available online: http://www.bar.gov.ph/organic-agriculture (accessed on 10 January 2015).
36. David, C.C. Agriculture. In *The Philippine Economy: Development, Policies, and Challenges*; Balisacan, A.M., Hill, H., Eds.; Oxford University Press: New York, NY, USA, 2003; pp. 175–218.
37. National Economic and Development Authority (NEDA). *Medium-Term Philippine Development Plan, 2004–2010*; NEDA: Manila, Philippines, 2004.
38. Herdt, R.W.; Barker, R.; Rose, B. *The Rice Economy of Asia*; International Rice Research Institute: Manila, Philippines, 1985; Volume 2.
39. Collado, P.M.G.; Tia, M.E.; del Prado, D.G.L.; Taguibolos, G.B.; Lipio, G.M., Jr. *Characteristics of Farm Holdings: Evidence from the Philippines' Census of Agriculture*; Southeast Asian Regional Center for Graduate Study and Research in Agriculture: Los Baños, Philippines, 2013.
40. Schmitt, L. Developing and applying a soil erosion model in a data-poor context to an island in the rural Philippines. *Environ. Dev. Sustain.* **2009**, *11*, 19–42.
41. Muramoto, J.; Hidaka, K.; Mineta, T. Japan: Finding opportunities in the current crisis. In *The Conversion to Sustainable Agriculture: Principles, Processes and Practices*; Gliessman, S.R., Rosemeyer, M., Eds.; CRC Press: Boca Raton, FL, USA, 2010; pp. 273–302.
42. Klonsky, K.; Greene, C. Widespread adoption of organic agriculture in the US: Are market-driven policies enough? In Proceedings of the American Agricultural Economics Association Annual Meeting, Providence, RI, USA, 24–27 July 2005.
43. D'Souza, G.; Cyphers, D.; Phipps, T. Factors affecting the adoption of sustainable agricultural practices. *Agric. Resour. Econ. Rev.* **1993**, *221*, 59–165.

44. Best, H. Environmental concern and the adoption of organic agriculture. *Soc. Nat. Resour.* **2010**, *23*, 451–468.
45. Mzoughi, N. Farmers adoption of integrated crop protection and organic farming: Do moral and social concerns matter? *Ecol. Econ.* **2011**, *7*, 1536–1545.
46. Cauwenbergh, N.V.; Biala, K.; Bielders, C.; Brouckaert, V.; Franchois, L.; Garcia Cidad, V.; Hermy, M.; Mathijs, E.; Muys, B.; Reijnders, J.; *et al.* SAFE—A hierarchical framework for assessing the sustainability of agricultural systems. *Agric. Ecosyst. Environ.* **2007**, *120*, 229–242.
47. Philippine Bureau of Agricultural Statistics. *Area Planted/Harvested of Crops*; Report 2007; Philippine Bureau of Agricultural Statistics: Manila, Philippines, 2007.
48. Hamill, L.; Gilbert, N. Social circles: A simple structure for agent-based social network models. *J. Artif. Soc. Soc. Simul.* **2009**, *12*, 3.
49. Datta, S.K.D. Improving nitrogen fertilizer efficiency in lowland rice in tropical Asia. *Fertil. Res.* **1986**, *91*, 71–186.
50. Schouten, M.; Verwaart, T.; Heijman, W. Comparing two sensitivity analysis approaches for two scenarios with a spatially explicit rural agent-based model. *Environ. Model. Softw.* **2014**, *54*, 196–210.
51. Crosetto, M.; Tarantola, S.; Saltelli, A. Sensitivity and uncertainty analysis in spatial modelling based on GIS. *Agric. Ecosyst. Environ.* **2000**, *81*, 71–79.
52. Gomez-Delgado, M.; Tarantola, S. GLOBAL sensitivity analysis, GIS and multi-criteria evaluation for a sustainable planning of a hazardous waste disposal site in Spain. *Int. J. Geogr. Inf. Sci.* **2006**, *20*, 449–466.
53. Chu-Agor, M.L.; Muñoz-Carpena, R.; Kiker, G.; Emanuelsson, A.; Linkov, I. Exploring vulnerability of coastal habitats to sea level rise through global sensitivity and uncertainty analyses. *Environ. Model. Softw.* **2011**, *26*, 593–604.
54. Saltelli, A.; Ratto, M.; Andres, T.; Campolongo, F.; Cariboni, J.; Gatelli, D.; Saisana, M.; Tarantola, S. *Global Sensitivity Analysis: The Primer*; Wiley-Interscience: Chichester, UK, 2008; p. 304.
55. Lilburne, L.; Tarantola, S. Sensitivity analysis of spatial models. *Int. J. Geogr. Inf. Sci.* **2009**, *23*, 151–168.
56. Homma, T.; Saltelli, A. Importance measures in global sensitivity analysis of nonlinear models. *Reliab. Eng. Syst. Saf.* **1996**, *52*, 1–17.
57. Saltelli, A.; Annoni, P.; Azzini, I.; Campolongo, F.; Ratto, M.; Tarantola, S. Variance based sensitivity analysis of model output. Design and estimator for the total sensitivity index. *Comput. Phys. Commun.* **2010**, *181*, 259–270.
58. Ligmann-Zielinska, A.; Sun, L. Applying time dependent variance-based global sensitivity analysis to represent the dynamics of an agent-based model of land use change. *Int. J. Geogr. Inf. Sci.* **2010**, *24*, 1829–1850.
59. Tarde, G. *The Laws of Imitation*; The Mershon Company Press: Rahway, NJ, USA, 1903.
60. Ryan, B.; Gross, N.C. The diffusion of hybrid seed corn in two Iowa communities. *R. Soc.* **1943**, *8*, 15–24.
61. Rogers, E.M. *Diffusion of Innovations*; Free Press: New York, NY, USA, 2003.

62. Heong, K.L.; Escalada, M.M. A comparative analysis of pest management practices of rice farmers in Asia. In *Pest Management of Rice Farmers in Asia*; Heong, K.L., Escalada, M.M., Eds.; International Rice Research Institute: Los Baños, Philippines, 1997; pp. 227–242.
63. Borron, S. *Building Resilience for an Unpredictable Future: How Organic Agriculture Can Help Farmers Adapt to Climate Change*; United Nations Food and Agriculture Organization: Rome, Italy, 2006.
64. Bankes, S. Exploratory modeling for policy analysis. *Oper. Res.* **1993**, *41*, 435–449.
65. Weaver, C.P.; Lempert, R.J.; Brown, C.; Hall, J.A.; Revell, D.; Sarewitz, D. Improving the contribution of climate model information to decision making: The value and demands of robust decision frameworks. *Wiley Interdiscip. Rev.: Clim. Change* **2013**, *4*, 39–60.
66. Bankes, S.; Lempert, R.; Popper, S. Making computational social science effective: Epistemology, methodology, and technology. *Soc. Sci. Comput. Rev.* **2002**, *20*, 377–388.
67. Downey, E.; Brill, J. The use of optimization models in public-sector planning. *Manag. Sci.* **1979**, *25*, 413–422.
68. Meadows, D. *Thinking in Systems: A Primer*; Chelsea Green: White River Junction, VT, USA, 2008.
69. Peck, S.L. Simulation as experiment: A philosophical reassessment for biological modeling. *Trends Ecol. Evol.* **2004**, *19*, 530–534.

8

Investigating Impacts of Alternative Crop Market Scenarios on Land Use Change with an Agent-Based Model

Deng Ding [1,2,*], David Bennett [1] and Silvia Secchi [3]

1 Department of Geographical and Sustainability Sciences, University of Iowa, 316 Jessup Hall, Iowa City, IA 52242, USA; E-Mail: david-bennett@uiowa.edu
2 Esri, 380 New York Street, Redlands, CA 92373, USA
3 Department of Geography and Environmental Resources, Southern Illinois University, Carbondale, IL 62901, USA; E-Mail: ssecchi@siu.edu

* Author to whom correspondence should be addressed; E-Mail: dding@esri.com

Academic Editors: James Millington and John Wainwright

Abstract: We developed an agent-based model (ABM) to simulate farmers' decisions on crop type and fertilizer application in response to commodity and biofuel crop prices. Farm profit maximization constrained by farmers' profit expectations for land committed to biofuel crop production was used as the decision rule. Empirical parameters characterizing farmers' profit expectations were derived from an agricultural landowners and operators survey and integrated in the ABM. The integration of crop production cost models and the survey information in the ABM is critical to producing simulations that can provide realistic insights into agricultural land use planning and policy making. Model simulations were run with historical market prices and alternative market scenarios for corn price, soybean to corn price ratio, switchgrass price, and switchgrass to corn stover ratio. The results of the comparison between simulated cropland percentage and crop rotations with satellite-based land cover data suggest that farmers may be underestimating the effects that continuous corn production has on yields. The simulation results for alternative market scenarios based on a survey of agricultural land owners and operators in the Clear Creek Watershed in eastern Iowa show that farmers see cellulosic biofuel feedstock production in the form of perennial grasses or corn stover as a more risky enterprise than their current crop production systems, likely because of market and production risks and lock in effects. As a result farmers do not follow a simple farm-profit maximization rule.

Keywords: agent-based model; land use change; agricultural land owners and operators survey; commodity and biofuel crop market scenarios

1. Introduction

Midwestern landscapes are dominated by commodity crops including corn, soybean, and wheat. Commodity prices and the policies affecting them are the key drivers of farmers' decisions about agricultural practices, such as crop rotations and fertilizer rates. Federal and state policies on renewable energy production interact with commodity markets to affect decisions about changes in crop type and land management. These decisions, in turn, have significant environmental impacts in terms of water quantity and quality, soil erosion, and carbon sequestration in the Midwest [1–4]. Farmers' involvement in the production of biofuel crops (e.g., corn stover, miscanthus, and switchgrass) could result in new land use patterns and, thus, altered environmental outcomes. Due to substantial risk associated with the adoption of novel agricultural practices, however, farmers may not follow the same maximum profit rules when considering biofuel crop production as they do when considering traditional crops. To achieve insight into the potential impact of biofuel crop markets on agriculture land use, empirical information about farmers' attitudes towards novel practices must be integrated into land use models.

Secchi *et al.* [5] investigated the potential water quality changes associated with market scenarios of decreasing soybean to corn price ratio in the Upper Mississippi river basin by integrating an economic-driven land use model and the Soil and Water Assessment Tool (SWAT), a surface water quality model. Their results showed that an increase in corn acreage by 14.4% could result in increase in N loadings to the watershed by 5.4% and P loadings by 4.1%. Another similar study was conducted to investigate the impacts of different corn price scenarios on crop rotation patterns and environmental consequences in Iowa, USA, by integrating an economic model and the edge of field environmental impact model EPIC (Environmental Policy Integrated Climate) [6]. The authors found that sustained high corn prices might result in continuous corn in crop rotation patterns on both current cropland and CRP (Conservation Reserve Program) land. This change in land management is associated with increased sediment and nitrogen loss from fields to surface water. Leaving crop residues such as corn stover in field can help recycle nutrients, control surface runoff and prevent water and wind erosion. Potential biofuel markets for corn stover feedstock could force further shifts from corn-soybean rotation to continuous corn and a significant reduction of crop residue, thus, creating further losses of soil, nitrogen (N) and phosphorus (P) [7]. As an alternative, switchgrass, a native perennial grass in the Midwest, has been extensively investigated as a potential feedstock for cellulosic biofuel production. Switchgrass is currently used in riparian buffer strips to reduce sediment, N and P in surface runoff [8], however, its production as a biofuel crop requires the use of fertilizers that can make their way to surface and groundwater [9].

Various approaches, such as statistical techniques, expert models (e.g., Bayesian probability), cellular models (cellular automata and Markov models), and hybrid models that combine multiple techniques have been applied in land use modeling [10]. These approaches, however, focus mainly on spatiotemporal patterns of land use change rather than the decision-making process of individual land managers, despite the fact that these managers are the essential driving force in the complex land use dynamics [11].

Agent-based model (ABM) is a computer simulation approach that can be used for land use modeling [10,12]. Agent-based land use models represent system complexity using a bottom-up approach that characterizes the decision making processes of heterogeneous agents as well as feedback processes among agents and between agents and the biophysical environment. Simulated agents, such as land owners, ranchers, farmers, and policy makers, act and interact following decision rules. They are characterized by sets of parameters that can be derived from theory or empirical data obtained by land use surveys, participatory observations, and field and laboratory experiments [13,14]. Due to the mechanistic nature of this approach and its unique perspective concerning agent-agent and agent-environment interactions, ABMs are often integrated with environment models to explore the impacts of socio-economic driving forces or alternative policies on land use and consequent environment outcomes [15–17]. ABM can be used to represent multiple types or levels of agents (e.g., institutions and individuals) and the co-evolution of and interaction between biophysical and human decision making processes in a coupled human-environment system [10,18–22].

In Midwestern agricultural landscapes, ABM has been applied in studies on how farmer decision making is impacted by agricultural policies, such as the CRP [23], potential market scenarios of biofuel crops [24], and price incentives for nitrogen and carbon abatement [25,26]. In the first two studies [23,24], the research is focused on how land use changes emerge from farmers' decisions under the assumption that the natural environment is static through time. In the latter two studies [25,26], the focus is on the impacts of farmers' management decisions on water quality. The authors integrated ABM with SWAT, and analyzed two-way interactions between farmer agents and the natural environment. The natural environment included two types of processes: biological and hydrological, both of which interacted with agents but at different levels. Agents' decisions on cropping practices not only influenced but also responded to crop yields. Agents' decisions affected watershed water quality, but there was no feedback from water quality into the agents' decisions. The study was considered as "semi-hypothetical" since no empirical information was used to parameterize agents' risk premiums towards novel agricultural practices.

The research objectives of this study are to: (1) develop an ABM to simulate farmers' decisions about land use practices given a risk-averse profit maximization rule; and (2) investigate the impacts of alternative market scenarios on land use change by integrating empirical information about farmers' attitudes into an ABM. Through this study, we address the following research questions: (1) is the current land cover pattern economically optimal given declines in crop yield resulting from continuous corn/soybean rotation? and (2) would biofuel crops be underutilized given farmers current attitudes towards biofuel crop production?

2. ABM of Agricultural Land Use

2.1. Mathematical Programming for Modeling Decision Making in ABM

Farm household modeling often relies on mathematical programming (MP) techniques to simulate land management decision of farmers and their responses to policy. This approach has several drawbacks [12,23]: (1) it generally assumes the maximization of a single objective (e.g., profit), which is often not appropriate when modeling the adoption of new technologies; (2) it ignores the social aspects of farm households such as communication and interaction among farmers in the same community; and

(3) it does not properly capture the heterogeneity of the social behaviors and responses of farmers. In ABMs of agricultural land use decision making, mathematical programming is generally applied at the farm level and combined with heuristic approaches and Bayesian inference or Bayesian probability networks [23,25–27]. Heuristic approaches, such as decision trees or rule-based models, assume limited human cognition, while optimization approaches, such as MP, assume that inefficiency in human decisions comes from external factors, such as the failure of institutions, imperfect markets, and lack of infrastructure or limited information [27]. Though the two approaches appear to be different in theory, Schreinemachers and Berger [27] argue that in practice they can be converted into each other. For example, decision rules about production and consumption in heuristic approaches can be incorporated into MP models as constraints. An optimization approach is even more appropriate for policy analysis and planning since it can quantitatively characterize the outcomes of alternative policies.

In the Midwestern U.S., Sengupta *et al.* [23] employed a hybrid approach to model land enrollment in an agricultural land set aside conservation program, CRP. Different types of farmers with distinct decision making rules were modeled in the Cache River watershed of southern Illinois. In another example of ABM for agricultural land use, Ng [25] and Ng *et al.* [26] combined a farm-level stochastic programming model with a Bayesian updating procedure to represent the optimization and adaption processes of farmers' decision in the Salt Creek watershed in Central Illinois. In Ng's ABM [25,26], the farmers are assumed to be economically rational with bounded information. Farmer agents learn new information through time with Bayesian updating. In this ABM, farmers interact with their geographic neighbors to exchange information about crop yields and costs.

2.2. Empirical Information for ABM

Empirically-based ABMs for land use simulation construct and parameterize decision making models from empirical information that characterizes macro-level patterns and micro-level processes. Information about micro-level processes could be directly utilized for developing and testing the structure of decision making models in ABM, while datasets about macro phenomena are usually applied in calibrating and validating the model [13]. In this study, we focus mainly on empirical information about micro-level processes and its use for the parameterization of the decision making model.

To characterize heterogeneous agents and their behavioral responses, empirical information needs to be analyzed using statistical methods (e.g., factor or cluster analysis, regression) and GIS techniques. Based on the statistical descriptions of agents characteristics derived from observation datasets, different populations of unique artificial agents can then be generated with Monte Carlo techniques [28].

Smajgl *et al.* [14] developed a framework for generating and calibrating the parameters that describe agent attributes and behavioral functions. In their framework, the parameterization process is composed of five steps: (1) create an agent typology based on behavioral differences; (2) specify the attribute values for each agent type; (3) specify the behaviors of each agent type by obtaining the parameter values for their behavioral functions; (4) develop agent types from agent attributes or behavior responses; and (5) generate a population(s) of agents based on the agent typology, and relevant attributes and parameters.

Empirical information can be collected using various methods [10,13,14]. These methods include role-playing games [29,30], sample surveys [31], participatory observations [32], field or laboratory experiments [33], and GIS and RS data collection [34]. Of these five methods, sample survey is the most quantitative approach and can be carried out at a relatively large spatial scale [13]. Though qualitative information about agents' interactions and feedbacks could be obtained using different methods, it is still challenging to empirically identify and quantify social, spatial and cross-scale interactions among agents and feedbacks between agents' decisions and the biophysical processes [35].

3. Study Area

The study area is the Clear Creek watershed located in Iowa and Johnson Counties in Eastern-Central Iowa, U.S. (Figure 1). It is a typical Midwestern agricultural watershed dominated by row crops (corn, soybean), with an area of about 267 km^2. The average annual temperature from 2000 to 2011 was 10 °C and average annual precipitation was 886 mm based on the weather records at the Iowa City Municipal Airport station. According to the United States Department of Agriculture National Agricultural Statistical Services (NASS) remote sensing Cropland Data Layer (CDL) images in 2008, the percentages of the predominant cover types were respectively 29.77% corn, 20.11% soybean, and 28.17% grassland/pasture. Additionally, 14.56% of the total watershed was in urban and developed uses and 6.9% was forest including deciduous, evergreen and mixed forest. Urban and forest areas are located mainly in the eastern part of the watershed. According to the Iowa Soil Properties and Interpretations Database (ISPAID), about 54% of the watershed is covered by soil with slope range of 2%–14%; while soils are distributed among several soil drainage type ranging from well drained soils (26%), moderate-well drained soils (23.5%), to poorly drained soils (11.9%). The dominant soil textures of the surface horizon in the region are silty clay loam (35.8%) and silt loam (35.1%); about 45% of the total land is highly erodible.

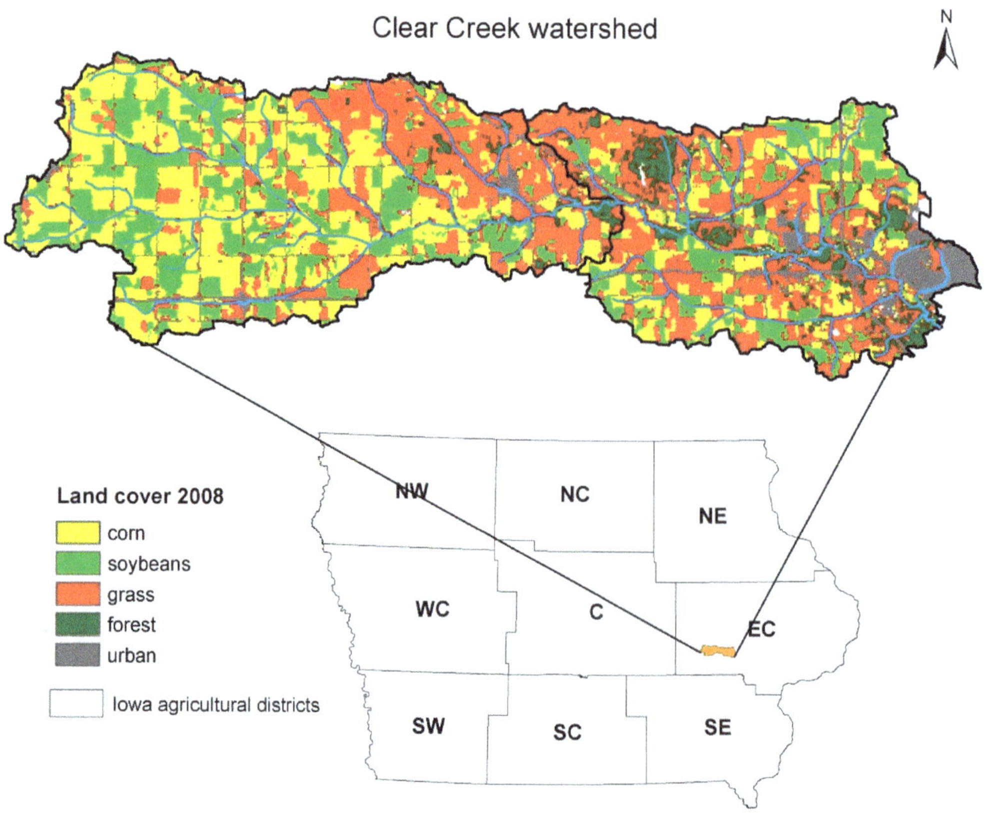

Figure 1. The Clear Creek Watershed in East-Central Iowa.

According to an agricultural land use survey conducted in 2010 [36,37], the dominant crop rotation type is corn-soybean and corn-corn-soybean with conventional tillage used for both corn and soybean. Some farmers adopt conventional tillage practices for corn but no-till or minimum-till practices for soybean. Specific information about actual application rates of inorganic N fertilizers in corn fields is not available for this watershed. According to Iowa State University Extension [38], the highest N rate is applied for corn after corn (168–224 kg/ha), followed by corn after soybean if no manure is applied (112–168 kg/ha), and corn on recently manured soils (0–101 kg/ha), given that all N is applied before corn planting or crop emergence.

4. Model Description

In this section, the model is described following the Overview, Design concept and Details (ODD) protocol [39].

4.1. Overview

4.1.1. Purpose

This model is used to simulate farmers' decisions on agricultural land use including crop type, tillage type, and fertilizer rates. We aim to run simulations to study farmers' decisions on biofuel crop productions in response to biofuel crop prices.

4.1.2. Entities, State Variables, and Scales

Agents in the model are farmers. Each farmer agent corresponds to a farm composed of one or more land parcels that are represented by command land units (CLUs) in this study. Farmer agents have attributes including farm size, and parameters about their attitudes towards biofuel crop production are summarized in Table 1.

We use eight parameters to characterize farmers' responses to the questions in Table A1 and thereby to describe their attitudes towards biofuel crop production and the associated risk premium. Descriptions of the eight parameters are included in Table 1. Additionally, there is a parameter *Area* representing the land acreage of the farm size of each farmer agent.

Each land parcel has attributes including area (in hectares), Corn Suitability Rate (CSR) (dimensionless) which is an index of land suitability for crop production, and potential yield of switchgrass (metric ton/ha).

Based on each land unit, farmer agents make CLU-specific decisions on crop type, tillage type, and fertilizer rates by changing the value of the CLU-specific (i as index) and year-specific (t as index) state variables: $X_{i,j,t}$, and j represents all available options for crop type, tillage type, and fertilizer rate levels. $X_{i,j,t}$ is a binary variable about crop type, tillage type, and N application rate level. Crop types include corn (with or without stover harvest), soybean, and switchgrass. Tillage types include conventional tillage, mulch tillage, and no tillage. Fertilizers include N, P, and K (potassium). For example, $X_{i,j=1,t} = 1$ means corn is planted in land parcel i at year t, $X_{i,j=2,t} = 1$ means soybean is planted in land parcel i at year t, and $X_{i,j=3,t} = 1$ means the land parcel i is fallow at year t. Similarly, $\sum_{j=4}^{6} X_{i,j,t} = 1$ and $j = 4, 5, 6$, respectively, mean conventional tillage, mulch tillage, and no tillage. When $j = 7, 8, ..., 16$, X_{ij} is the decision variable of the N application level from the lowest to the highest. In the biofuel crop market scenarios, crop type decision variables also have two extra options, switchgrass, and corn with stover harvested.

Table 1. Parameters characterizing farmers' attitudes towards biofuel crop production.

Parameter	Value	Agent Characteristics
Type	0	Not interested in either marketing corn stover or planting switchgrass
	1	Only interested in marketing corn stover
	2	Only interested in planting switchgrass
	3	Interested in both
Profit SWG	Numeric	Profit rate ($/ha) required by the farmer to plant, harvest and market switchgrass *
Percent1 SWG	Numeric	Percent of farm acreage on which the farmer would plant switchgrass if the Profit SWG can be achieved
Percent2 SWG	Numeric	Percent of farm acreage on which the farmer would plant switchgrass if 1.5 times the Profit SWG can be achieved
Profit Stvr	Numeric	Profit rate ($/ha) required by the farmer to harvest and market corn stover *
Percent1 Stvr	Numeric	Percent of farm acreage on which the farmer would consider harvesting and marketing corn stover if the *Profit Stvr* can be achieved
Percent2 Stvr	Numeric	Percent of farm acreage on which the farmer would harvest, and market corn stover if 1.5 times the *Profit Stvr* can be achieved
Portion Stvr	Numeric	Portion of corn stover that the farmer would harvest

* Given that in 2010 when the survey was conducted, the average price for corn was $196.85/metric ton and for soybean was $404.14/metric ton).

One simulation step represents one year. Farmer agents make decisions on each land parcel (CLU) of their farms every year. CLUs are the smallest land units with common crop choice and management, a continuous boundary and the same owner and operator. Thus, in the ABM, land parcels represented as CLUs are the smallest simulation units for which both the physical conditions (soil fertility, slope, *etc.*) and farmer's land use management practices are assumed to be homogeneous. Generally, a farm is composed of multiple CLUs and each farm corresponds to a single land use decision maker represented as a farmer agent in the ABM. Data for CLU boundaries in Iowa are available to the public through the Natural Resources Geographic Information Systems (NRGIS) Library. However, the information about farm boundaries is confidential and thus cannot be publically displayed. In this study, we generate pseudo farm boundaries by utilizing the statistical distribution of farm sizes, and the spatial locations of farm centroids based on real information of farm boundaries.

4.1.3. Process Overview and Scheduling

Each year, farmer agents estimate commodity prices and fuel prices, calculate crop production costs and returns, and make decisions on crop type, tillage type, and fertilizer rates. To make these decisions, farmer agents try to maximize farm profit given the constraints of crop yield drag effect, and their risk aversions towards biofuel crop production.

More specifically, at each time step, for each farmer agent, the land use decision making process is implemented in two parts in the model. In Part I, linear programming is used to look for land use practices that maximize farm profit with planning horizon of 1 year given the crop choices of corn without stover, soybean and fallow. The 1-year planning horizon is used in biofuel crop market scenarios. In model verification, 1-year, 2-year, and 3-year planning horizons are used. In Part II, a decision-tree based algorithm is designed for comparing the profit rates determined from Part I with the profit rate from corn with stover at the simulation step, and with the average yearly profit rate from corn with stover and switchgrass calculated within a 10-year planning horizon (discount rate = 0.05). If the risk premium for biofuel crops is met and expected profits exceed those of traditional crops, the profit maximizing biofuel crop will be selected and planted. Total biofuel production is limited by farm specific acreage constraints. For next time step, the same two processes are implemented except that the land parcel that was previously enrolled in switchgrass is be excluded from the farm profit maximization until it is out of the 10-year enrollment limit.

4.2. Design Concepts

4.2.1. Principles

We assume that farmers are risk averse profit maximizers, and that there is little risk associated with the choice between corn and soybeans, since they are both well-established annual crops. Therefore, in the case of these crops the decision rule collapses to profit maximization. However, there is substantial risk associated with new markets and crops. Further, if these crops are perennials, they lock in farmers for substantial periods of time, thereby increasing both market and production risk.

In all cases, farm profit depends on yield potential for corn, soybean, and switchgrass specific to each land parcel. Corn and soybean yields are lower if the crops are not rotated, and corn yield responds to different levels of N rates. Because of the additional risk associated with cellulosic ethanol markets, farmer agents might require higher profits from biofuel crops than they do from traditional crops and limit the total acreage allocated to biofuel crops.

4.2.2. Emergence

Watershed-scale agricultural land use acreages and fertilizer application rates are modeled as an emergent property of farmer agent decisions. For model results, we expect to observe insensitivity of cropland switched to switchgrass and corn stover harvest in response to the price change due to farmers' risk aversions towards biofuel crop production.

4.3. Details

4.3.1. Initialization

We use NASS CDL data from 2001 to 2011 for initializing the land cover in the first two years of simulation and compare with the simulation results in the remaining years. NASS CDL is a raster data set at 30 meter spatial resolution. It contains land cover information of specific crop types including corn and soybean. We use the majority zonal operator in ArcGIS 10.0 to aggregate the pixel-specific crop type into the CLU level.

4.3.2. Input Data

The ABM inputs include prices and land parcel information. The price file contains annual time series of prices for crops (corn, soybean, switchgrass, corn stover), fertilizers (N, P, K), and fuels (diesel, LPG-liquid petroleum gas). The parcel information file contains land parcel specific data about acreage, CSR, yield potential for switchgrass, and land cover types for the previous two years. These data were used to construct net returns for each field in the watershed for each crop rotation [6]. In addition, a parameter file characterizing farmer agents' attitudes towards biofuel crop production (as in Table 1) is a necessary input into this model. The price data was based on United States Department of Agriculture's (USDA's) Economic Research Service, U.S. Energy Information Administration and Iowa State University Extension's budgets [38,40,41], the CSR data was obtained from the soil database ISPAID, the historical landcover was constructed using the CDL, and the attitudes of farmers were parameterized from a land use and attitudes survey conducted in 2010 [36,37].

4.3.3. Submodels

As mentioned, there are two parts of the decision making processes. In Part I, for current iteration year, the farm based objective function is:

$$maximize\ P = \sum_{i=1}^{n} (S_i - C_i) \tag{1}$$

where i is the land parcel index, n is the total number of land parcels within a farm, P is the profit, S is revenue from crop sale, and C is crop production cost. Equation (1) illustrates that the farm agents' objective is to maximize farm profit each year. With decision variables included, Equation (1) can be expressed as:

$$maximize\ P = \sum_{i=1}^{n} \sum_{j=1}^{m} f(c_{ij} X_{ij}) \tag{2}$$

where c_{ij} is the objective function coefficient corresponding to the jth variable for land parcel i, and X_{ij} is the jth decision variable for land parcel i, n is the total number of land parcels within a farm, and m is the total number of decision variables. As illustrated in Equation (2), the farm profit maximization for corn and soybean is a linear programming problem, which is scripted in the mathematical programming software (AIMMS) and implemented as an ABM using C++.

After the implementation of Part I, the decision variable values and the resulting profit rates from the AIMMS optimization component are passed to the decision-tree based algorithm in Part II. In this part, profit rates for biofuel crops (corn with stover harvest, and switchgrass) are evaluated with the agent-specific risk premiums and compared to commodity crop profit rates obtained from the Part I algorithm. The biofuel crop is chosen for a land parcel when the risk premium for biofuel crops is met and expected profits exceed those of traditional crops. Empirical parameters characterizing farmers' risk premiums about biofuel crops are incorporated in the decision-tree based algorithm. The parameters are derived from the land use survey and applied in the land use decision rule for switchgrass and corn stover production. The parameters and the sampling strategy of parameter values derived from the land use survey database are described in Table 1 and the Appendix B.

5. Data and Simulation Settings

5.1. Agricultural Landowners and Operators Survey

A survey of agricultural landowners and operators in the Clear Creek watershed was conducted in 2010 to obtain information about current farming practices and farmers' willingness to participate in conservation practices and biofuel crop production [36,37]. The survey questions were sent by mail to all non-urban landowners and agricultural operators (about 998) within the watershed. Responses were received from 397 of them (response rate 41.1%). A survey database was developed based on those responses. The database covers multiple topics including personal information, farm characteristics, and farmers' information sources and their attitudes towards watershed conservation and growing biofuel crops. For this study, we are interested in questions about (1) farm size; (2) minimum net profit rates that farmers require from biofuel crop production; (3) the acreage on which farmers would plant, harvest, and market corn stover and/or switchgrass; and (4) the proportion of corn stover harvested if such profits were realized (Table A1 in the Appendix B).

5.2. Price Scenarios and Simulation Settings

Price inputs to the model include commodity prices of corn, soybean, corn stover, switchgrass, fertilizers (N, P, K) and fuels (LPG and diesel). In this study, we run the ABM simulations with four different input datasets for commodity and biofuel crop prices to investigate: (1) the land use pattern resulting from profit maximization (Simulation Set I); (2) the sensitivity of land use pattern to commodity crop prices (Simulation Set II); (3) the impacts of biofuel crop prices on land use patterns given the risk associated with biofuel crop production (Simulation Set III, and Simulation Set IV).

5.2.1. Historical Market Prices

Simulation Set I is based on historical market prices from 2003 to 2011 (Table 2). The corn and soybean prices are the average prices in the calendar year (ISU cash corn and soybean prices) [42]. From 2003 to 2006, corn prices are replaced with the target prices of the counter-cyclical payments for corn. The fertilizer prices are the average U.S. farm prices of selected fertilizers in March or April published by USDA's Economic Research Service. The LPG price is the weekly Iowa propane residential price in the middle of March and the diesel price is the price of Midwest No. 2 diesel retail sales by all sellers. The data is published by the U.S. Energy Information Administration.

5.2.2. Price Scenarios for Corn and Soybean

Simulation Set II (Table 3) is based on price scenarios for 360 combinations of corn prices and soybean prices. By confining the price ranges within the historical records and the ten year baseline projections for U.S. agricultural markets by the Food and Agricultural Policy Research Institute, University of Missouri (FAPRI-MU) [43], we determine 24 levels of corn prices and 15 levels of soybean to corn price ratios (Table 3). The 24 levels of corn price start from 360 (24 × 15) combinations of corn price and the price ratio are used for generating price inputs into the model. For the 360 price input files, the other prices including switchgrass, corn stover, fertilizers (N, P, and K), and fuels (diesel and LPG) are held constant. Switchgrass and corn stover are not taken into account and so their prices are considered as zero. Prices of fertilizers and fuels are the average values of FAPRI-MU projections.

Table 2. Historical market prices (corn, soybean switchgrass and corn stover prices in \$/metric ton, fertilizer prices in \$/kg, fuel prices in \$/liter, soybean to corn price ratio is dimensionless).

Year	Corn	Soybean	Switch Grass	Corn Stover	N	P	K	Diesel	LPG	Soybean to Corn Ratio
2003	102.36 *	223.38	0	0	0.41	0.27	0.18	0.38	0.29	2.18
2004	103.54 *	281.06	0	0	0.42	0.29	0.20	0.44	0.28	2.71
2005	103.54 *	216.03	0	0	0.46	0.33	0.27	0.56	0.33	2.09
2006	103.54 *	203.91	0	0	0.57	0.36	0.30	0.75	0.39	1.97
2007	132.68	285.84	0	0	0.58	0.46	0.31	0.73	0.40	2.15
2008	188.19	417.00	0	0	0.83	0.88	0.62	1.16	0.56	2.22
2009	150.00	369.60	0	0	0.75	0.70	0.94	0.57	0.43	2.46
2010	151.97	362.26	0	0	0.55	0.56	0.56	0.80	0.46	2.38
2011	234.65	458.88	0	0	0.83	0.70	0.66	1.06	0.49	1.96

* Corn prices are replaced with the target prices of the counter-cyclical payments according to the Farm Security and Rural Investment Act of 2002.

Table 3. Price levels of alternative scenarios (corn and soybean prices in \$/metric ton, switchgrass and corn stover prices in \$/metric ton, fertilizer prices in \$/kg, fuel prices in \$/L, price ratios are dimensionless).

Simulation Set	Number of Combinations	Corn	Soybean	Soybean: Corn Price Ratio	Switchgrass	Corn Stover	SWG: Stvr Price Ratio	N	P	K	Diesel	LPG
II	360	Start = 102.36 To = 283.46 Step = 7.9 (24 levels)	Corn price × SC price ratio (15 levels)	Start = 1.8 To = 3.2 Step = 0.1 (15 levels)	0	0		0.89	0.80	0.76	1.03	0.48
III	16	186.6	419.94	2.25	Start = 58.41, To = 223.71, Step = 11 (16 levels)	SWG price/ SS price ratio (16 levels)	1.31	0.89	0.80	0.76	1.03	0.48
IV	96	186.6	419.94	2.25	Start = 58.41, To = 223.71, Step = 11 (16 levels)	SWG price/ SS price ratio (6 levels)	Start = 1.1, To = 2.1, Step = 0.2 (6 levels)	0.89	0.80	0.76	1.03	0.48

5.2.3. Price Scenarios for Switchgrass and Corn Stover

Simulation Set III (Table 3) is based on 16 levels of switchgrass prices. By referring to the projected prices of warm season grasses and corn stover for biofuel markets by FAPRI-MU [44], we determine 16 levels of switchgrass prices from 58 \$/metric ton to 224 \$/metric ton with an interval of 11 \$/metric ton. The price ratio of switchgrass to corn stover is fixed as 1.31 according to the average value in FAPRI-MU (2011). The other price inputs for corn, soybean, fertilizers and fuels are fixed as the average prices for Year 2012 to 2021 in the FAPRI-MU projections. In Simulation Set III, simulations are run for each of the 16 price levels with the 30 samples of the empirical parameter set. The model is run for a total of 480 (16 × 30) simulations. The last set of simulations (Simulation Set IV) (Table 3) uses the same price settings as Simulation Set III except that the price ratio of switchgrass to corn stover varies from 1.1 to 2.1 with interval of 0.2 (a total of 6 levels) Therefore, in Simulation Set IV, price scenarios are 96 (16 × 6) combinations of switchgrass and corn stover prices. Similarly to Simulation Set IV, simulations are run for each of the 96 combinations with 10 samples of the empirical parameter set. The model is run for totally 960 (96 × 10) simulations.

6. Results and Discussions

6.1. Model Verification

The yield drag incorporated in the ABM model on the basis of agronomic data effectively precludes the choice of continuous corn for the simulated farmers. Thus, farmers are *de facto* constrained to think in terms of corn-soybeans or corn-corn-soybeans in the model. In practice farmers may not behave as if the yield drag matters, which is verified by the crop rotation pattern comparison below.

6.1.1. Crop Rotation Pattern

Figure 2 compares the simulated crop rotations with the NASS CDL based crop rotations using graphs that summarize the land area percentage and the CLU count percentage corresponding to the maximum number of continuous crop years during the 11-year time period. In Figure 2a, we see that about 74% of the simulated land is in corn-corn-soybean and about 26% of land is in corn-soybean during any of the 11 years. In reality, however, less land is in corn-corn-soybean (about 39%) while about the same percentage of land (25%) is in corn-soybean; and there are some fields (about 36%) which were in continuous corn for three or more years. For soybean (Figure 2b), the simulated and real landscapes are very similar: Single-year soybean is predominant (more than 80%). This indicates that in reality farmers do not necessarily follow the rotation patterns of corn-corn-soybean or corn-soybean, and that the continuous corn years could be longer than two, but soybean is grown almost always only in rotation with corn.

Assuming the NASS CDL is reality, we can conclude the inclusion of yield drag in the model and the likely underestimate of the effect of yield drag by actual farmers causes the model to underestimate the amount of continuous corn, because simulated farmers are in practice restricted in their options. For example, after two years of corn a simulated farmer will plant soybeans even if relative crop prices suggest that corn is the more profitable crop because of the effects of yield drag on productivity. Yield drag is, however, hard for actual farmers to ascertain because they do not run long term controlled experiments in their fields, many factors change simultaneously from year to year, and yield drag is affected by tillage and weather. Unfortunately, though there is anecdotal evidence [45,46] that farmers may underestimate yield drag, there are no peer-reviewed studies that compare perception and reality on this issue.

Overall though, the model gives a reasonable approximation of the behavior of farmers in the watershed for the study period. For example, the average acreage of cropland in corn according to the CDL dataset was 56.7% and the simulated average using the one year planning horizon was 52.2%. Again, the underestimation of corn acreage is related to the yield drag issue.

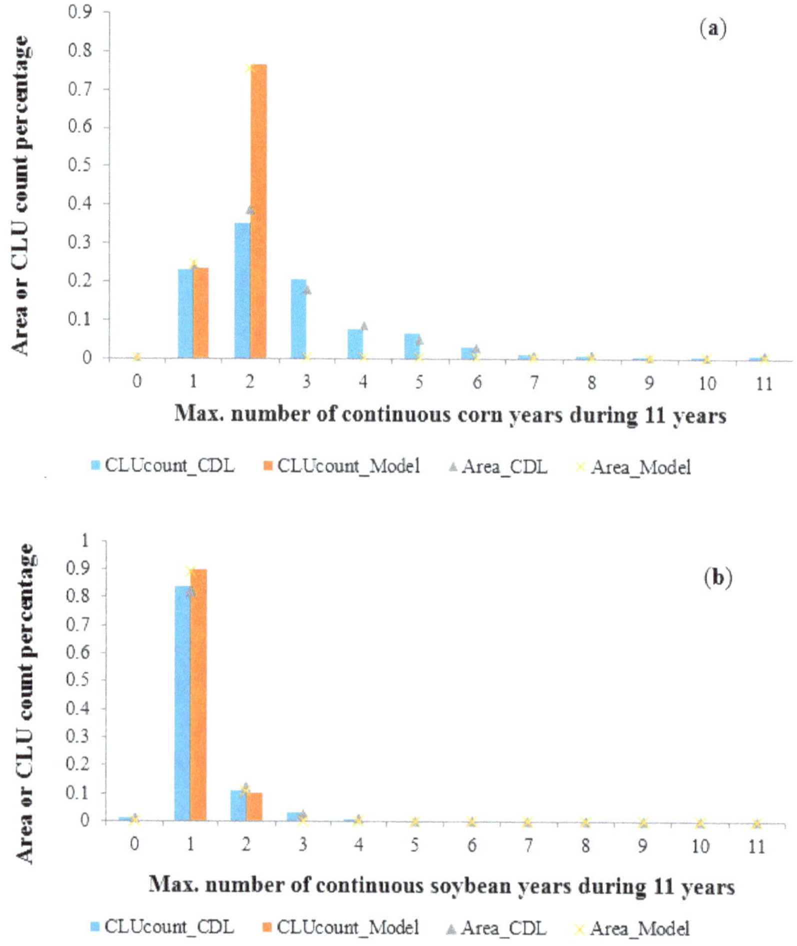

Figure 2. Maximum number of continuous corn (**a**); and soybean (**b**) years during an 11-year time period.

6.1.2. Field and Farm Scale Statistics

At the field scale, box plots are made for the cost and profit rates of corn and soybean throughout the 11 simulation years (Figure 3a–d). Generally, the simulated cost and profit rates increase through the years, and corn (Figure 3a,c) has higher cost and profit rates than soybean (Figure 3b,d). The box heights are generally larger in Figure 3c than Figure 3d, which indicates that the spatial variability (across fields) of the simulated cost rate of corn is much higher than soybean. The reason is that nitrogen fertilizer is only applied for corn and the N application rate varies across different corn fields depending on the physical characteristics of land (e.g., yield potential). At the farm scale, a box plot is made for the simulated profit rate over all simulation years (Figure 3e). The simulated average net farm income rates from 2008 to 2011 is about 694 \$/ha, which is comparable to the overall average of typical cash rents from 2008 to 2011 for corn and soybean fields (respectively 460, 477, 484, and 541 \$/ha) in the corresponding agricultural district (District 6) in Iowa [47]. The lower cash rent is likely due to the fact that cash rent is generally built as the average of recent past net returns and therefore lags.

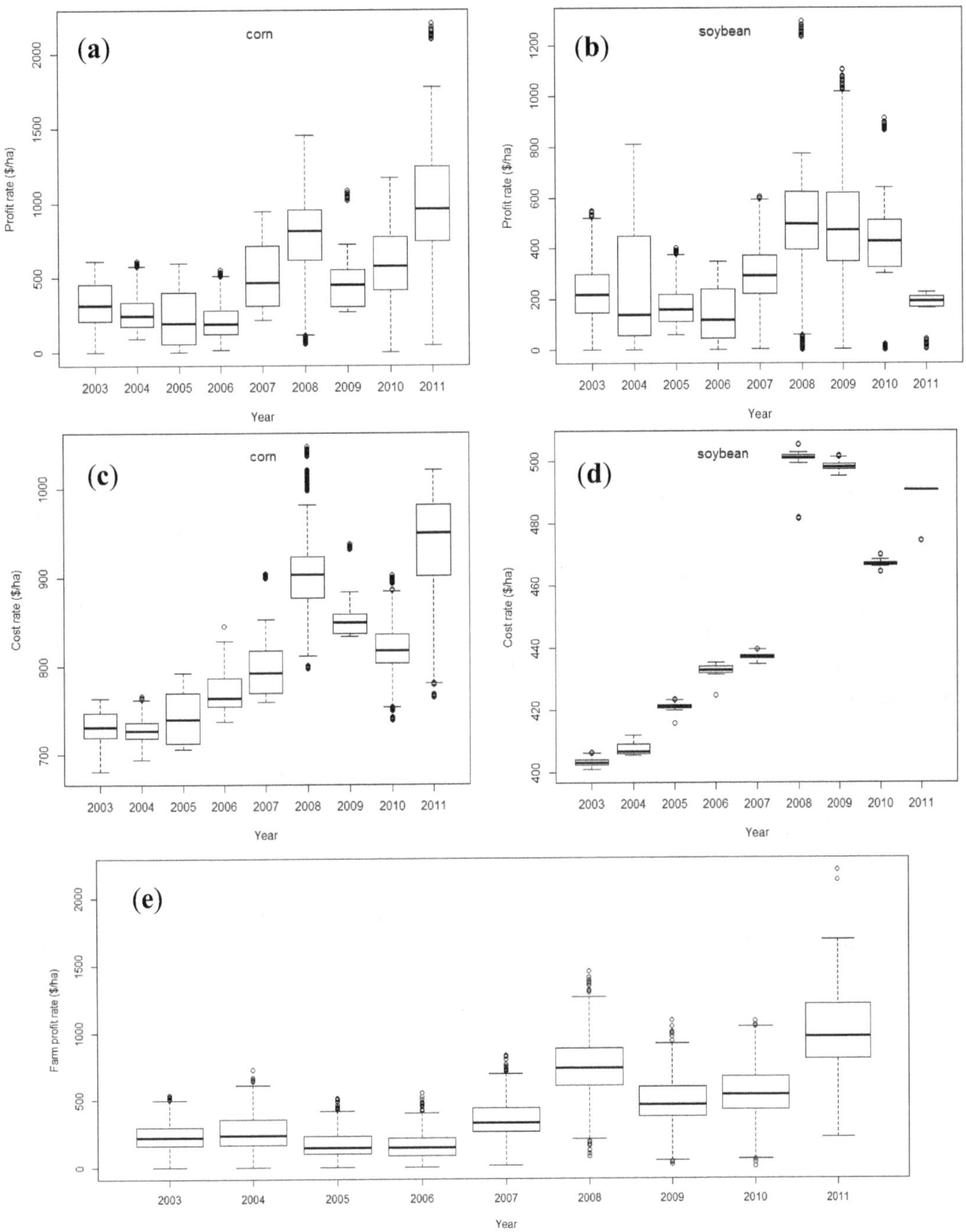

Figure 3. Box plots of simulated profit rate ((**a**) corn; (**b**) soybean); and cost rate ((**c**) corn; (**d**) soybean); and of farm-level profit rate (**e**).6.2. Model Results: Corn and Soybean Price Scenarios (Simulation Set II).

With the second set of simulation results, we plotted watershed-scale crop area percentages (Figure 4a,b) and fertilizer rates (Figure 4c,d) against changes in corn price and soybean to corn price ratio. As illustrated by Figure 4a, the percent of land in corn increases as corn price increases and/or as the ratio of soybean to corn price decreases. The percent of land in soybean changes with the opposite trends (Figure 4b). Corn and soybean lands stabilize at 50/50 within the corn price range of about 181.1–283.5 \$/metric ton and the range of soybean to corn price ratio of about 2.8–3.2. Beyond this range, the percent of the landscape in corn and soybean is very sensitive to changes in corn price and/or soybean to corn price ratio.

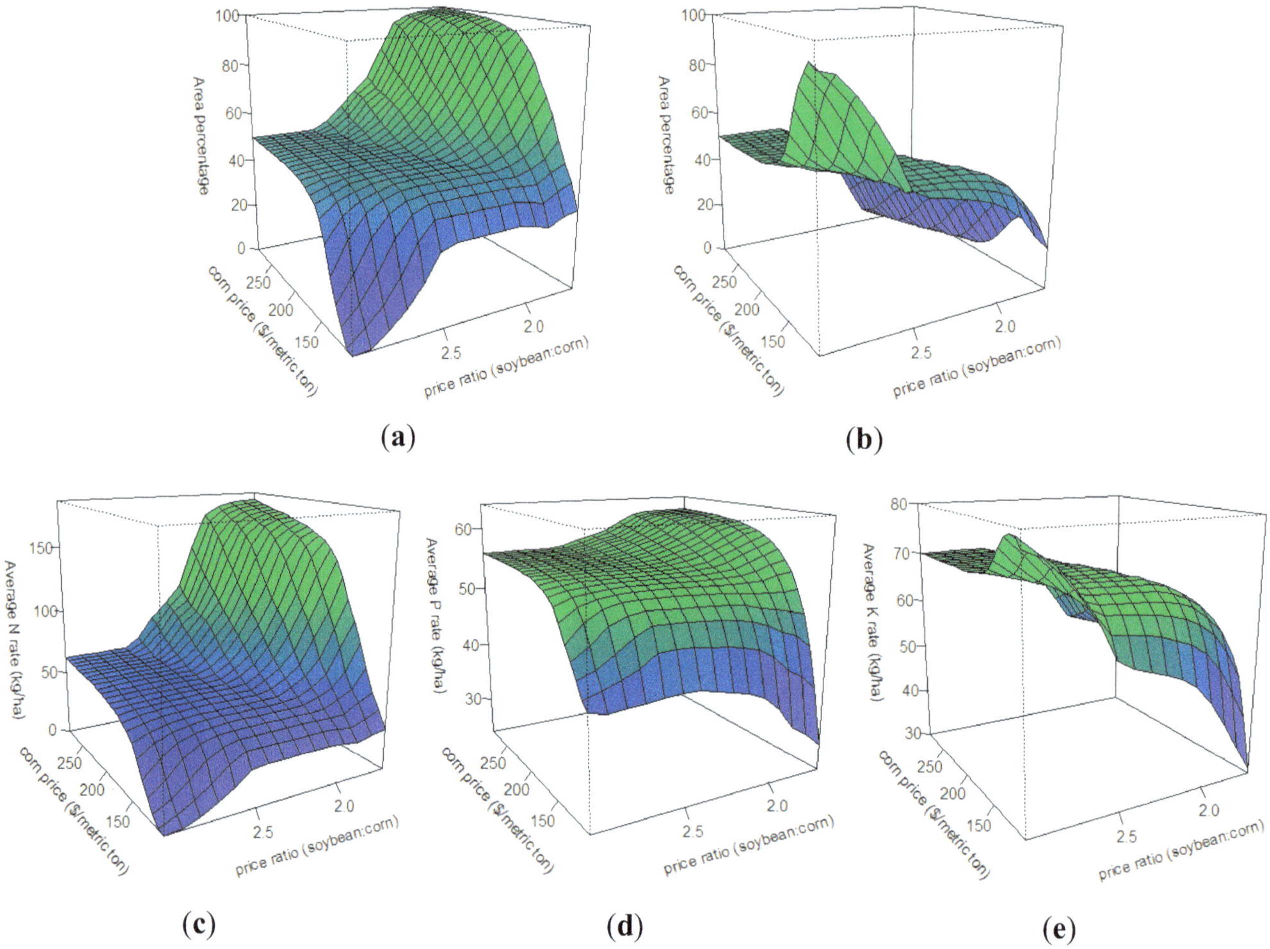

Figure 4. Simulated crop area percentage ((**a**) corn; (**b**) soybean); and fertilizer rates averaged over the whole watershed ((**c**) N; (**d**) P; (**e**) K) for 360 combinations of corn prices and soybean to corn price ratios.

According to Figure 4c, nitrogen responds to the changes in corn price and soybean to corn price ratio in a very similar way to corn acreage since N fertilizer is only applied for corn. The shape of the P surface is more similar to the corn land percentage surface than to the soybean, except that the lowest value of the average P application rate occurs when corn price is at the lowest (181.1 \$/metric ton) and soybean to corn price ratio is at the lowest (2.8). This is because the average P application rate is related to both corn and soybean planting but dominated by corn since corn requires more P input than soybean. Figure 4e characterizes the response of K application rate to changes in corn price and the soybean to corn price ratio. The K surface has a shape more similar to that of the percent of the land in soybean than to corn, except that the lowest value of the average K application rate occurs at the same position as P. This is reasonable because the average K application rate is related to both corn and soybean planting but dominated by soybean since soybean requires more K input than corn. Within similar ranges

(181.1–283.5 \$/metric ton corn prices, and 2.8–3.2 of soybean to corn price ratio), the average N, P, and K rates stabilize respectively at around 65, 56, and 71 kg/ha.

6.2. Model Results: Switchgrass and Corn Stover Price Scenarios

6.2.1. Simulations with *vs.* without Land Use Survey Information Included (Simulation Set III)

We run the model with positive switchgrass prices as defined in Simulation Set III under two scenarios:

Scenario 1. The agricultural landowners and operators survey information is not included in the model, and agents do not appreciate that there is additional risk in planting perennial crops for a new market. In this scenario farmer agents follow the same rule of farm profit maximization as they do in Simulation Set I (profit maximization unaffected by farmer risk perception);

Scenario 2. Information from the farmer survey is incorporated into the model and used to parameterize the perceived risks associated with planting a new perennial crop. In this scenario, agent-specific constraints are incorporated into the model which stipulate the minimum expected economic return required before a farmer would consider biofuel crops and the maximum percent of their farm they would allocate to biofuel crop production if such returns could be realized.

As illustrated in Figure 5, the incorporation of farmer attitudes about risk has a significant impact on land use. Considerably more land is allocated to switchgrass and corn stover in scenario 1 (profit maximization, triangles in Figure 5) than in scenario 2 (constrained by perceived risk, circles in Figure 5). Even with 30 simulations for each of switchgrass price levels, the variations are low (in Figure 5, the dispersion of 30 circles is small compared to the discrepancy between the circles and triangles). Figure 5a–d shows that when perceived risk is modeled, the simulated crop composition starts to change at switchgrass price of \$88.16/metric ton (switchgrass to corn stover ratio fixed at 1.31) and above: Corn (no stover) (Figure 5a) and soybean (Figure 5b) acreages start to decrease and switchgrass (Figure 5c) and corn (stover) (Figure 5d) start to increase. The crop area percentage curves approach a stabilization stage at switchgrass prices of about 132–165 \$/metric ton. The crop composition stabilizes at about 32% corn (no stover), 38% soybean, 22% switchgrass, and 8% corn (stover).

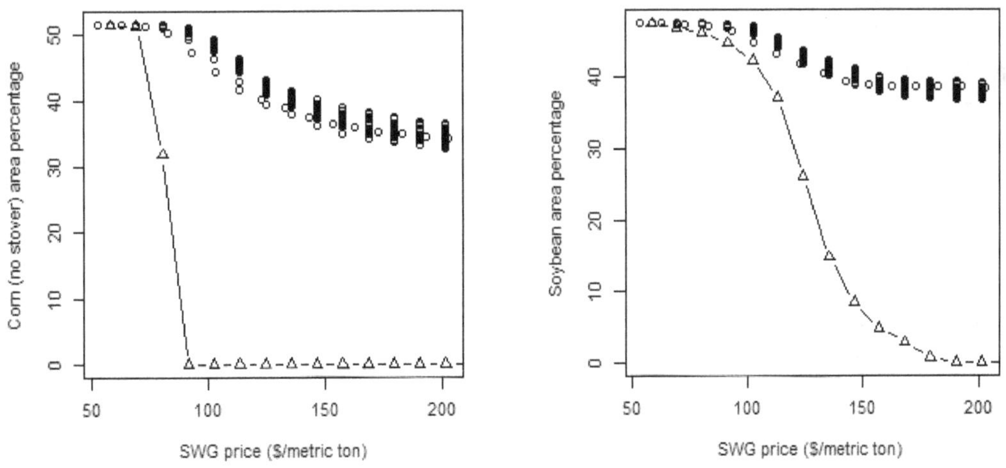

(a) **(b)**

Figure 5. *Cont.*

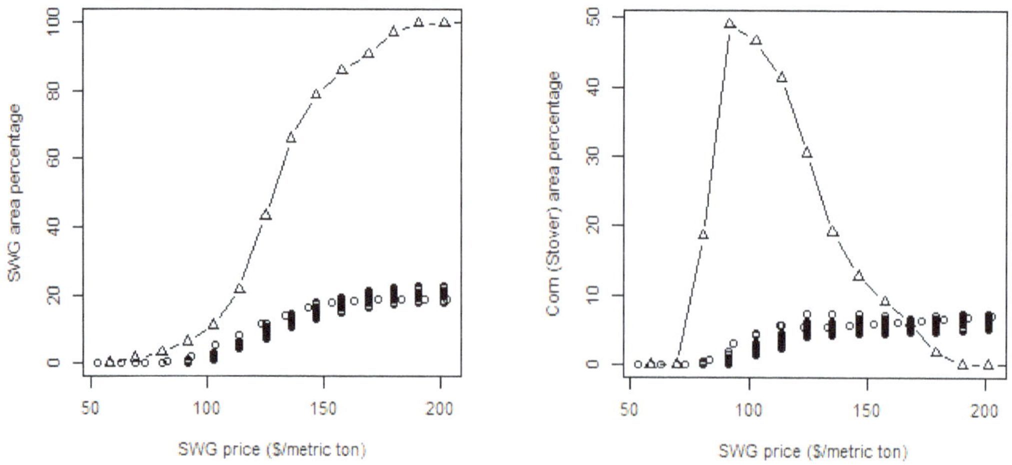

(c) (d)

Figure 5. Crop area percentages ((**a**) corn—no stover; (**b**) soybean; (**c**) switchgrass; (**d**) corn—stover) from 36 simulations (36 SWG price levels and fixed switchgrass to corn stover price ratio of 1.31) on the basis of with (dots) *vs.* without (triangles in dashed line) land use survey information included.

Similarly, there is a large discrepancy in simulated watershed level fertilizer application rates between scenario 1 and 2 (Figure 6). The variation of the fertilizer rates with scenario 2 given 30 random samples from the empirical distribution is small. Under scenario 2 the simulated N and K curves are sigmoid shaped while the P curve mirrors the sigmoid shape. As the switchgrass price increases within a certain range, the average N rate and K rate increase while the average P rate decreases. This is reasonable because P rate for switchgrass is relatively low compared to corn (stover or no stover) and soybean.

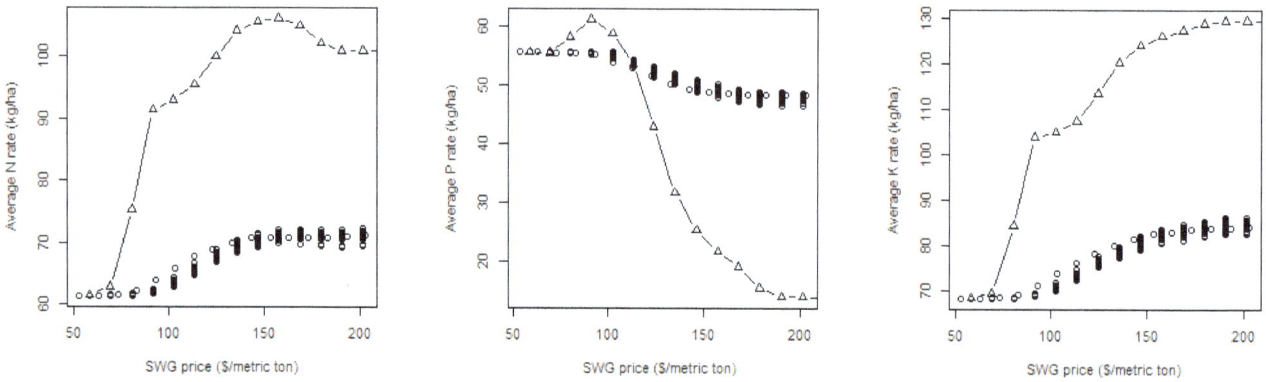

(a) (b) (c)

Figure 6. Watershed-averaged fertilizer rates ((**a**) N; (**b**) P; and (**c**) K) from 36 simulations (36 SWG price levels and fixed switchgrass to corn stover price ratio of 1.31) on the basis of with (dots) *vs.* without (triangles in dashed line) land use survey information included.

The two runs in Simulation Set II correspond to the commodity crop price settings in Simulation Set III, corn price is 186.6 $/metric ton and the price ratio of soybean to corn is 2.25. The two runs are (1) corn price is 181.1 $/metric ton, and the soybean to corn price ratio is 2.4; and (2) corn price is 188.98 $/metric ton, and the soybean to corn price ratio is 2.24. Considering the two runs in Simulation Set II as baseline scenarios, the watershed-average N rate in Simulation Set III (ranging from 62.77 to 71.14 kg/ha) is higher than in the baseline (between 60.53 and 62.77 kg/ha). So is the watershed-average K rate in Simulation III (ranging from 67.25 to 84.07 kg/ha) compared to the baseline (68.37 kg/ha). The watershed-average P rate in Simulation III (ranging from 47.08 to 56.04 kg/ha) is lower than in the baseline (56.04 kg/ha). These results show that biofuel crop markets may potentially result in more N and K inputs and less P inputs into the watershed. This is understandable because (1) in the baseline scenario, corn-soybean rotation only needs N input in corn years, while once enrolled in switchgrass planting, it requires N input every year; (2) switchgrass requires more K input than corn and soybean do. Since switchgrass potentially has beneficial effects of reducing nutrient and sediment runoff [48,49] while commodity crops management and harvesting corn stover may have adverse effects [50], it would therefore be very interesting to investigate the impacts of potential biofuel crop markets on watershed water quality in a further study.

6.2.2. Impacts of Corn Stover and Switchgrass Price (Simulation Set IV)

In Simulation Set IV, both the switchgrass price and the price ratio of switchgrass to corn stover vary. Switchgrass price varies from 58.41 $/metric ton to 223.71 $/metric ton, and the price ratio of switchgrass to corn stover varies from 1.1 to 2.1. For each combination of the two variables, we run the simulation 10 times and plotted the average of the 10 as a response "surface" (Figures 7 and 8).

Figure 7c,d shows that within the price ranges in Simulation Set IV, switchgrass occupies at most about 22%–24% of the watershed and corn stover occupies at most 10%. These numbers are consistent with those in Simulation Set III. In Figure 7a–d, at any fixed ratio of switchgrass: Corn stover (1.1–2.1), the crop area percent responds similarly to the curves represented by circles in Figure 6a–d except for corn (stover). At any fixed ratios greater than about 1.4, the corn (stover) area percentage curve does not stabilize within the switchgrass price range from 58.41 to 223.71 $/metric ton. Instead, it still keeps increasing even at the very high switchgrass price of 223.71 $/metric ton.

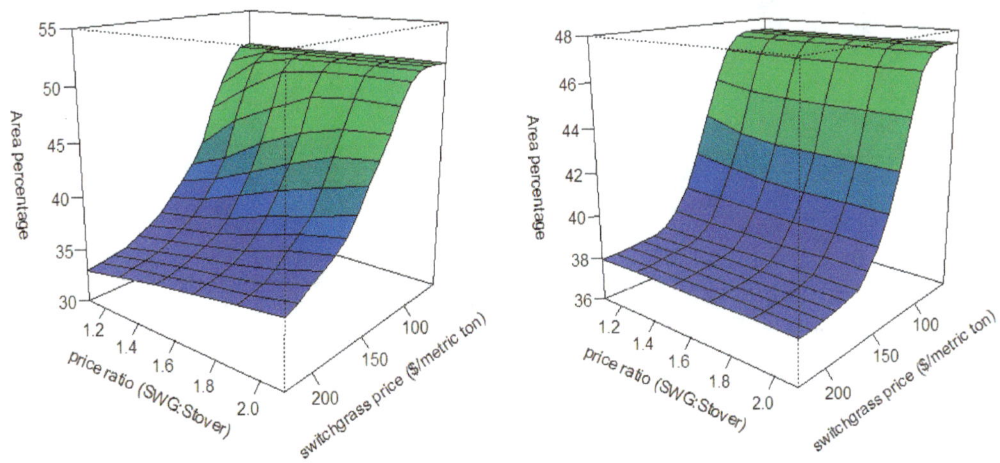

(a) (b)

Figure 7. *Cont.*

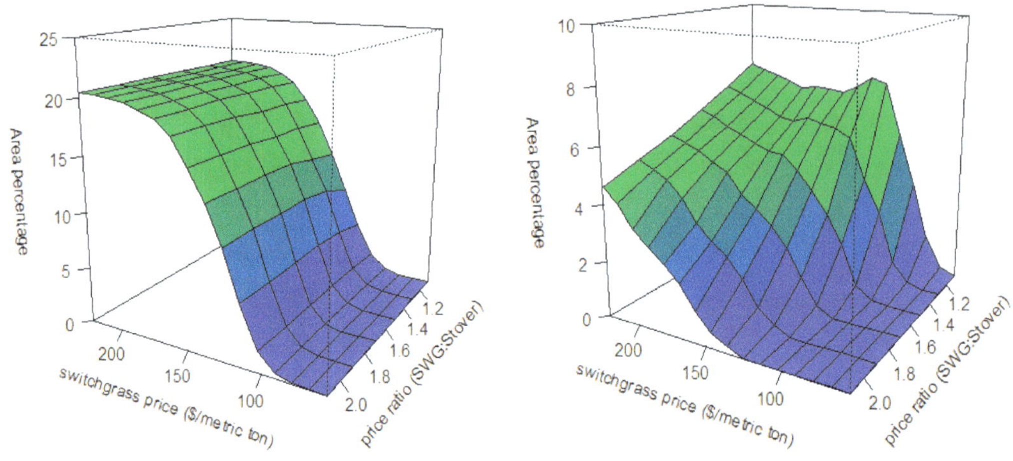

(c) (d)

Figure 7. Crop area percentages ((**a**) corn—no stover; (**b**) soybean; (**c**) switchgrass; (**d**) corn—stover) from 960 simulations (16 levels of switchgrass price by 6 levels of switchgrass to corn stover price ratio, average of the 10 random samples based simulations for each combination).

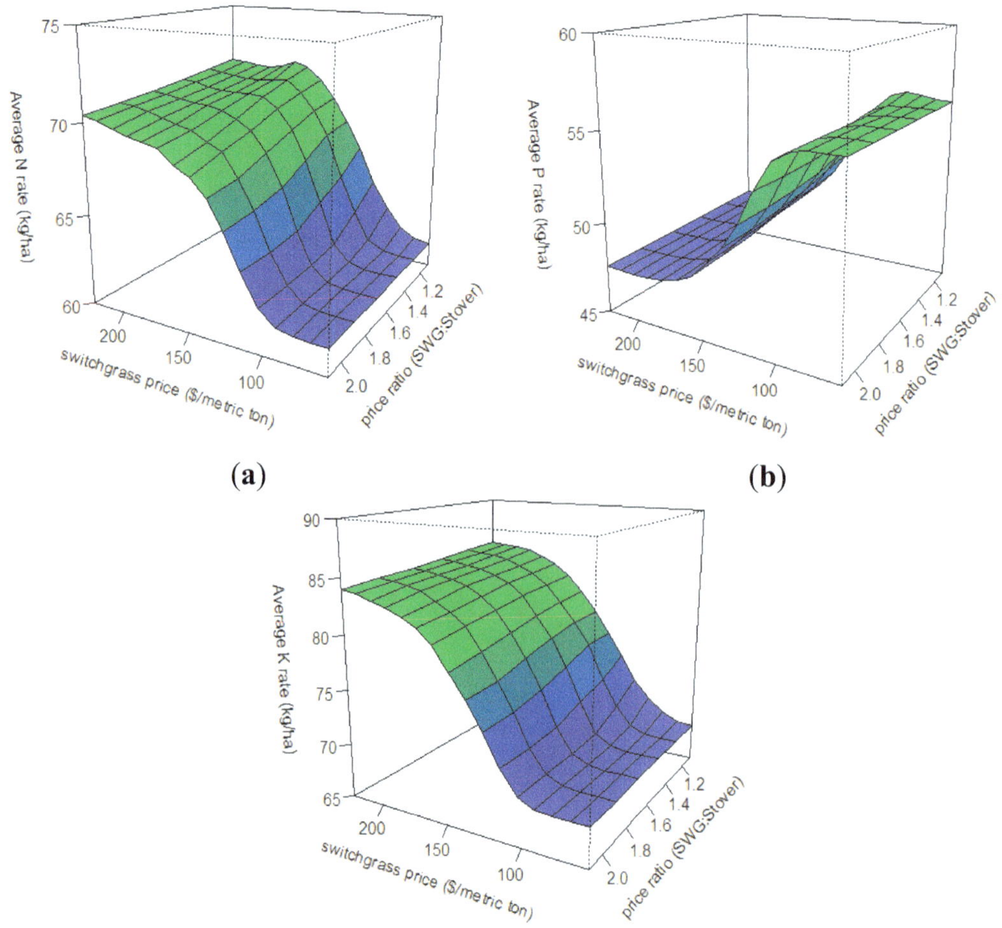

(c)

Figure 8. Watershed-averaged fertilizer rates ((**a**) N; (**b**) P; and (**c**) K) from 960 simulations (16 levels of SWG price by 6 levels of switchgrass to corn stover price ratio, average of the 10 random samples based simulations for each combination)

Overall, the crop area percentage in Figure 4a,b (corn and soybean scenarios) responds more rapidly to the change of price ratio than in Figure 7a–d (switchgrass and corn stover scenarios). We may conclude that the price ratio of switchgrass to corn stover does not influence the crop composition as much as the price ratio of soybean to corn.

In Figure 8a–c, we plotted watershed-averaged N, P, and K rates in response to the change of switchgrass prices and switchgrass to corn stover price ratios. Similarly to Figure 6a–c, the watershed-averaged N and K rates increase as switchgrass prices increase but the watershed-averaged P rate decreases as switchgrass prices increase. N, P, and K rates stabilize, respectively, at 67.25–73.98 kg/ha, 44.83–49.32 kg/ha, and 81.82–87.43 kg/ha when switchgrass price reaches 132.24–165.3 $/metric ton and higher. The fertilizer rates in the switchgrass-corn stover scenarios (Figure 8a–c) respond more rapidly to the change of switchgrass price than to the change of price ratio. At fixed level of switchgrass price, fertilizer rates hardly respond to the change of switchgrass to corn stover ratio within the range of 1.1 to 2.1.

7. Conclusions

The major findings in this research include: (1) discrepancies exist between simulated and satellite-derived land acreages and crop rotation patterns, with implications in terms of farmers' estimation on crop yield drag effects; (2) the simulated biofuel crop land acreage response surface starts to plateau at prices beyond $150/ton for switchgrass; and (3) simulated biofuel crop land acreages and fertilizer application rates in response to alternative crop market scenarios differ significantly depending on whether or not the risk averse behavior of farmers towards biofuel crop production is considered. These findings help define the decision space for future biofuel production, and provide insights for different stakeholders, such as agricultural policy makers who are concerned about commodity and biofuel crop land use and inputs of fertilizer into the agricultural system. Since currently no crop insurance is available for biofuel crops, our results also indicate the importance of creating such a program if biofuel crops are to be promoted.

The simulated corn and soybean land percentages respond to market prices (commodity crop, fertilizers, and fuel) more strongly than would be suggested by satellite imagery. The simulations strictly follow the rotation pattern of corn-soybean or corn-corn-soybean while according to the satellite data, about 36% of the crop land were in more than two-year continuous corn. The discrepancy between the simulations and satellite data suggests that farmers are underestimating the yield drag associated with continuous corn. The topic of farmers' yield drag perception and estimation *versus* agronomic evidence is worth further investigation and verification since misperceptions can result in non-optimal economic and environmental consequences.

The simulated corn and soybean land acreage and fertilizer rates responses to corn price change and/or soybean to corn price ratio changes are realistic. Those responses stabilize within 181.10–283.46 $/metric ton and the range of soybean to corn price ratio of about 2.61–2.99: Corn and soybean area percentages stay 50/50; average N, P, and K rates stabilize at respectively about 65.01, 56.04, and 70.61 kg/ha.

Given farmers' attitudes towards biofuel crop production, large scale cellulosic biofuel crop production is likely to require some mechanism to reduce risk for farmers. Our results indicate that if risk for planting switchgrass is eliminated and farmers only follow a farm-profit maximization rule, given a switchgrass to corn stover price ratio of 1.31, switchgrass would occupy the whole watershed when its price is about 179.63 $/metric ton and higher. However, in the perceived risk case, switchgrass would occupy at most about 22%–24% of the watershed when switchgrass price is about 132.24–165.3 $/metric ton or higher, and corn stover would occupy at most about 10%. Correspondingly, average N, P, and K rates stabilize respectively at 67.25–73.98 kg/ha, 44.83–49.32 kg/ha, and 81.82–87.43 kg/ha when switchgrass price reaches 132.24–165.3 $/metric ton and higher. The 2014 Farm bill eliminates crop subsidies, though temporary establishment subsidies for dedicated biomass feedstocks were maintained in the Energy title of the bill under the Biomass Crop Assistance Program (up to five years for a maximum of $111.2/ha). However, this is unlikely to suffice. Some type of subsidized crop insurance would also have to be developed.

Acknowledgments

This study was supported by the National Science Foundation Cyber-Enabled Discovery and Innovation Program (award 0835607), the National Science Foundation Dynamics of Coupled Natural and Human Systems Program (award 1114978), and the IIHR-Hydroscience and Engineering at the University of Iowa.

Author Contributions

Deng Ding designed the ABM, wrote the model scripts, analyzed the data and wrote the manuscript. David Bennett supervised the research, participated in designing the model, reviewed the manuscript, and provided suggestions and edits. Silvia Secchi provided data and citation sources about the economic models adapted in the model, reviewed the manuscript and provided suggestions and edits.

Conflicts of Interest

The authors declare no conflict of interest.

Appendix

A. Crop Yield Drag Coefficients and Fertilizer Rates

The effects of crop rotation, tillage type, and nitrogen fertilizer rates (corn only) are accounted for when calculating crop yields. Crop yield drag coefficients that capture the rates at which the crop productivity declines as a function of continuous planting, are derived from corn and soybean yield functions [7] and represented in Figure A1. Given the impact of the previous year's crop, continuous corn (CC) corresponds to lower yield than corn in annual rotation with soybean (SC). Soybean yields in the present year are influenced by the previous two years crops. Among the crop rotation series (corn-corn-soybean: CCS, corn-soybean-soybean: CSS, soybean-corn-soybean: SCS), CCS corresponds to the highest soybean yield, followed by SCS, and then CSS. The impacts of tillage types on crop yields are taken into account for both corn and soybean. Among the three tillage types (conventional tillage, mulch tillage, and no tillage), the crop with conventional tillage has the highest yield and no till has the lowest. Instead of using a quadratic function [7] to characterize corn yield response to nitrogen, ten different levels of nitrogen rates and corresponding corn yield drag coefficients are sampled from the response curve. This implementation reduces the optimization problem from a nonlinear into a linear programming problem.

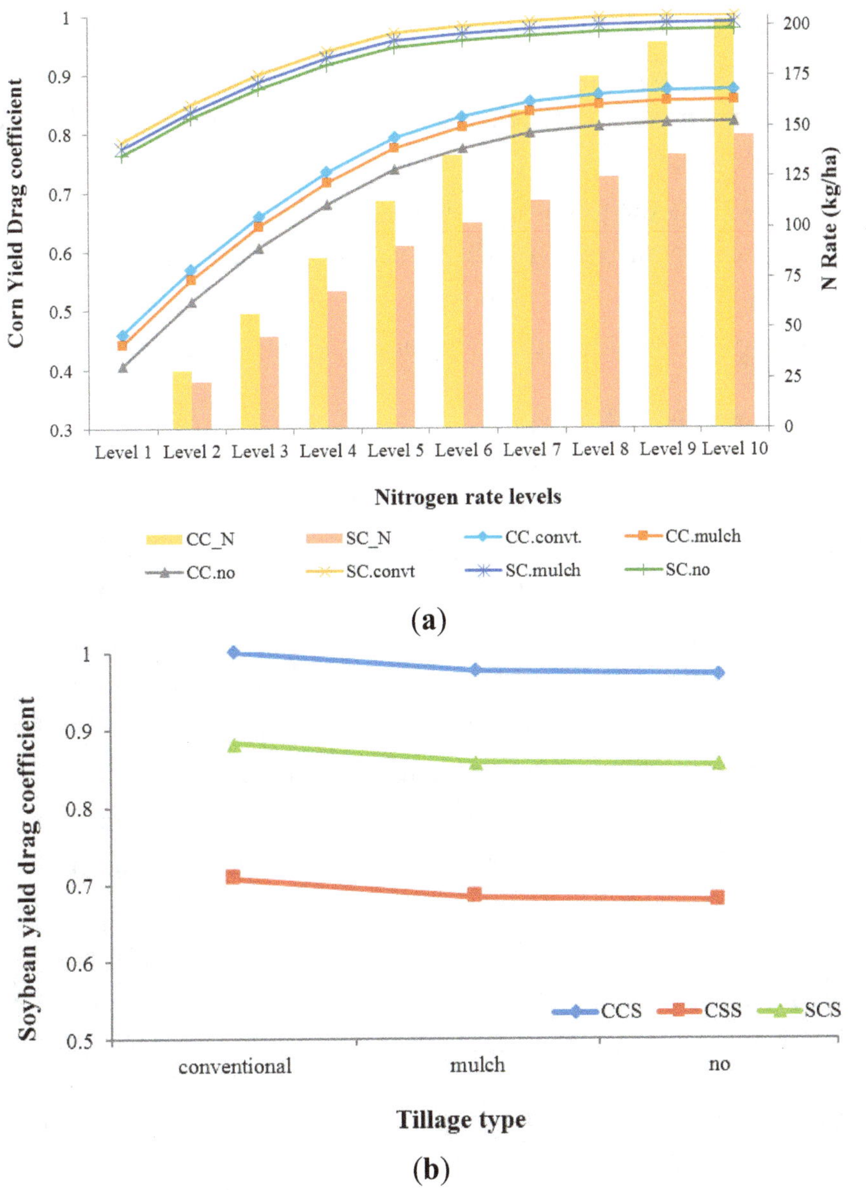

Figure A1. Corn yield drag coefficients (**a**) for different rotation, tillage, and N fertilizer rate levels, and soybean yield drag coefficients; (**b**) for different rotation types: Corn is denoted as "C" and soybean as "S", the last letter in a rotation pattern means the current crop (e.g., CCS means corn-corn-soybean, and soybean is in the current year).

In the model, it is assumed that soybean does not need nitrogen fertilizer. The fertilizer rates of phosphorus (P) and potassium (K) are fixed for soybean: Respectively, 44.83 kg/ha and 84.07 kg/ha. According to Iowa State University Extension [38,51], corn after corn requires lower fertilizer input of phosphorus (P) and potassium (K) than corn after soybean. For corn after corn, the P rate is 61.65 kg/ha and the K rate is 50.44 kg/ha; for corn after soybean, the rates are 67.25 kg/ha and 56.04 kg/ha.

Switchgrass is a perennial with an establishment period. Therefore, we assume that the switchgrass starts to be harvested from the second year. Switchgrass yields depend on soil type. We scale the tall introduced grasses yield (TIGRSYLD) in the Iowa Soil Properties and Interpretations Database (ISPAID) into the data range from 4.48 to 14.35 metric ton/ha to represent the switchgrass yield for different soil types. According to Duffy (2008) [40], in the first year when switchgrass is established nitrogen fertilizer is not required, but 33.63 kg/ha of P and 44.83 kg/ha of K are applied. 112.09 kg/ha of N is applied from the second year on to compensate for the harvest of switchgrass. The amount of P and K applied depends on the amount of switchgrass biomass harvested. For each ton of switchgrass that is harvested, 0.88 kg of P and 10.34 kg of K are applied.

Corn stover yield is estimated as the product of the portion of corn stover harvested and corn yield (we assume that dry weight of corn grain is equal to corn stalk weight). Additional amounts of N, P, and K fertilizers are applied to compensate for the nutrients removed from the soil system by corn stover harvest. 9.07 kg of N, 2.68 kg of P, and 11.34 kg of K are applied per ton of dry matter stover harvest [41].

B. Empirical Parameters and Parameterization

To generate samples from the agricultural landowners and operators survey database for the eight parameters above, we performed the following analysis on the survey responses. Firstly, we considered the respondents who self-identified as farmers. Concerning the parameter *Type*, the farmer respondents were classified into the four types based on whether they responded to Question 48 and/or 56 (Table A1). Secondly, we performed one-way ANOVA test on the farm sizes (*Area*) of the four types of farmers. The results showed that there is sufficient evidence (significance level: 0.001) to reject the hypothesis that the farm size means are all equal among the four farmer types. Thirdly, we analyzed the correlation among the other parameters (Table A2). The results do not support the conclusion that *Percent1 SWG* and *Profit SWG* are correlated for Type 2 and 3 farmers. Neither is there evidence to suggest that *Percent1 Stvr* and *Profit Stvr* are correlated for *Type* 1 and 3 farmers. But for both *Type* 2 and 3 farmers, *Percent1 SWG* and *Percent2 SWG* are significantly correlated. Similarly, *Percent1 Stvr* and *Percent2 Stvr* are significantly correlated for *Type* 1 and 3 farmers. For *Type* 3 farmers, *Profit SWG* and *Profit Stvr* are significantly correlated as are *Percent1 SWG* and *Percent1 Stvr*.

Table A1. Land use survey questions adopted for model parameterization.

Question #	Question
3	Do you consider yourself a farmer?
18c	Number of acres you own that you farmed in 2009
18d	Number of acres you leased or rented to other people to farm in 2009
18e	Number of acres you rented from others to farm in 2009
18f	Number of acres in the Conservation Reserve Program (CRP) in 2009
48	What is the minimum net profit per acre you would need to get in order to consider marketing corn stover?
49	If you could get that profit per acre, how many acres of corn stover would you consider harvesting?
50	If you supplied corn stover to a bio-refinery, would you prefer to harvest 30%, 50%, or 70% of the corn stover in your fields?
51	If you could get a net profit 50% higher than what you indicated in question 48, how many acres of corn stover in total would you consider harvesting?
56	What is the minimum net profit per acre you would need to get in order to consider growing switchgrass?
57	If you could get that profit per acre, how many acres of switchgrass you consider planting?
58	If you could get a net profit 50% higher than what you indicated in question 55, how many acres of switchgrass in total would you consider growing?

Based on the results of this analysis, we used the following strategy to empirically draw 30 samples using the bootstrap sampling method. First, farmer respondents and farmer agents are respectively classified into five groups based on the land acreages they reported (≤40.47, 40.47–121.41, 121.41–202.34, 202.34–303.51, and >303.51 ha). Second, for all agents in a certain group, the sample of parameter *Type* is drawn from the empirical distribution of *Type* for farmers in the corresponding group. Third, the remaining parameter values are generated for farmer agents in each group and of each type. For *Type* 1 agents, parameter values of *Profit Stvr*, *Percent1 Stvr* and *Portion Stvr* are independently generated; parameter values of *Percent2 Stvr* are associated with *Percent1 Stvr*. For *Type* 2 agents, parameter values of *Profit SWG*, *Percent1 SWG* are independently generated; parameter values of *Percent2 SWG* are associated with *Percent1 SWG*. For *Type* 3 agents, parameter values for *Profit Stvr*, *Percent1 Stvr*, and *Portion Stvr* are independently generated; parameter values of *Profit SWG* are associated with *Profit Stvr*; *Percent2 Stvr*, *Percent1 SWG* and *Percent2 SWG* are associated with *Percent1 Stvr*.

Table A2. Pearson's r (*p*-value) between the parameters characterizing farmers' attitudes.

	Type	*Profit SWG*	*Percent1 SWG*	*Percent2 SWG*	*Profit Stvr*	*Percent1 Stvr*	*Percent2 Stvr*	*Portion Stvr*
Profit SWG	1 or 2	1	0.4145 (0.1407)	na *	na	na	na	na
	3	1	−0.2488 (0.1560)	na	0.8471 (2.03×10^{-11})	na	na	na
Percent1 SWG	1 or 2		1	0.9687 (4.18×10^{-11})	na	na	na	na
	3		1	0.8885 (1.06×10^{-12})	na	0.6822 (3.31×10^{-6})	na	na
Percent2 SWG	1 or 2			1	na	na	na	na
	3			1	na	na	na	na
Profit Stvr	1 or 2				1	0.1918 (0.4768)	na	−0.0525 (0.8526)
	3				1	−0.1714 (0.3248)	na	−0.0648 (0.7033)
Percent1 Stvr	1 or 2					1	0.8897 (9.01×10^{-6})	na
	3					1	0.8952 (1.76×10^{-13})	na
Percent2 Stvr	1 or 2						1	na
	3						1	na
Portion Stvr	1 or 2							1
	3							1

* No Data.

References

1. O'Neal, M.R.; Nearing, M.A.; Vining, R.C.; Southworth, J.; Pfeifer, R.A. Climate change impacts on soil erosion in Midwest United States with changes in crop management. *Catena* **2005**, *61*, 165–184.

2. Lettenmaier, D.P.; Hooper, E.R.; Wagoner, C.; Faris, K.B. Trends in stream quality in the continental United States, 1978–1987. *Water Resour. Res.* **1991**, *27*, 327–339.
3. Yadav, V.; Malanson, G.P.; Bekele, E.; Lant, C. Modeling watershed-scale sequestration of soil organic carbon for carbon credit programs. *Appl. Geogr.* **2009**, *29*, 488–500.
4. Bekele, E.G.; Lant, C.L.; Soman, S.; Misgna, G. The evolution and empirical estimation of ecological-economic production possibilities frontiers. *Ecol. Econ.* **2013**, *90*, 1–9.
5. Secchi, S.; Gassman, P.W.; Jha, M.; Kurkalova, L.; Kling, C.L. Potential water quality changes due to corn expansion in the Upper Mississippi River Basin. *Ecol. Appl.* **2011**, *21*, 1068–1084.
6. Secchi, S.; Kurkalova, L.; Gassman, P.W.; Hart, C. Land use change in a biofuels hotspot: The case of Iowa, USA. *Biomass Bioenergy* **2011**, *35*, 2391–2400.
7. Kurkalova, L.A.; Secchi, S.; Gassman, P.W. Corn stover harvesting: Potential supply and water quality implications. In *Handbook of Bioenergy Economics and Policy*; Khanna, M., Scheffran, J., Zilberman, D., Eds.; Springer-Verlag: New York, NY, USA, 2010; pp. 307–323.
8. Lee, K.H.; Isenhart, T.M.; Schultz, R.C.; Mickelson, S.K. Nutrient and sediment removal by switchgrass and cool-season grass filter strips in central Iowa, USA. *Agrofor. Syst.* **1998**, *44*, 121–132.
9. Sarkar, S.; Miller, S.A.; Frederick, J.R.; Chamberlain, J.F. Modeling nitrogen loss from switchgrass agricultural systems. *Biomass Bioenergy* **2011**, *35*, 4381–4389.
10. Parker, D.C.; Manson, S.M.; Janssen, M.A.; Hoffmann, M.J.; Deadman, P. Multi-agent systems for the simulation of land-use and land-cover change: A review. *Ann. Assoc. Am. Geogr.* **2003**, *93*, 314–337.
11. Bennett, D.A.; Tang, W.; Wang, S. Toward an understanding of provenance in complex land use dynamics. *J. Land Use Sci.* **2011**, *6*, 211–230.
12. Matthews, R.; Gilbert, N.G.; Roach, A.; Polhill, J.; Gotts, N.M. Agent-based land-use models: A review of applications. *Landsc. Ecol.* **2007**, *22*, 1447–1459.
13. Robinson, D.T.; Brown, D.G.; Parker, D.C.; Schreinemachers, P.; Janssen, M.A.; Huigen, M.G.A.; Wittmer, H.; Gotts, N.M.; Promburom, P.; Irwin, E.G.; *et al.* Comparison of empirical methods for building agent-based models in land use science. *J. Land Use Sci.* **2007**, *2*, 31–55.
14. Smajgl, A.; Brown, D.G.; Valbuena, D.; Huigen, M.G.A. Empirical characterisation of agent behaviours in socio-ecological systems. *Environ. Model. Softw.* **2011**, *26*, 837–844.
15. Manson, S.M. Challenges in evaluating models of geographic complexity. *Environ. Plan. B Plan. Design* **2007**, *34*, 245–260.
16. Matthews, R. The people and landscape model (palm): Towards full integration of human decision-making and biophysical simulation models. *Ecol. Model.* **2006**, *194*, 329–343.
17. Matthews, R.; Selman, P. Landscape as a focus for integrating human and environmental processes. *J. Agric. Econ.* **2006**, *57*, 199–212.
18. Epstein, J.M. *Generative Social Science: Studies in Agent-Based Computational Modeling*; Princeton University Press: Princeton, NJ, USA, 2006.
19. Lansing, J.S.; Kremer, J.N. Emergent properties of Balinese water temple networks: Coadaptation on a rugged fitness landscape. *Am. Anthropol.* **1993**, *95*, 97–114.

20. Le, Q.B.; Park, S.J.; Vlek, P.L.G. Land Use Dynamic Simulator (LUDAS): A multi-agent system model for simulating spatio-temporal dynamics of coupled human-landscape system 2. Scenario-based application for impact assessment of land-use policies. *Ecol. Inform.* **2010**, *5*, 203–221.
21. Le, Q.B.; Park, S.J.; Vlek, P.L.G.; Cremers, A.B. Land-Use Dynamic Simulator (LUDAS): A multi-agent system model for simulating spatio-temporal dynamics of coupled human-landscape system. I. Structure and theoretical specification. *Ecol. Inform.* **2008**, *3*, 135–153.
22. Bousquet, F.; Le Page, C. Multi-agent simulations and ecosystem management: A review. *Ecol. Model.* **2004**, *176*, 313–332.
23. Sengupta, R.; Lant, C.; Kraft, S.; Beaulieu, J.; Peterson, W.; Loftus, T. Modeling enrollment in the Conservation Reserve Program by using agents within spatial decision support systems: An example from southern Illinois. *Environ. Plan. B Plan. Des.* **2005**, *32*, 821–834.
24. Scheffran, J.; BenDor, T. Bioenergy and land use: A spatial-agent dynamic model of energy crop production in Illinois. *Int. J. Environ. Pollut.* **2009**, *39*, 4–27.
25. Ng, T. Response of Farmers' Decisions and Stream Water Quality to Price Incentives for Nitrogen Reduction, Carbon Abatement, and Miscanthus Cultivation: Preditions Based on Agent-Based Modeling Coupled with Water Quality Modeling. Ph.D. Dissertation, University of Illinois at Urbana-Champaign, Urbana-Champaign, IL, USA, 2010.
26. Ng, T.; Eheart, J.W.; Cai, X.M.; Braden, J.B. An agent-based model of farmer decision-making and water quality impacts at the watershed scale under markets for carbon allowances and a second-generation biofuel crop. *Water Resour. Res.* **2011**, *47*, doi:10.1029/2011WR010399.
27. Schreinemachers, P.; Berger, T. Land use decisions in developing countries and their representation in multi-agent systems. *J. Land Use Sci.* **2006**, *1*, 29–44.
28. Berger, T.; Schreinemachers, P. Creating agents and landscapes for multiagent systems from random samples. *Ecol. Soc.* **2006**, *11*, 19.
29. Castella, J.C.; Verburg, P.H. Combination of process-oriented and pattern-oriented models of land-use change in a mountain area of Vietnam. *Ecol. Model.* **2007**, *202*, 410–420.
30. Torii, D.; Bousquet, F.; Ishida, T.; Trébuil, G.; Vejpas, C. *Using Classification Learning in Companion Modeling Multi-Agent Systems for Society*; Lukose, D., Shi, Z., Eds.; Springer: Berlin/Heidelberg, Germany, 2009; Volume 4078, pp. 255–269.
31. Brown, D.G.; Robinson, D.T. Effects of heterogeneity in residential preferences on an agent-based model of urban sprawl. *Ecol. Soc.* **2006**, *11*, 46.
32. Huigen, M.G.A. First principles of the mameluke multi-actor modelling framework for land use change, illustrated with a Philippine case study. *J. Environ. Manag.* **2004**, *72*, 5–21.
33. Castillo, D.; Saysel, A.K. Simulation of common pool resource field experiments: A behavioral model of collective action. *Ecol. Econ.* **2005**, *55*, 420–436.
34. Irwin, E.G.; Bockstael, N.E. Interacting agents, spatial externalities and the evolution of residential land use patterns. *J. Econ. Geogr.* **2002**, *2*, 31–54.
35. Brown, D.G.; Robinson, D.T.; An, L.; Nassauer, J.I.; Zellner, M.; Rand, W.; Riolo, R.; Page, S.E.; Low, B.; Wang, Z.F. Exurbia from the bottom-up: Confronting empirical challenges to characterizing a complex system. *Geoforum* **2008**, *39*, 805–818.

36. Druschke, C.G.; Secchi, S. The impact of gender on agricultural conservation knowledge and attitudes in an Iowa watershed. *J. Soil Water Conserv.* **2014**, *69*, 12.
37. Varble, S.; Druschke, C.G.; Secchi, S. An examination of growing trends in land tenure and conservation practice adoption: Results from a farmer survey in Iowa. *Environ. Manag.* **2015**, doi:10.1007/s00267-015-0619-5.
38. Iowa State University Extension. *Nitrogen Fertilizer Recommendations for Corn in Iowa*; Leopold Center: Ames, IA, USA, 1997. Available Online: http://www.iasoybeans.com/advancenewsletter/PDF/ADV15_0611_4_PM1714.pdf (accessed on 1 October 2015).
39. Grimm, V.; Berger, U.; DeAngelis, D.L.; Polhill, J.G.; Giske, J.; Railsback, S.F. The ODD protocol: A review and first update. *Ecol. Model.* **2010**, *221*, 2760–2768.
40. Duffy, M. *Estimated Costs for Production, Storage and Transportation of Switchgrass*; Iowa State University Extension: Ames, IA, USA, 2008. Available Online: https://www.extension.iastate.edu/agdm/crops/pdf/a1-22.pdf (accessed on 1 October 2015).
41. Edwards, W. *Estimating a Value for Corn Stover*; Iowa State University Extension: Ames, IA, USA, 2014. Available Online: https://www.extension.iastate.edu/agdm/crops/pdf/a1-70.pdf (accessed on 1 October 2015).
42. Johanns, A.M. *Iowa Cash Corn and Soybean Prices*; Iowa State University Extension and Outreach: Ames, IA, USA, 2015. Available Online: http://www.extension.iastate.edu/agdm/crops/pdf/a2-11.pdf (accessed on 1 October 2015).
43. Food and Agricultural Policy Research Institute. *U.S. Baseline Briefing Book: Projections for Agricultural and Biofuel Markets*; University of Missouri: Columbia, MO, USA, 2015. Available online: http://www.fapri.missouri.edu/wp-content/uploads/2015/03/FAPRI-MU-Report-01-15.pdf (accessed on 1 October 2015).
44. Food and Agricultural Policy Research Institute. *Competition for Biomass among Renewable Energy Policies: Liquid Fuels Mandate versus Renewable Electricity Mandate*; University of Missouri: Columbia, MO, USA, 2011. Available Online: http://www.fapri.missouri.edu/wp-content/uploads/2015/02/FAPRI-MU-Report-11-11.pdf (accessed on 1 October 2015).
45. Davidson, D. *Corn on Corn: How Much Yield is Enough*; Progressive Farmer: Birmingham, AL, USA, 2007.
46. Smith, P. *The Pros and Cons of Going corn-on-Corn*; Fields of Facts: Brandon, MS, USA, 2013. Available Online: http://agfax.com/2013/09/23/the-pros-and-cons-of-going-corn-on-corn/ (accessed on 1 October 2015).
47. Edwards, W.; Johanns, A.M. *Cash Rental Rates for Iowa 2012 Survey*; Iowa State University Extension and Outreach: Ames, IA, USA, 2013. Available Online: https://www.extension.iastate.edu/agdm/wholefarm/pdf/c2–10_2012.pdf (accessed on 1 October 2015).
48. McLaughlin, S.B.; Walsh, M.E. Evaluating environmental consequences of producing herbaceous crops for bioenergy. *Biomass Bioenergy* **1998**, *14*, 317–324.
49. Lee, K.H.; Isenhart, T.M.; Schultz, R.C. Sediment and nutrient removal in an established multi-species riparian buffer. *J. Soil Water Conserv.* **2003**, *58*, 1–8.
50. Mann, L.; Tolbert, V.; Cushman, J. Potential environmental effects of corn (*Zea mays* L.) stover removal with emphasis on soil organic matter and erosion. *Agric. Ecosyst. Environ.* **2002**, *89*, 149–166.

51. Iowa State University Extension. *A General Guide for Crop Nutrient and Limestone Recommendations in Iowa*; Iowa State University Extension and Outreach: Ames, IA, USA, 2013. Available online: http://store.extension.iastate.edu/Product/A-General-Guide-for-Crop-Nutrient-and-Limestone-Recommendations-in-Iowa-PDF (accessed on 1 October 2015).

9

Spatial Distribution of Estimated Wind-Power Royalties in West Texas

Christian Brannstrom *, Mary Tilton, Andrew Klein and Wendy Jepson

Department of Geography, Texas A&M University, 810 O&M Building, College Station, TX 77843, USA; E-Mails: m.tilton83@gmail.com (M.T.); klein@geog.tamu.edu (A.K.); wjepson@geos.tamu.edu (W.J.)

* Author to whom correspondence should be addressed; E-Mail: cbrannst@geos.tamu.edu

Academic Editors: Claudia A. Radel and Jacqueline M. Vadjunec

---

Abstract: Wind-power development in the U.S. occurs primarily on private land, producing royalties for landowners through private contracts with wind-farm operators. Texas, the U.S. leader in wind-power production with well-documented support for wind power, has virtually all of its ~12 GW of wind capacity sited on private lands. Determining the spatial distribution of royalty payments from wind energy is a crucial first step to understanding how renewable power may alter land-based livelihoods of some landowners, and, as a result, possibly encourage land-use changes. We located ~1700 wind turbines (~2.7 GW) on 241 landholdings in Nolan and Taylor counties, Texas, a major wind-development region. We estimated total royalties to be ~$11.5 million per year, with mean annual royalty received per landowner per year of $47,879 but with significant differences among quintiles and between two sub-regions. Unequal distribution of royalties results from land-tenure patterns established before wind-power development because of a "property advantage," defined as the pre-existing land-tenure patterns that benefit the fraction of rural landowners who receive wind turbines. A "royalty paradox" describes the observation that royalties flow to a small fraction of landowners even though support for wind power exceeds 70 percent.

Keywords: wind power; royalty; property; land use; income; Texas

---

1. Introduction

Explaining how biofuels affect land change is a key research theme in hybrid land change science that aims to incorporate insights from political ecology [1], but linkages between land change and utility-scale renewable energy are poorly understood. The rapid expansion of wind-power development in the U.S., primarily on private landholdings in a region from Texas to Minnesota, has produced unknown outcomes for livelihoods and land uses through the production of royalties for landowners who make private contracts with wind-farm operators. In dozens of rural communities, wind power has produced obvious impacts on land cover from construction of wind turbines and maintenance roads; however, these plainly visible land changes contrast with far less visible effects on landowner livelihoods and land cover, which have multiple potential behavioral, financial, and environmental determinants.

More than 97 percent of the ~60 GW of installed wind power capacity in the U.S. has been built on private land [2,3]. With 12,214 MW of installed capacity, Texas is the leading U.S. state, where 10 percent of its electricity comes from wind farms—virtually all sited on private lands—in the state's western region [4]. Wind turbines represent a land use that farmers and ranchers view as compatible with livestock grazing, hunting, or agriculture, and social acceptance is high [5–7]. Knowledge of the spatial distribution of royalty payments from wind energy is a crucial first step to understanding how income from renewable power may alter land-based livelihoods and may encourage land-use changes.

Here we use GIS-based methods and a royalty formula to estimate the amount and spatial distribution of wind-power royalties in Nolan and Taylor counties (Figure 1), a core area of windfarms in Texas. We show that royalty distribution among landowners is unequal, especially between two sub-regions, because of pre-existing "property advantage." We also identify a "royalty paradox:" a tiny fraction of the overall rural population receives royalties, but over 70 percent of the population support wind power. Spatial patterns of royalties should inform studies of land change in areas of energy development and studies of livelihoods possibly altered by energy royalties. Studies of wind power support or opposition should consider the flow and distribution of financial resources to landowners when attempting to understand the social acceptance of wind power.

Figure 1. Nolan and Taylor Counties, Texas.

1.1. The Property Advantage

One important spatial and land-based aspect of royalties is the "property advantage," which refers to unequal land-tenure patterns that may encourage unequal distribution of turbines and royalties. The distribution of royalties is important because it affects household wealth through increased land value [8]. More broadly, the spatial distribution of ownership of energy assets has attracted little critical attention, apart from recent work on shale gas in Denton, Texas [9]. Although Sowers included royalties into a "please-in-my-backyard" explanation for high apparent acceptance of wind power in rural Iowa [10], he did not estimate spatial distribution. Scholars of social, economic, and political aspects of wind-power expansion in North America have focused on other economic benefits, such as direct employment and indirect economic activity, arising from wind-power development, but livelihood implications are unknown.

More critical work reports possible tensions between landowners. In Ontario, Canada, wind turbines sited on private land generate income of at least $8000 per year, but neighbors without turbines receive no compensation. Walker *et al.* [11] report that wealthy farmers were getting wealthier because of lease payments and other farmers argued that lease payments helped sustain their farm operations. One-third of respondents to their study agreed strongly with the claim that "residents living near wind turbines, but without one on their land should also receive financial benefits" ([11]; p. 738). This finding resonates with a warning that energy payments could "undermine social capital" in U.S. rural communities because so few landowners received income from energy ([8]; p. 22). The energy revenue stream is highly concentrated because of the small number of rural households that own land.

Royalty payments are absent from studies regarding community support or opposition to wind power. Swofford and Slattery [5] measured public attitudes to wind power near a 100 MW wind farm in Texas, but their questionnaire did not prompt respondents to report whether they received royalties or lease payments. Abbott [12] found that public comments regarding wind farms in Washington state were positive with respect to impacts on agriculture, but it is not clear whether material benefits formed part of the view of wind power as "a new phase for extraction economies" ([12]; p. 981). Mulvaney *et al.* [13] reported that support for Indiana wind farms included the claim that turbines would help preserve farming or prevent farms from being turned into housing subdivisions, but they did not mention leases or royalties. Kahn [14] reported improvements in quality of life in places with wind farm development, specifically lower property taxes and increased school quality, but did not analyze royalty payments to landowners. Brown *et al.* [15] found an increase in county-level income and employment resulting from wind power, but did not indicate the specific impact of royalties and land leases on the estimated $11,000 increase in annual personal income per MW of wind capacity installed.

One possible reason for the focus on economic and employment aspects is reliance on the Jobs and Economic Development Impacts (JEDI) model, developed by the National Renewable Energy Laboratory, which includes a measure of landowner lease payments between $3000 and $7000 per MW per turbine per year (default = $6000). Several research teams have relied on this input-output model to estimate direct employment in construction and operational phases and indirect economic activity [16,17]. Greene and Geisken [18] applied JEDI to a case in Oklahoma, estimating nearly $400,000 in land lease payments for a 147 MW wind farm, but they did not report the number of landowners affected in a county nor the spatial distribution of the wind turbines on properties. Indeed, JEDI estimates are not spatially explicit, focusing at county or state levels rather than on individual landowners. Moreover, lease payments are considered synonymous with royalties, while in practice, leases for roads and other infrastructure represent a separate revenue stream from royalties, which are based on variable electricity output per wind turbine and electricity price.

1.2. The Royalty Paradox

Following from the "property advantage" relating to land distribution is the broader social and political significance of royalties. We describe this as the "royalty paradox," which other scholars have suggested. For example, Slattery *et al.* [6] reported high social support for wind power in the west Texas and Iowa, with more than 70 percent of respondents to a mail-out survey indicating that wind energy improved employment and the overall economy. But they noted that, ironically, only "relatively few residents" ([6]; p. 3699) received direct benefits in the form of royalty income from wind power. Wind turbines were "seen as the vehicle that will reverse economic decline" plaguing many rural communities ([6]; p. 3699). Economic benefits, in terms of royalty payments, may help explain why people in some parts of the U.S. are highly supportive of wind power; however, we do not know how these payments are distributed spatially.

Financial aspects of the "royalty paradox" have been suggested by many journalists and wind-power promoters. Just as owners of mineral rights receive royalties based on the extraction of hydrocarbon resources, owners of surface land benefit from wind power generated from turbines sited on their land. Landowners with turbines receive royalties estimated at between $4000 and $10,000 per wind turbine sited on their land per year, depending upon electricity output and contract terms [19,20]. Sowers [10] reported that landowners in Iowa received between $2000 and $2400 per turbine per year under 30-year contracts for turbines ranging from 0.75–0.90 MW capacity. However, scholars have not yet examined royalties from wind-power generation in spatial detail. Weber *et al.* [8] report that approximately 8 percent of farms in the U.S. Plains received an energy payment for wind or natural gas. They concluded that energy payments on a national basis reached 3.4 percent of farms and totaled $2.3 billion in 2011 compared to $11.8 billion in annual government payments to farmers, although important spatial and temporal uncertainties persist.

More generally, financial benefits are part of community support for wind power in cases where public input is highly limited and large landowners have influence on siting decisions. Sharing or distributing financial benefits helps predict support for wind farms [11,21,22]. Bohn and Lant [23] predicted that "local investiture" and "local control over siting" wind-energy projects will lead to strong local and community support for wind power, but it was not clear whether "local investiture" includes royalty payments made to individual landowners ([23]; p. 97). This view echoed Pasqualetti's [24] prediction that support for wind will increase as economic returns and property values increase, but the specific influence of royalties in economic returns was not elaborated. Walker *et al.* [11] indicate that determining how financial benefits from wind farms are distributed should demand attention of researchers and policy makers.

The property advantage and royalty paradox represent new ways to analyze how utility-scale renewable energy may alter land-based livelihoods of some landowners, and possibly encourage land-use changes. Determining the spatial distribution of royalty payments from wind energy and estimating their total value are important first steps toward this objective.

2. Research Design

2.1. Study Region

Our findings focus on Nolan and Taylor counties, a microcosm of a wider process that has made Texas the leading U.S. state in wind power. Nolan County (population = 14,700) had a median household income of $37,500 in 2009. Taylor County, where Abilene is the county seat, has 126,600 residents and a median household income of $40,400 [6]. Rural landowners in both counties are at the eastern margins of the Permian Basin oil fields that developed in the 1920s and 1930s [25,26] and have some familiarity with energy payments and leases, mainly through family history and legal expertise, but no published data or analysis on oil and gas royalties in the study region exist.

Writing in the 1960s, Abilene journalist A.C. Greene ([25]; p. 149) wrote that wind was the "greatest natural resource in West Texas" but it was also the cause, or the excuse, for most of what is "uncomfortable, abominable, and harsh about the country," creating difficult conditions for ranching and farming. This steady wind would later become characterized more optimistically as favorable wind climatology, which, coupled with buoyant demand for electricity, little political opposition to wind-farm construction, and the country's lowest permitting requirements, made west Texas the North American leader for wind power. Nolan and Taylor counties, Texas (Figure 1), received large wind-power investments dating from the 150 MW Trent Mesa site, built in 2001 by Enron's wind division. By 2009, the counties and adjacent areas had approximately 3000 MW of installed wind power. Installation of new turbines in Nolan and Taylor counties slowed after 2009, but completion of new transmission lines around 2011 helped increase efficiencies and provide incentives for future development. Sweetwater, Nolan County's largest city, aims to brand itself as the center of wind-energy development by promoting trade shows, encouraging manufacturing related to wind power, and hosting a wind technician training program at a local community college (Figure 2).

Figure 2. Wind turbine blade on Interstate 20 in Sweetwater suggesting the importance of wind energy to the city's identity and economy.

The large and rapid investments in Texas wind power rely upon legislation promoting renewable energy standards and electricity market deregulation, authorized in 1999, just a few days after then-governor George W. Bush declared his presidential candidacy. This statute resulted from Bush's desire to encourage wind—he told his top utility appointee to "go get smart on wind" ([7]; p. 121) in 1996—and Enron chief Kenneth Lay's desire for a deregulated electricity market in Texas, which helped create conditions for Enron's construction of the Trent Mesa wind farm, signaling the start of the wind boom in Nolan and Taylor counties [7]. Complementing Bush's stance on renewables, Governor Rick Perry in 2005 authorized Competitive Renewable Energy Zones (CREZ), which linked rural regions with the greatest wind energy potential to Texas's growing urban areas through a $7 billion transmission lines project capable of moving 18,500 MW [27]. These policies, combined with the federal Production Tax Credit (PTC) for renewable power generation [28] and the fact that Texas offers simple permitting of wind farms, requiring no public consultation or deliberation [23,29], created powerful economic and political incentives for wind farms.

However, the rapid growth of wind power in Texas inspired work critical of subsidies that were "compensating the market for unfavorable production locations" for wind power and had resulted in negative electricity prices ([30]; p. 30). Zarnikau ([31]; p. 3912) noted that wind-generated power cost three times more than producing electricity from natural gas, adding that wind power in the Electricity Reliability Council of Texas (ERCOT) produced "adverse impacts on system reliability." Recent reports indicate that the negative prices reported in 2011 and 2012 disappeared in 2013 and 2014 because of ERCOT's switch to a nodal system and increased transmission capacity through the CREZ program [27,32].

At the county level, property tax abatements and a "wind welcomer" posture by public officials and landowners attracted wind farm construction to the region [33,34]. Many political elites view tax abatements as essential to attracting firms to establish wind farms, although some officials are skeptical of the long-term benefits on public finances and infrastructure, suggesting an impending "day of reckoning" ([33]; p. 847). County officials justify tax abatements to constituents by arguing that the positive economic benefits— increased employment and economic activity—are greater than property tax forgone through abatements. Economic and political elites in the study area confirm the influence of county, state, and federal policies for attracting wind-energy firms, but they also note the importance of the under-utilized 345 kV transmission line that carries power from gas-powered electricity generators in the Odessa and Midland through Nolan and Taylor counties to Dallas and Fort Worth [7].

Wind farms in the study region are sited exclusively on private properties. Firms establish land-use agreements with landowners, who usually pursue legal counsel as they seek to maximize their return on land leased for locating turbines, roads, substations, and maintenance facilities. During our previous research, key actors (large landowners, elected officials, business owners) told us that royalties were between $8000 and $10,000 per turbine per year [33]. Our earlier research led us to agree with Kahn [14] and Brown *et al.* [15], viewing the lease as a voluntary transaction for a land use that does not replace a more lucrative land use from the perspective of the land owner, and that the firm offering the contract prefers the site to other possible locations. The firm, of course, is influenced by incentives such as the federal PTC, which encourages rapid construction, and county-approved property tax abatements.

2.2. Data Sources

Location of wind turbines was determined by the point locations for individual turbines obtained in 2010 from the Federal Aviation Agency (FAA), which we confirmed by checking data from the American Wind Energy Association and the Texas Public Utilities Commission (PUC). Wind turbines must be registered with the FAA because they are considered to be a flight obstruction; they must also be registered with the PUC because they provide electricity to ERCOT, the grid operator. In addition, we visually compared the mapped turbine locations against aerial photography because wind turbines are obvious features of the landscape. Each turbine has a known "nameplate" or potential output measured in megawatts. We omitted turbines outside Nolan and Taylor counties and a small number of turbines within the counties owned by state agencies. Assigning turbines to land parcels in our two-county study region yielded 1701 turbines with ~2700 MW capacity.

Next, GIS was used to assign wind turbines to land parcels to identify which landowners had wind turbines. We obtained geo-located parcel data accurate to the 2010 tax year from Nolan and Taylor County Appraisal Districts. We eliminated urban properties using a separate GIS layer identifying the boundaries of Sweetwater, Abilene, and other cities, which resulted in 17,500 total non-urban parcels in the two-county region. Next, we consolidated parcels according to tax address where parcels of land having the same tax address were combined into a single landholding. This procedure yielded 8370 landholdings with unique tax addresses. A spatial join was used to connect wind turbines to tax parcels, which yielded 241 private landholdings with turbines in the two-county area.

2.3. Royalty Estimation

To estimate royalties generated by wind turbines we employed electricity output, wholesale electricity price, and royalty rate. The Texas grid operator, ERCOT, reports electricity production by wind farm in 15-minute intervals. However, use of these electricity production data would not materially improve our royalty estimates because electricity sales prices and the royalty rates are confidential between landowner and wind farm operator. Neither of the latter figures is available in precise terms. Moreover, our goal was to determine the spatial distribution of estimated wind-energy royalties rather than establish a precise figure for royalties, which is not feasible because of the private nature of contracts governing transactions between wind-farm operator and landowner.

Instead of actual electricity output we employ capacity factor (CF), which is the efficiency of the wind turbine expressed as a percentage. CF is a product of engineering characteristics (hub height, rotor characteristics, scheduled maintenance, and sweep area of rotor blades, among others), wind characteristics (air density, wind velocity), transmission capacity, and electricity price. Ironically, CF data are made available in real time to wind farm technicians and managers, but are not reported transparently to the public; rather, CF data are notoriously shrouded in secrecy and often over-estimated by wind farm operators [35].

We considered various CF estimates appropriate to our study region. A Nolan County based group, the West Texas Wind Energy Consortium, commissioned a study of economic impacts of wind power that relied on a 35 percent CF estimate as the "accepted norm for Rolling Plains wind energy projects" ([36]; p. 7). The Texas Comptroller General [37] reported that the mean CF for Texas wind farms installed in 2004 and 2005 was

39 percent, and cited at least one specific wind farm in the study region, Sweetwater-2, as having a CF between 35 to 40 percent. Bolinger and Wiser [28], based on data from 24 projects with installed capacity of 2275 MW, reported the U.S. average CF at between 33 and 35 percent for projects installed after 2003; however, Boccard [35] compared US installed capacity with output, arriving at 25.7 percent CF. A recent global estimate of wind power potential used a 20 percent CF [38]. The US average CF has remained at about 32 percent since 2000 [3].

In the Texas grid, transmission capacity lagged so far behind installed capacity that NextEra, a wind-farm developer, built a 230-mile transmission line that boosted capacity factor of a west Texas wind farm from 20 to 29 percent. Lack of transmission capacity resulted in curtailment, which was most severe in 2009, immediately prior to construction of transmission lines through the CREZ [39]. The trend is for capacity factor to remain at approximately 32 percent in west Texas because new wind farms are being built in areas with relatively low wind quality [3]. Even with the construction of new transmission lines, wind farm managers in the region reported informally to the first author the occasional use of "economic curtailment," a situation in which the wind farm operator orders turbines into neutral position because the price of electricity does not compensate the cost of operating the turbine.

The second and third variables (wholesale electricity price and royalty rates) used to estimate royalties may have less variability but they are still cloaked in secrecy because landowners negotiate confidential contracts with wind-power firms. The wholesale electricity rate is estimated at approximately $40 per MWh [36], although the figure of $35 per MWh for Texas was reported recently [3], which is likely to include a Renewable Energy Credit (REC) [28]. It is known that several wind farms have long-term contracts with utilities, while others sell on the spot market, so any single figure is unlikely to match actual prices received. However, we consider our figures to be representative of the most likely prices that are used to compute royalties. Moreover, imprecision on electricity price does not influence the spatial distribution of estimated royalties.

Nolan County landowners receive a royalty, normally 4 percent or a minimum guaranteed payment, on the basis of the wholesale price of electricity produced [36,37]. Key actors in wind power in Nolan County noted that the Trent Mesa wind farm was on a 1 percent royalty rate while contracts signed in 2009 had reached 5 percent [40,41]. Landowners with turbines report differences between "good" and "bad" contracts, in that "bad" contracts have a low royalty and do not carry a guaranteed minimum payment, and often were negotiated during the early 2000s, the "wild west" period, according to one landowner. Ranchers who were among the earliest to host wind farms obtained royalty rates and payment terms inferior to landowners who signed contracts later. One key actor in wind power refers to discussion about contracts as "Dairy Queen talk," the sort of discussion among rural landowners over lunch specials at Dairy Queen, a fast-food chain popular among farmers and ranchers in rural Texas. The reference to "Dairy Queen talk" suggests that variability exists among royalty rates, but that is not possible to determine actual rates without gaining trust among landowners. Therefore, we rely on the 4 percent royalty rate in our calculations.

To estimate royalty rate (per turbine per year) from wind power we created an equation relating MW capacity (CF), wholesale electricity price, and royalty rate:

$$\text{Estimated royalty (per turbine per year)} = \text{CF (MW)} * \text{Price (MW/h)} * \text{hours/year} * \text{royalty rate} \quad (1)$$

Using Equation (1) we computed a low royalty estimate (CF = 25%, wholesale electricity price = \$35/MWh, and royalty rate = 4%), a high estimate (CF = 35%, wholesale electricity price = \$45/MWh, and royalty rate = 4%). We calculated a mean royalty estimate per turbine from high and low estimates, then we applied this to the entire wind farm and aggregated to the landholding on which the turbines were located. Table 1 reports wind farms in the study region, their estimated electrical production and royalty rate per turbine per year. For example, the 67 turbines at Buffalo Gap would produce ~264,000 MWh per year at capacity factor of 25 percent, yielding gross revenues of \$9.2 million at \$35 per MWh. But at 35 percent capacity factor, and \$45 per MWh, gross revenues would increase to \$16.9 million per year. A fixed royalty of 4 percent would yield a mean, between high and low estimates, of \$7726 per turbine per year (Table 1).

We used distance as a proxy for absentee ownership and land rental. Specifically, we determined driving distance from the *situs* (property) address, defined as the latitudinal and longitudinal location of the centroid of a single land parcel or aggregated parcels, to the tax appraisal (mailing) address. We assumed that the address listed for mailing the property tax statement is the same as the residential address, although it is possible that an attorney or financial advisor receives the tax correspondence for some landowners. The latitude and longitude coordinates of the parcel were entered into a Google Maps directions form, yielding turn-by-turn driving directions. To determine the driving distance for all 241 landowners with turbines, we wrote a C# script to query Google Maps for driving distance and time between two points excluding the turn-by-turn instructions. These distances were then added as landowner attributes.

Table 1. Summary of wind farms, capacity, and estimated royalty.

Wind Farm Name	No. Turbines	MW per Turbine	Total MW	Estimated Royalty per Turbine per Year (\$)
Buffalo Gap	67	1.8	120.6	7726
Buffalo Gap 3	73	2.3	167.9	9873
Buffalo Gap II	139	1.5	208.5	6439
Callahan Divide	76	1.5	114	6439
Champion (Roscoe II)	54	2.3	124.2	9873
Horse Hollow II	125	2.3	287.5	9873
Horse Hollow	292	1.5	438	6439
Inandale (Roscoe IV)	143	1.0	143	4292
Roscoe	100	1.0	100	4292
South Trent Mesa	43	2.3	98.9	9873
Sweetwater	171	1.5	257	6439
Sweetwater 4a	124	1.0	124	4292
Sweetwater 4b	46	2.3	105.8	9873
Sweetwater 5	34	2.3	78.2	9873
Trent Mesa	101	1.5	151.5	6439
Turkey Track	113	1.5	169.5	6439
Total	1701		2688.1	

We interpret results based on our knowledge of the study region, obtained through key informant interviews conducted in 2010 and subsequent annual two-day field trips with undergraduate students enrolled in a "Geography of Energy" course taught by the first author. During these field trips, which included wind farm site visits and panel discussions, students inquired about royalties. Answers from key informant interviews and panelists, who included county commissioners and landowners, helped us determine the values for formula (1) and interpret results.

3. Results and Discussion

Total royalties for the two-county study region were estimated at ~$11.5 million per year, slightly less than the $12.264 million per year estimate of royalties accrued from 2500 MW installed capacity, which assumed a 35 percent CF, $40/MWh wholesale electricity price, and 4 percent royalty rate [36]. The mean annual royalty received per landowner per year was estimated at $47,879 (for 7 turbines on 274 hectares). The median annual royalty was $25,756 (for 4 turbines on 129 hectares), less than median household income in either county. Mean driving time from landholding to tax address was slightly more than 2 hours, although the median driving time was 38 minutes. The mean landholding for the sample was 274 hectares (Table 2).

Only 241 landowners out of ~8370 rural landowners in the study region have wind turbines on their land. This small fraction, less than 3 percent, of rural landowners shows how pre-existing land-tenure patterns, the "property advantage," enabled some landowners to obtain royalties from wind farms because of factors relating to size or relative location of their land before the development of wind farms. A similar low fraction of participation in energy assets has been reported for shale gas underneath Denton, Texas, where 2.4 percent of subsurface mineral (gas) value was owned by family trusts and individuals with Denton tax addresses [9].

Table 2. Summary statistics for estimated royalty, distance between landholding and tax address, landholding area, and number of turbines per landholding (n = 241).

	No. Turbines	Estimated Royalty Per Landholding ($)	Driving Time (minutes) to Tax Address	Landholding Area (hectares)
Total	1701	11,538,769	33,198	66,117
Mean	7	47,879	138	274
Median	4	25,756	38	129
Mode	2	29,619	21	65
SD	11	70,850	260	494

The total area of the 8370 landholdings was 452,800 hectares (mean = 54.1 hectares), and median size of 2.4 hectares; this compares to the ~66,000 hectares of the 241 landholdings that have turbines located on their land. Some landowners without turbines may have refused to sign contracts with wind-power developers, but many others may have had land parcels too small for siting turbines, or land with unsuitable topography, low wind potential, or poor road access.

Distribution of royalties by quintile is shown in Table 3. The top quintile concentrates 60 percent of royalties, with a mean of $146,189 (median = $100,444) per landholding per year. This group (n = 48) has a mean landholding area of 842 hectares. By contrast, the fifth quintile captures less than 4 percent of total royalties (mean = $7656, median = $8584), with 1.3 turbines per landholding on average.

Table 3. Landholdings with turbines divided into quintiles for number of turbines and estimated royalty per landholding per year.

Quintile and Metric	No. Turbines	Royalty per Landholding per Year ($)	Area per Landholding (ha)
Q1 Total (n=48)	1020 (60.8%)	7,017,091 (60.0%)	40,425 (61.1%)
Mean	21.3	146,189	842.2
Median	14	100,444	497.5
Standard Dev.	17.9	111,288	883.9
Q2 Total (n=48)	328 (18.5%)	2,129,078 (19.3%)	11,442 (17.3%)
Mean	6.8	44,356	238.4
Median	6	39,492	211.5
Standard Dev.	1.7	10,932	152.4
Q3 Total (n=48)	173 (10.8%)	1,249,539 (10.2%)	5334 (8.1%)
Mean	3.6	26,032	111.1
Median	4	27,473	86.5
Standard Dev.	0.5	4134	87.0
Q4 Total (n=48)	117 (6.7%)	767,913 (6.9%)	5307 (8.0%)
Mean	2.4	15,998	110.6
Median	2	17,168	70.0
Standard Dev.	0.5	3340	84.5
Q5 Total (n=49)	63 (3.3%)	375,148 (3.7%)	3609 (5.5%)
Mean	1.3	7656	73.7
Median	1	8584	65.0
Standard Dev.	0.5	2067	60.6

Regarding the spatial distribution of royalties per landholding, there is a clear divide between relatively large ranches (used for livestock grazing and hunting leases) south of Sweetwater and small farms, mainly growing cotton without irrigation, near Roscoe (Figure 3). We found significant differences ($p < 0.05$) for royalty per landholding per year by landholding area (Table 4). Roscoe landholdings had a mean annual royalty of $19,310, while Sweetwater landholdings had a mean royalty of $63,000 per year.

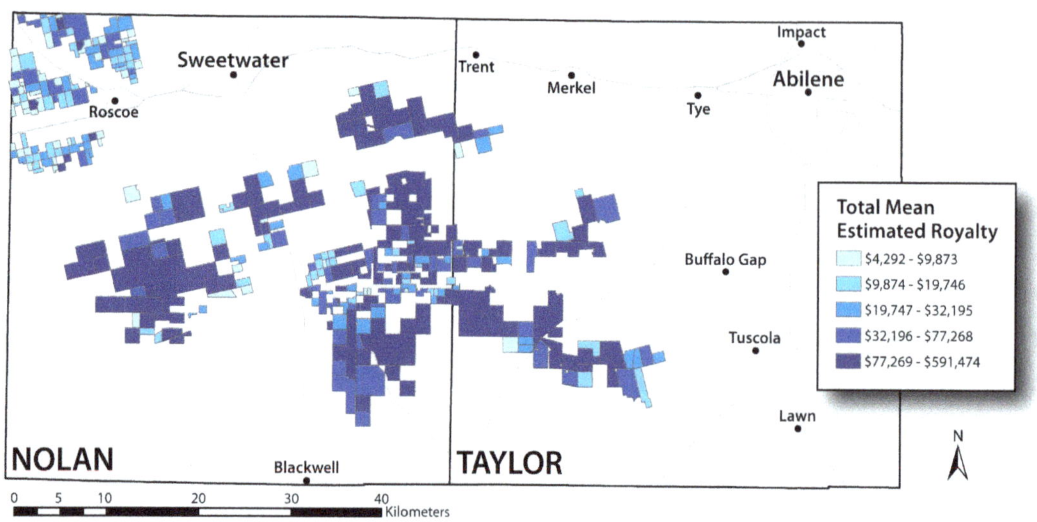

Figure 3. Spatial distribution of estimated royalties per landholder per year in Nolan and Taylor Counties.

The distribution of Roscoe royalties may be different than our estimates indicate. Around 2005, an entrepreneurial leader organized farmers as a "unitized" group of 120 landowners covering ~13,000 hectares [41]. Roscoe's many small landholdings, mainly devoted to rainfed cotton, were considered unattractive to wind developers. Unitization, a well-established practice in oil and gas production whereby small fields are grouped into one, and royalties are divided among members [42], helped make the region attractive to firms because they would negotiate one contract with a group, rather than with many small owners of rights. Royalties to all participants in the "unitized" group would include landowners without turbines, perhaps with land parcels too small or not well suited to the designs of the wind-farm developer, therefore slightly increasing the number of rural landowners receiving royalty payments.

Table 4. Descriptive statistics for landholdings with turbines in the Roscoe and Sweetwater areas. Royalty column refers to estimated royalty per landholding per year.

	Roscoe (n = 85)			Sweetwater (n = 156)		
	No. Turbines	**Royalty**	**Hectares**	**No. Turbines**	**Royalty**	**Hectares**
Mean	3.6	19,310	115.6	8.9	63,307	361.3
Median	3.0	12,876	65.0	4.0	32,195	182.5
Mode	2.0	8584	65.0	3.0	29,619	259.0
SD	3.1	15,857	97.6	12.9	83,351	593.2

We also determined the driving distance from landholding to tax address (Table 5) as a proxy for land rental and absentee ownership. We separated landowners with royalties into two groups, using 45 minutes driving time as a threshold, but found no significant difference in estimated royalty. Considering the tax address, we found that approximately 60 percent of total royalties accrued to landowners residing within Nolan and Taylor counties. It is possible that some of these landowners rent their land to ranch operators, but it is also possible that most landowners residing in Nolan and Taylor counties are the operators of their farm or ranch land. Around 30 percent of total royalties went to landowners residing elsewhere in Texas, and 7 percent of total royalties accrued to landowners residing outside Texas.

Table 5. Landholdings, turbines and estimated royalty per landholding per year according to location of tax address.

Tax Address	No. Landholdings	No. Turbines	Royalty ($)
Nolan County	105 (43.6%)	739 (43.4%)	4,867,684 (42.2%)
Taylor County	42 (17.4%)	312 (18.3%)	2,219,266 (19.2%)
Other Texas	79 (32.8%)	538 (31.6%)	3,652,962 (31.7%)
Other U.S.	15 (6.2%)	112 (6.6%)	798,857 (6.9%)
Total	241	1701	11,538,769

Estimating livelihood and land-use impacts is not possible with our findings, but theoretical claims and qualitative evidence offer several suggestions. First, Weber *et al.* [8] rely on a combination of life-cycle and behavioral frameworks [43] to argue that energy payments, such as wind royalties, may be consumed, saved, or invested. They assume that energy amenities or disamenities will affect land value, and, therefore, household wealth. Wind turbines are considered amenities and therefore add to land value. They argue that increased land value accounts for approximately 75 percent of total estimated private wealth from energy royalties. By comparison, consumption and farm investment are low, approximately 4.2 cents of marginal consumption for every dollar of energy payment.

Qualitative evidence from fieldwork offers other suggestions. Landholders offer many answers—except specific ones regarding actual investments—when approached on the topic of royalties. Previous work has suggested the existence of a particular moral economy in which wind power keeps farms and ranches in families, while landowners remain environmentally skeptical [33,34]. “Pay my notes (farm debt)” was the response of a cotton farmer in Roscoe when asked what he did with his wind-power royalties. For this farmer, landowners with turbines on cotton fields lost between 3 and 5 percent of farm area to pad site and access roads, making turbines attractive [41]. Similarly, a rancher outside Sweetwater joked that royalties allowed him to “maintain my hobby,” by which he meant ranching in difficult climatic times. Farming or ranching is a “pretty tough gig” in the Nolan-Taylor region. Royalties help “pay bills and put kids through college.” Landowners with turbines indicated that royalties helped families maintain their land, which had been in precarious state because of the declining cattle economy and lack of rainfall. But wind might also be just another type of energy income. Wind, according to one landowner, was “strictly money,” echoing findings reported elsewhere [34].

It is not known whether life-cycle and behavioral frameworks [43] should be modified to particular conditions of west Texas. Perhaps the only published observation is that of Greene ([25]; pp. 24–25), who argued that west Texas landowners viewed income from cattle as more noble than income from oil, recounting the story of a rancher who “hated the idea that none of this new wealth [from oil] was his doing, as his cattle and his land deals had been.” Greene also drew contrasts between the region’s Christian morality (“God is the biggest industry in west Texas”) and the “sinful humanity” and oilmen (a “villain, seen as a creature of lawyers and bankers”) attracted by oil to west Texas ([25]; pp. 130, 163). It is unclear how broadly these assertions, if accurate, apply to current landowners, because as one landowner told us, “you do not gripe about the money that is coming in from oil wells, and you shouldn’t be concerned about money coming in from the wind farms” ([34]; p. 859). It is also unknown whether a form of energy payment that is held in contempt, as Greene [25] suggested, is any different from wind royalties, which are viewed positively, in terms of consumption, savings, or investment. Whether households use royalties produced in this context any differently than models suggest is an open empirical question with implications for the vast areas of wind-power development.

Royalties from wind-power production accrue to hundreds of landowners in a vast area of central North America that hosts much of U.S. wind-power capacity. This area is spatially coincident with high reported social acceptance of wind power. The “property advantage” describes unequal distribution in royalties predicated upon land-tenure inequalities established before wind-power development. Because of their “property advantage,” and without “unitization,” some landowners will benefit significantly more than other landowners. The vast majority of people in rural areas who do not own land are excluded from royalties, but may benefit from wind power in other ways [14–16]. Future studies of social acceptance or rejection of wind farms should test for royalty effects, especially in rural areas where drought or poor commodity prices adversely affect farming and ranching, paying attention to particular moral economies that may not conform to models.

We also identify a “royalty paradox,” the situation in which estimated royalties accrue to only a small fraction of rural landowners, even though support for wind power apparently extends into the general population, far broader than the small sub-population of landowners with turbines. We do not know how large payments to a small number of landowners translate into apparent high social acceptance. But it is likely that the landowners who benefit the most from royalties are prominent spokespersons for the wind economy, occupying key positions in organizations such as local school boards and other local civic groups that help influence opinion. They may be the "wind welcomers" previously described [33]. It is beyond the scope of this study to understand the behavior of local elites, but it is likely that landowners with royalties influence local discourse and opinion through varied means.

The influence of royalties on livelihoods and, by extension, land-change patterns, is not known. However, understanding the spatial distribution of royalties is an important starting point. A next step is to consider how landowners treat wind royalties, perhaps informed by a particular moral economy, and whether actual behaviors conform to the more well studied government payments that are independent of farm production decision making [43]. Even if a moral economy for wind royalties does not exist, it is possible that landowners at the lower quintiles treat royalties no differently than other sources of farm income—perhaps they just "pay notes," as one farmer told us. But at the upper quintiles, landowners may take different decisions with their royalty income, using royalties to fund capital expenses, invest in education of their children, or support their "hobby." Behavioral models may prove useful to offer estimates based on amount of royalty income and existing land cover (cotton or scrub vegetation), for example. An increase in land value among landholdings with energy payments, which Weber *et al.* [8] predict, may have numerous direct and indirect consequences that future research may determine. Certainly, the property value effect of wind power as an amenity may become apparent in property tax datasets and will figure highly in household wealth.

Our estimates of royalties are conservative because they do not include land leases for roads or maintenance facilities; nor does our estimated royalty include the signing bonus, which varies substantially among firms and landowner contracts. In addition, actual royalties could be higher because nameplate capacity may have been increased through software and hardware upgrades, either under manufacturer warranty or in the after-market period beyond warranty.

We cannot validate our royalty estimates because royalty rates are confidential between wind farm and landowner, and because wholesale electricity prices may vary considerably depending on contract terms between wind farms and utility. Future research aiming to model wholesale electricity price, coupled with analysis of actual electricity output per wind farm, could improve the accuracy of our estimates, but those estimates would still not have accurate estimates of the royalty rate.

The flow of wind-power royalties is contingent, of course, on the price of wholesale electricity, the output of turbines, and contract terms. Sales of land with wind turbines may slow, as farmers are reluctant to part with their royalty payments; however, some observers report the development of contract terms that allow land sellers to retain their royalties after selling the land. It is not yet known how landowners will respond when contracts approach the end of the projected lifetime of wind turbines, and whether re-negotiation opportunities may exist, especially as wind farms change ownership.

At a county level, future research could compare royalties that accrue privately with the cost of subsidies, in terms of property tax abatements, that are borne by the public in terms of property taxes assessed overall. County subsidies in the form of tax abatements may lower overall property tax rates, and help county officials politically, but they also encourage the flow of considerable benefits to a tiny minority of landowners.

Our findings indicate some challenges in conducting research aiming to link energy with land change and livelihoods in North America. The development of renewable energy in central North America, from Texas to Minnesota, is relatively recent, and while apparent impacts on land cover from the construction of wind turbines are plainly visible, possible effects on land use and landowner livelihoods are far more difficult to discern. The fact that wind power is located almost entirely on private landholdings means that landowners are "visible" in terms of tax addresses, but potentially high non-response bias will confront researchers aiming to determine the use of royalties through mail-out or online surveys. Qualitative and ethnographic methods may reveal important insights into the ways that renewable energy influences land uses through royalties, but require high levels of trust and long periods of field work.

4. Conclusions

Estimated royalties from wind power are unequally distributed among a small fraction of rural landowners who use the income in unknown ways to support their livelihoods. Landowners may try to overcome initial land inequality though cooperative agreements, known as "unitization," borrowed from oil and gas institutions. Royalties may help explain reported high support for wind power in the rural U.S., but livelihood and land-change outcomes of royalties are not yet known. We identified the "property advantage," the land-tenure patterns established before wind-power development that determine the unequal distribution of royalties. The "royalty paradox" refers to how royalty payments to a small number of landowners help produce high social acceptance, probably through the discursive and political actions of this small group of landowners. Future research should determine behavioral aspects of wind-royalty income and impacts on property value, which could influence land use and land cover.

Acknowledgments

We thank the Wind Initiative at Texas Christian University for supplying datasets and funding earlier research, and Nicole Persons for assisting in data collection in early phases of the research. Comments by Jackie Vadjunec, Claudia Radel, and anonymous reviewers improved earlier drafts.

Author Contributions

Christian Brannstrom designed the research, conducted research, analyzed the data, and wrote the paper; Mary Tilton conducted research; Andrew Klein conducted research and analyzed the data; Wendy Jepson designed the research and conducted research.

Conflicts of Interest

The authors declare no conflict of interest. The founding sponsors had no role in the design of the study; in the collection, analyses, or interpretation of data; in the writing of the manuscript, and in the decision to publish the results.

References and Notes

1. Brannstrom, C.; Vadjunec, J. Notes for avoiding a missed opportunity in sustainability science: Integrating land change science and political ecology. In *Land Change Science, Political Ecology, and Sustainability: Synergies and Divergences*; Brannstrom, C., Vadjunec, J., Eds.; Earthscan from Routledge: Abingdon, UK, 2013; pp 1–23.
2. Public Land and Wind Energy. Available online: http://www.awea.org/Issues/Content.aspx?ItemNumber=858 (accessed on 25 July 2014).
3. Wiser, R.; Bolinger, M. *2012 Wind Technologies Market Report*; U.S. Department of Energy: Washington, DC, USA, 2013.
4. Wind Generates More than 10% of Texas Electricity in 2014. Available online: http://www.eia.gov/todayinenergy/detail.cfm?id=20051 (accessed on 7 June 2015).
5. Swofford, J.; Slattery, M. Public attitudes of wind energy in Texas: Local communities in close proximity to wind farms and their effect on decision-making. *Energy Policy* **2010**, *38*, 2508–2519.
6. Slattery, M.C.; Johnson, B.L.; Swofford, J.A.; Pasqualetti, M.J. The predominance of economic development in the support for large-scale wind farms in the U.S. Great Plains. *Renew. Sustain. Energy Rev.* **2012**, *16*, 3690–3701.
7. Galbraith, K.; Price, A. *The Great Texas Wind Rush: How George Bush, Ann Richards, and a Bunch of Tinkerers Helped the Oil State Win the Race to Wind Power*; University of Texas Press: Austin, TX, USA, 2013.
8. Weber, J.G.; Brown, J.P.; Pender, J. *Rural Wealth Creation and Emerging Energy Industries: Lease and Royalty Payments to Farm Households and Businesses*; Federal Reserve Bank of Kansas City: Kansas City, MO, USA, 2013.
9. Fry, M.; Briggle, A.; Kincaid, C. Fracking and environmental (in) justice in a Texas City. *Ecol. Econ.* **2015**, *117*, 97–107.
10. Sowers, J. Fields of opportunity wind machines return to the plains. *Gt. Plains Q.* **2006**, *26*, 99–112.
11. Walker, C.; Baxter, J.; Ouellette, D. Beyond rhetoric to understand determinants of wind turbine support and conflict in two Ontario, Canada communities. *Environ. Plan. A* **2013**, *46*, 730–745.
12. Abbott, J.A. The localized and scaled discourse of conservation for wind power in Kittitas County, Washington. *Soc. Nat. Resour.* **2010**, *23*, 969–985.
13. Mulvaney, K.K.; Woodson, P.; Prokopy, L.S. A tale of three counties: Understanding wind development in the rural midwestern United States. *Energy Policy* **2013**, *56*, 322–330.
14. Kahn, M.E. Local non-market quality of life dynamics in new wind farm communities. *Energy Policy* **2013**, *59*, 800–807.
15. Brown, J.P.; Pender, J.; Wiser, R.; Lantz, E.; Hoen, B. Ex post analysis of economic impacts from wind power development in U.S. counties. *Energy Econ.* **2012**, *34*, 1743–1754.
16. Slattery, M.; Lantz, E.; Johnson, B.L. State and local impacts from wind energy projects: Texas case study. *Energy Policy* **2011**, *39*, 7930–7940.
17. Lantz, E.; Tegen, S. Variables affecting economic development of wind energy. Presented at WINDPOWER 2008, Houston, TX, USA, 1–4 June 2008.
18. Greene, J.S.; Geisken, M. Socioeconomic impacts of wind farm development: A case study of Weatherford, Oklahoma. *Energy Sustain. Soc.* **2013**, doi:10.1186/2192-0567-3-2.

19. BBC News. The world's largest wind farms. *BBC News* 25 November 2010.
20. Shlachter, B. Rancher's windfall. *Ft. Worth Star-Telegram* 29 January 2006.
21. Baxter, J.; Morzaria, R.; Hirsch, R. A case-control study of support/opposition to wind turbines: Perceptions of health risk, economic benefits, and community conflict. *Energy Policy* **2013**, *61*, 931–943.
22. Warren, C.R.; McFayden, M. Does community ownership affect public attitudes to wind energy? A case study from south-west Scotland. *Land Use Policy* **2010**, *27*, 204–213.
23. Bohn, C.; Lant, C. Welcoming the wind? Determinants of wind power development among U.S. States. *Prof. Geogr.* **2009**, *61*, 87–100.
24. Pasqualetti, M.J. Opposing wind energy landscapes: A search for common cause. *Ann. Assoc. Am. Geogr.* **2011**, *101*, 907–917.
25. Greene, A.C. *A Personal Country*; Texas A&M University Press: College Station, TX, USA, 1979 (1969).
26. Moore, R.R. *West Texas after the Discovery of Oil: A Modern Frontier*; Jenkins Publishing and Pemberton Press: Austin, TX, USA, 1971.
27. Fewer Wind Curtailments and Negative Power Prices Seen in Texas after Major Grid Expansion. Available online: http://www.eia.gov/todayinenergy/detail.cfm?id=16831 (accessed on 7 June 2015).
28. Bolinger, M.; Wiser, R. Wind power price trends in the United States: Struggling to remain competitive in the face of strong growth. *Energy Policy* **2009**, *37*, 1061–1071.
29. Parker, B.D. *Capturing the Wind: The Challenges of a New Energy Source in Texas*; House Research Organization: Austin, TX, USA, 2008.
30. Baldick, R. Wind and energy markets: A case study of Texas. *IEEE Syst. J.* **2012**, *6*, 27–34.
31. Zarnikau, J. Successful renewable energy development in a competitive electricity market: A Texas case study. *Energy Policy* **2011**, *39*, 3906–3913.
32. Tao, H.Y.S.; Srivastava, A.K.; Pineda, R.L.; Mandal, P. Wind power generation impact on electricity price in ERCOT. In Proceedings of the IEEE Power and Energy Society General Meeting, San Diego, CA, USA, 22–26 July 2012; pp. 1–7.
33. Brannstrom, C.; Jepson, W.; Persons, N. Social perspectives on wind-power development in west Texas. *Ann. Assoc. Am. Geogr.* **2011**, *101*, 839–851.
34. Jepson, W.; Brannstrom, C.; Persons, N. "We don't take the pledge": Environmentality and environmental skepticism at the epicenter of US wind energy development. *Geoforum* **2012**, *43*, 851–863.
35. Boccard, N. Capacity factor of wind power realized values *vs.* estimates. *Energy Policy* **2009**, *37*, 2679–2688.
36. New Amsterdam Wind Source. Nolan County: Case study of wind energy economic impacts in Texas. 2008. Available online: http://www.moakcasey.com/articles/viewarticledoc.aspx/Nolan%20County%20Case%20Study.pdf?AID=168&DID=288 (accessed on 2 March 2009).
37. Combs, S. *The Energy Report*; Texas Comptroller of Public Accounts: Austin, TX, USA, 2008.
38. Lu, X.; McElroy, M.B.; Kiviluoma, J. Global potential for wind-generated electricity. *Proc. Natl. Acad. Sci. USA* **2009**, *109*, 10933–10938.
39. Wiser, R.; Bolinger, M. *2008 Wind Technologies Market Report*; Department of Energy: Washington, DC, USA, 2009.

40. Confidential interview with respondent 17, 20 May 2009.
41. Confidential interview with respondent 25, 2 October 2010.
42. Weaver, J.L. *Unitization of Oil and Gas Fields in Texas: A Study of Legislative, Administrative, and Judicial Policies*. Resources for the Future: Washington, DC, USA, 1986.
43. Whitaker, J.B. The varying impacts of agricultural support programs on U.S. farm household consumption. *Am. J. Agric. Econ.* **2009**, *91*, 569–580.

10

Entity-Based Landscape Modelling to Assess the Impacts of Different Incentives Mechanisms on Argan Forest Dynamics

Farid El Wahidi [1,*], Julien Radoux [2], Quentin Ponette [2] and Pierre Defourny [2]

[1] Centre de la Recherche Forestière, 10050 Rabat-Agdal, Marocco

[2] Earth and Life Institute, Université catholique de Louvain, 1348 Louvain-la-Neuve, Belgium; E-Mails: julien.radoux@uclouvain.be (J.R.); quentin.ponette@uclouvain.be (Q.P.); pierre.defourny@uclouvain.be (P.D.)

* Author to whom correspondence should be addressed; E-Mail: elwahidifarid@hotmail.com

Academic Editor: Andrew Millington

Abstract: Illegal occupation of argan forest parcels by local households is a new phenomenon in South West Morocco. This is primarily due to the weakening of traditional common control systems and to the boom of the argan oil price. The scope of this work is to develop a decision support system based on dynamic spatial modelling, allowing to anticipate the land tenure dynamics and their impact on forest stand degradation under different policy scenarios. The model simulates the change of land possession by locals and the forest stand degradation levels. The methodological approach combines a Markov chain analysis (MCA) with stakeholders' preferences for land tenure. First, parcels' transition probabilities are computed using the MCA. Second, the acquiring suitability map is derived from multi-criteria evaluation procedure (AHP) using biophysical and socio-economic data. Finally, uncertainty is introduced in the simulation based on probabilistic analysis for supporting socio-economic diversity and non-mechanistic human behavior. The modelling approach was successfully used to compare three scenarios: business as usual (continuation of illegal acquiring), total disengagement of the population and private/public partnership with incentives for restoring argan parcel. The model yields geographic information about (*i*) the magnitude of the on-going process; (*ii*) the potential occurrence of land use conflicts induced by new policies; and (*iii*) the location of land conservation or degradation hot-spots. The outcomes of the "business as usual" and of the "total disengagement" models were similar over a 30-year simulation period: in both cases, the proportion of "highly degraded" parcels was doubled and the number of "quite degraded" parcels was increased by 50%. On the other hand, should the private/public partnership effectively work, about 40% of the parcels could be restored to a sustainable level.

Keywords: spatial modelling; multicriteria evaluation; Markov chain analysis; entity based; stakeholder's preferences; incentives mechanisms; argan forest dynamics

1. Introduction

Argan (*Argania spinosa*) is an endemic tree species in southern Morocco. Argan forests are linked with complex landscapes where human society and the natural environment strongly interact. Recent observations show that argan forests undergo a very slow but steady decrease of tree density and a rapid qualitative degradation [1,2]. In parallel, local stakeholders are abandoning traditional usage rights to privatize forest lands as a result of the argan oil boom and socio-economic mutations.

Recently, the scientific community produced a large set of applied models that can be used to predict or explore possible land use/cover change (LUCC) trajectories [3–8], vegetation dynamics [9], crop sequences and bio-economic interactions [10], land suitability evaluation [11] and stakeholder preferences for land conservation [12]. In addition to being a learning tool, LUCC models play an important role in exploring possible future developments in the landscape system. These exploratory and projective capacities allow models to be used as a communication and learning environment for stakeholders involved in land use decision making processes [13,14]. Through "*what if*" scenarios analysis, land use models can also support land use planning and policy [15,16].

A variety of modelling approaches are used for LUCC simulation [17]. These include: (*1*) *Equation based models and system models:* The most common mathematical formulation used for these models is sets of equations based on economic theory [18], linear programming [10], heuristic solution techniques such as simulated annealing for forest landscape management [19] or theories of population growth and diffusion over time [20]. A major drawback of such models is that a numerical or analytical solution of the system of equations must be obtained, limiting the level of complexity that may practically be built into such models [21]; (*2*) *Empirical and statistical techniques* include stochastic methods (e.g., Monte Carlo techniques) and a variety of regression techniques such as generalized linear models [3], logistic models applied to space and more tailored spatial statistical methods [22,23].They sometimes incorporate expert judgment with probability techniques, such as Bayesian probability or Dempster-Schaefer theory [24]. Artificial neural networks approaches can also be classified in this category. (*3*) *Cellular models* (*CM*) include cellular automata (*CA*) and Markov models [25]. CM methods have recently supported many LUCC models [4,5,16], but they hardly incorporate human decision making. (*4*) *Agent-Based Models* (*ABM*) focus on human decision making materialized by agents [26]. Agents must act according to some model of cognition that links their autonomous goals to the environment through their behaviors. Such models are very data intensive and the specification of realistic agent behavior and diversity is very challenging, especially when it comes to the behavior of the agent at higher levels of organization such as institutions [5,16].

In spatially explicit models, geographic space is typically represented as a grid of regular cells. The modelling results may vary according to the cell size [27], the neighborhood configuration [28,29], the stability of the model and its ability to reproduce spatial patterns [25]. To cope with this scale sensitivity, some authors have used, instead of a regular grid, an irregular tessellation to represent space layer such as Voronoi diagrams [30] or Delaunay triangle network [31]. Irregular tessellation has been used with success in CM. One limitation of these approaches, however, is that the polygons do not match the real-world entities composing the landscape as determined by stakeholders' perceptions. In addition, using a pixel-based approach results in an increased patchiness of the landscape compared to entity-based simulations using land management units such as parcels [16]. The entity-based landscape models avoid these disadvantages by splitting the landscape into meaningful objects with semantic features.

Land use modelling rarely reflects the uncertainties in stakeholders' decision making. Deterministic modelling approaches indeed take a single value for each parameter and yield a single prediction of the system response. One way to incorporate uncertainty is probabilistic modelling (e.g., using the Monte Carlo technique), where uncertain input and output parameters are described by probability density functions [32].

This study aims at developing a decision support system to assess the impact of different incentive mechanisms on the argan forest degradation within a co-management framework. The model simulates the expected change for the next 30 years of land possession by locals and forest stand degradation levels according to three scenarios based on land tenure policy and economic incentives. The three scenarios are compared for their effects on argan forest degradation measured by tree density and qualitative stand degradation level. The proposed modelling approach relies on two keys elements: (*i*) the use of an *entity-based model* to represent real-world parcels making the landscape; and (*ii*) the implementation of stochastic processes to simulate uncertainty and document the variability among model repetitions for the simulated land possession and stand degradation dynamics.

2. Challenges for Argan Forest Conservation

2.1. Brief History of Forest Management in Morocco

Until the beginning of the 20th century, the Moroccan forest policy was non-existent. At that time, the community management (*common property management*) based on Islamic and traditional rights completely controlled the access and use of various goods and services of the forest. It is only in 1917 that the forest administration was created and that forest and private domains were distinguished. Although usufruct rights were granted to the local population under the forest domain, this led to a deep mistrust between local communities and forest officers and to persistent conflicts between the parties. In the 90s, following the recommendations of the Rio World Summit on Sustainable Development, forest management plans started to be elaborated in collaboration with the local population. This participation was, however, limited to discussions about the infrastructure and social facilities to be developed. In general, the Moroccan forest policy is still empowered by the central administration according to a "top-down" scheme but includes progressively more and more social realities. Local communities are now considered as stakeholders in the land management and forest conservation measures. Nevertheless, the involvement of the local population is not sufficient to stop forest degradation.

2.2. Argan Forest Agro-Sylvo-Pastoral System

Only 10% of the total argan area is currently covered by forest management plans. The lack of natural regrowth and the failure of reforestation are the main challenges for the administration, mostly due to the difficulty of regulating the grazing activity or enforcing closure periods. Most of the time, the actions of the forest officers generate the hostility of the population and local authorities.

Private parcels in the argan region are usually exploited as farmland. On the other hand, the majority of argan parcels are public properties under one of the two traditional users' management strategies, *i.e.*, "*agdal*" and "*mouchaa*": (i) Trees in *agdal* areas are assigned to particular users through traditional arrangements within the community that exclusively holds the rights of nut harvesting. *Agdal* areas are precisely delimited and the related rights are transmitted to heirs; (ii) *Mouchaa* land is a common forested area, basically a buffer for other categories, where the trees are not assigned to individuals.

Recently, a new kind of *agdal*, named "*managed agdal*", appeared in the region. It can be considered as semi-private zones with permanent restrictions imposed on the grazing rights of the other users. In addition to being illicit possession, this proliferation of "*managed agdal*" demonstrates the weakening of the community power and a trend towards a more individual management of argan stands resulting from the increasing profits from argan oil.

While user rights are officially recognized, the State remains the legal owner of the trees, which is arguably discouraging users' investments [33]. Today, several authors contest the current legislation and claim that more efforts have to be made to gain the trust between users and forest officers [34], to reform the law towards co-administration of natural resources [35], rethinking the actual boundaries of the forest domain [36,37], to return to the state before delimitation (before the protectorate period) [38] or to empower local institutions for management [39].

In short, the argan forest is increasingly seen as an agro-ecosystem where users must be involved in any conservation management strategy. It should be fully managed like an agro-ecosystem, taking into account its peculiarities of combining fruit trees and a pastoral area.

2.3. Experiences of Public/Private Partnership in Natural Resource Management

Incentive and market-based mechanisms (IMBM) have been considered as likely solutions to support the farmers, communities, private companies and other managers of the natural resources to adopt sustainable management practices [40]. Nevertheless, incentives solely based on market valuation (labels, geographic areas (IGP—Indication Géographique Protégée) or certifications), give most advantage to stakeholders at the tail of the marketing chain compared to the local farmers, who are the effective users of the natural resources. In many cases, these IMBM therefore remain insufficient to encourage the population to invest and be involved in sustainable management practices. The efficiency of these mechanisms could be improved if used in combination with financial incentives that stakeholders profit from directly such as through payment for ecosystem services (PES) [41,42]. This however raises some operational issues: how to proceed with the economic incentives? At which level (household or community) and at which land unit (parcel or forest) should the sustainable actions be executed and controlled? How could these incentives insure the involvement of all (beneficiaries and non-beneficiaries)?

The traditional management practices linking local inhabitants to argan trees identify the spatial units (*agdal* parcels) to successfully set up economic incentives. Nevertheless, because of the deeply embedded mistrust between locals and forest officers, such economic incentives should be coupled with a land tenure policy offering more warranties to locals. Combining economic incentive with ownership recognition could make the difference in adopting conservation and afforestation actions in argan forests.

Public/Private Partnerships (PPP) allowing more flexibility for forest co-management yielded promising results for protecting and developing natural resources. For instance, China's wastelands policy [40] and the joint forest management in India [43] achieved their sustainable objectives. In Morocco, households' plantations of almond trees on collective domains proved to be successful in the Anti-Atlas Mountain (Tafraout, Ighrem) and the Green Morocco Plan (GMP) promoting the PPP for agricultural lands.

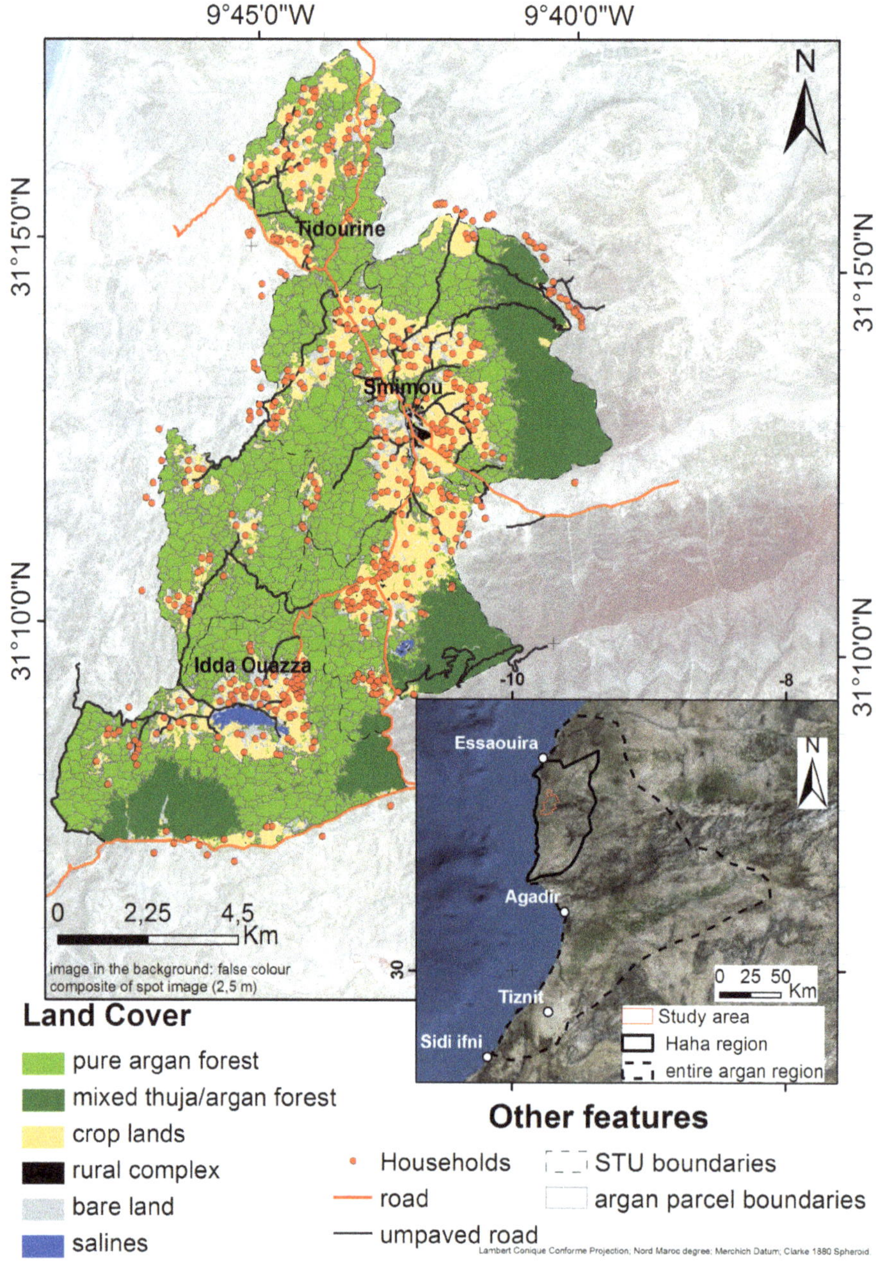

Figure 1. Study area showing land cover, socio-territorial units (STU) and argan parcels boundaries.

3. Study Area and Related Data

The argan (*Argania spinosa*) forest covers about 950,000 ha [44] bordering the Sahara desert, in Southern Morocco. Argan trees have provided multiple ecosystem goods and services, argan oil being the special flagship product. It became famous for its cosmetic virtues and has been exported at prices up to several hundred dollars a liter to Europe, Japan and the United States for the last two decades.

The study area covers approximately 12,500 ha (Figure 1) and includes 11 villages corresponding to two rural districts (Smimou and Idda ou Azza). These are located on the Haha plateau, in the northern part of the argan region. The argan tree is the keystone species within the study area and two natural vegetation types (argan and thuja forests) can be distinguished based on their socio-economic roles. Thuja forests (*Tetraclinis articulate sp.*) have closed canopies and dominate the upper level of the mountains (Amsiten, Aguirar). They are the main supply for fuel-wood consumption. The pure argan woodlands are agro-sylvo-pastoral systems located at lower elevations. They are generally scattered throughout the undulating landscape with a tree density that varies from 30 to 80 trees/ha.

The study area can be divided into six socio-territorial units (STU) (Tidourine, Ida Ou Talilt, Sidi Brouzekri, Tadart, Id M'barek ou Ahmed, Tourhnist; Figure 1). A STU is defined as a geographic region exploited by the same community (fraction of the local population linked by relations of solidarity, common interests and specific organization for the exploitation of agro-sylvo-pastoral resources). Benomar [45] delineated the STUs' boundaries from village surveys based on ethnic affiliation and herd movements.

4. Methods

4.1. Outline

The proposed model (Figure 2) computes the fate of argan parcels over a 30-year simulation period. Each parcel is initially characterized by a set of attributes related to environmental variables, socio-economic variables and forest stand degradation. At each step, the attributes of the parcels are updated according to two interrelated modules: (i) the argan parcels' possession; and (ii) the stand degradation. Three scenarios were compared using the model in order to simulate two possible types of land possession dynamics with the current process of illicit possession:

- *The Business As Usual scenario* (*BAU scenario*) predicts the results of the current dynamic of illicit parcels possession inferred from the "*managed agdal*". Farmers protect "their" parcels but are not involved in any kind of afforestation process. Possession and degradation are yearly updated.
- *The Public/Private Partnership scenario* (*PPP scenario*) is an optimistic scenario that simulates the impact of coupled economic incentive/rights recognition that encourages farmers to plant and protect trees in the newly acquired parcels. The possession of a parcel becomes legal upon condition that it is sustainably managed. Possession and degradation are updated at each step, but afforestation is also included in the forest degradation module.
- *The Total Disengagement of the Population for forest protection* (*TDP scenario*) corresponds to the pessimistic scenario when no possession (even illegal) occurs. This total disengagement

of the local population could happen if the argan trees' income is significantly low compared to non-farm activities. In this case, only the forest degradation will be updated.

The modules include a stochastic component. The model is therefore repeated 30 times (30 repetitions) and the outputs are summarized using the mean and standard deviation.

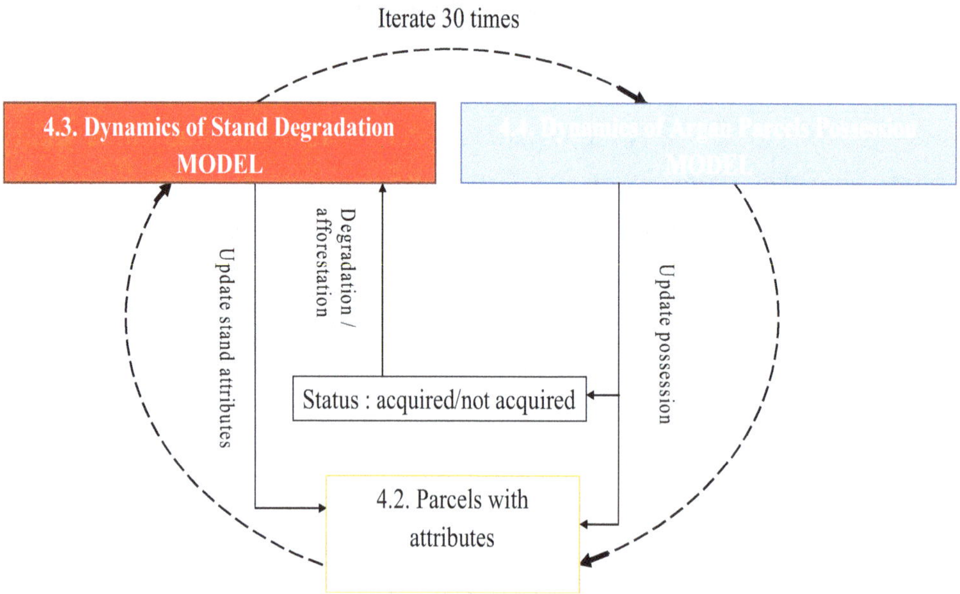

Figure 2. General overview of the core model structure. The modules are described in the next sections (Sections 4.2, 4.3 and 4.4). The forest degradation module is common to all scenarios, while the land possession dynamics is absent from the TDP scenario.

4.2. Parcel Extraction and Description

Argan parcels boundaries have been automatically generated using an image segmentation algorithm (eCognition software®) applied to the SPOT-5 scene of July 2008. The parcels boundaries were delineated with a minimum area of 1000 m^2. Despite a small over-segmentation, the extracted boundaries accurately matched the reference boundaries. The attributes of the parcel were then derived from remote sensing and GIS analysis:

- A land cover map depicting six classes (*i.e.*, pure argan forest, mixed thuja/argan forest, agricultural lands, bare land, build-up areas and salines) was obtained using object-oriented image classification of the SPOT-5 image. The overall thematic accuracy estimated from a field survey of 200 points was 80%. This information is used to screen out all parcels without argan tree from the analysis.
- The stand degradation level in the parcels was defined by two criteria, *i.e*, tree density and qualitative degradation level. Argan tree density was estimated based on an automated tree crown detection method and validated with a sample of field plots (R^2= 0.85). A discrete variable was then used to characterize the forest Stand Degradation Level (SDL). Three SDL (slightly degraded, quite degraded and highly degraded) have been derived using the method described in [46]. The overall accuracy was estimated at 76% using 127 survey plots. All

parcels in the region were at least slightly degraded, but a restored class has been added to account for positive change in the future.

- Topographic variables (slope, slope exposure and position in the valley) were computed from the GDEM ASTER with 30m ground resolution and ±9.8 m of locally estimated geometric precision [47].
- *Household density* was derived from a spatial layer showing the point location of the isolated houses (weight = 1) or grouped houses (hamlet) materialized by a point with a so-called weight of habitations (number of houses by hamlet). The layer of house location is digitized from a Quick Bird image with a 0.6 m spatial resolution acquired in 2006. The distance of argan parcels to the nearest household was computed with the same dataset.
- *Argan parcel scarcity* corresponded to the surface of argan forest available within a radius of 1 km from each household
- *Distance-based variables* (*i.e.*, to the farm land, road, and unpaved trail) were obtained from the land cover map and digitized layers of roads and unpaved trails from the satellite image.
- Two types of Traditional Land Management (TLM), *mouchaa* and *agdal*, were derived from a set of spatial variables (tree density, slope and distance to the household). The accuracy was 67% as estimated from the 127 field observation plots. Another attribute defined the actual possession status (legal or not) by local households.

4.3. Dynamics of Argan Parcel Degradation

Based on field observations, natural regeneration is completely absent in the study area and is not expected to occur as long as the grazing pressure does not decrease. Hence, the tree density will keep decreasing in the absence of plantation. Its evolution was therefore simulated using the rate of change (−2.04% in 22 years) estimated from remote sensing data in the study area [48]. This density change rate was assumed constant over the whole period and applied to all parcels.

Qualitative degradation dynamics occurs in parcels located in the common domain. Transition matrices for *agdal* and *mouchaa* were computed from the actual distribution of the stand degradation level (Appendix A). At each time step, parcels that degrade were selected among those with the upper vulnerability level. The transition probabilities for the *m* steps were computed using Markov transition analysis.

The argan parcel vulnerability was estimated using AHP (Analytic Hierarchy Process)[49] method (Figure 3). It depends on (i) the parcel ecological potential of resilience; and (ii) the risk of over-grazing. The ecological potential of the site was computed from the argan stand conditions (SDL and tree density) and from the topographic variables. The over-grazing risk was estimated using the household's density and the forage availability in the parcel environment within a radius of 1000 m. The forage availability was in turn derived from Equation (1) (adapted from [2,34,50]).

$$\text{Forage availability} = 0.55 \times \text{SD} + 0.36 \times \text{QD} + 0.09 \times \text{HD} \quad (1)$$

where SD, QD and HD are the total area of slightly, quite and highly degraded parcels, respectively.

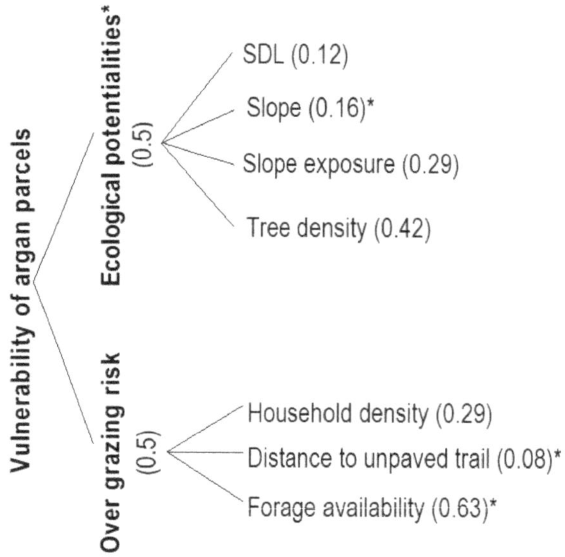

Figure 3. Components and weights (between brackets) used to estimate the vulnerability of argan parcels to further degradation. All variables are standardized (between 0 and 1). Variables with * are negatively correlated to the vulnerability.

In the special case of the *PPP* scenario, we assumed that legal possession was a sufficient incentive for undertaking a restoration process by plantation of argan trees. The planting success rate was set at 10% of the actual number of trees per hectare every three years. A parcel was allowed to move to the "restored" level after reaching a threshold tree density of 80 trees/ha (the maximum density currently observed) as long as the associated density change was higher than 10 trees/ha. The latter condition was introduced to prevent the denser parcels (acquired and having a density close to 80 trees/ha) from changing their status (restored) during the first five years of simulation. A minimum period of six years was deemed necessary to remove qualitative degradation.

4.4. Dynamics of Argan Parcel Possession

The possession dynamics was simulated using the suitability of argan parcels for the household possession and Markov chain analysis. The Markov chain process controls the temporal change among the parcel land tenure types (acquired/not acquired) based on transition matrices defined from the original transition probability matrix of land tenure type [4]. The suitability function controls the spatial pattern of change through decision rules simulating the farmers' rationale for choosing parcels. Furthermore, the final assignment incorporates local rules (required suitability or required rate of successful planting) as well as uncertainty related to farmer behavior.

The dynamics of the parcel possession are different for each of the three scenarios:

- There is no change in possession for the TDP scenario.
- The proportion of converted parcel possession is assumed to be constant over the next 30 years for *scenario BAU*. The annual rate of parcels recently enclosed was derived by photointerpretation of very high resolution images of 2006 (QuickBird) and 2009 (Google Earth) for each STU.

- The transition matrix in the PPP scenario is established by considering, in the initial stage, that up to one parcel is allowed for each household. The same household is not entitled to a second parcel before a period of three years, a period deemed necessary to ensure the survival of argan seedlings. The forecast transition proportions of acquired parcels (f_m^p) are estimated using transition probability matrix (P) and Markov process. The Markov process has two properties: (i) the status of a parcel in the *m* step depends only on its status in the ($m - 1$) *step*; and (ii) the probability of a "not acquired" parcel to be "acquired" in *m* steps, is the *m*th power of the transition matrix.

The converted parcels are selected according to a stochastic process, with uniform randomization, by favoring the parcels with a high possession suitability score while allowing other parcels to be included. While being driven by suitability scores, it also introduces a random effect in the household decision, accounting for considerations not included in the model (*i.e.*, historical, patrimonial or psychological interests). Appendix B shows the theoretical development of the approach.

In addition to the above, the possession process was allowed to change over time as a function of parcels' scarcity and vulnerability level to more closely mimic household decisions in the field.

Due to the risk of parcel withdrawal as a penalty, the local stakeholder is expected to invest more in the planting effort with an increasing scarcity of parcels under the *PPP* scenario; this is the case despite the fact that the conditions for planting success are becoming increasingly hard. The constraint of the required rate of planting success (T_i) was empirically derived from the scarcity level and compared to the vulnerability level (V_i; $0 \leq V_i \leq 1$). The number of parcels that satisfy this constraint ($1 - V_i \geq T_i$) corresponds with the real number (f_m^r; $f_m^r < f_m^p$) that will be chosen by households.

For the BAU scenario, the minimum required suitability (S_i') was lowered along with time as a result of increasing parcel scarcity, and less suitable parcels were thus progressively included in the range of regarded parcels.

The two constraints considered above (S_i';T_i) have been useful to ensure spatial contiguity of the selected parcels by raising the parcels chances to be selected in the neighborhood of the previously selected parcels, as was observed in the field.

Parcels' possession suitability attempts to simulate the decision making of the farmers. The selection of a parcel is influenced by biophysical and socio-economic conditions within as well as in the neighbourhood of the argan parcel. Three components were considered in the suitability function, namely Forest and Pasture Aptitude (FPA), the Exploitation Convenience (EC), and the Social Attractiveness Level (SAL).

- FPA translates the biophysical consistency of the forest stand regarding its tree density, its SDL and its position in the toposequence.
- The parcel exploitability expressed by EC is computed based on the parcel area and the distance-based variables.
- SAL reports the social pressure in the parcel environment and the competition between farmers for exploiting the same area. It reflects the scarcity of argan resources.

Parcels possession suitability was evaluated using the AHP method. In the model development, a two-stage AHP was performed (Figure 4) with FPA, EC and SAL playing the role of intermediate variables with equal weighting.

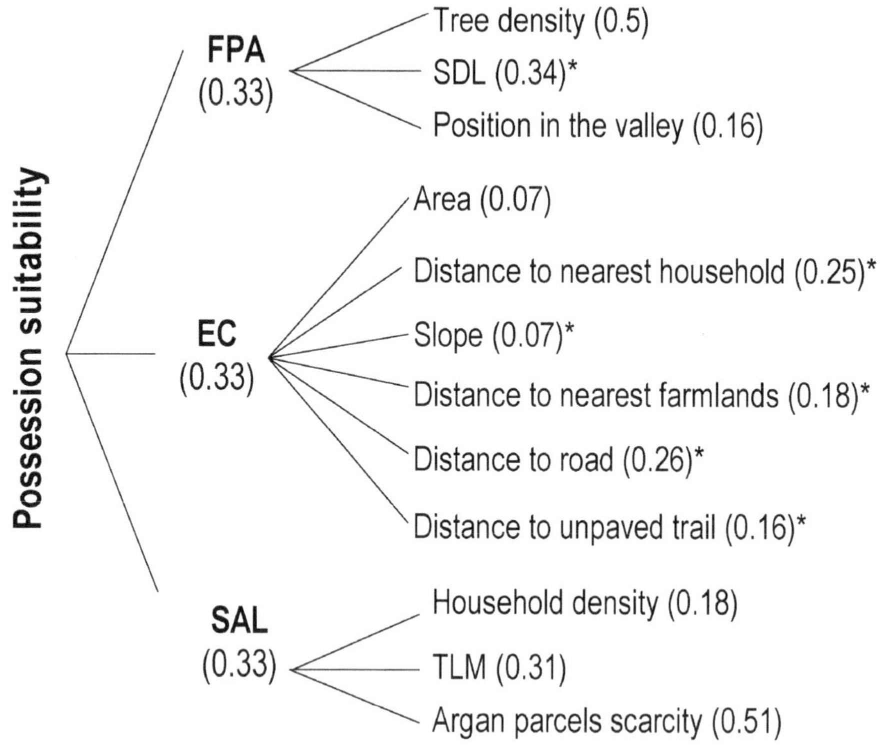

Figure 4. Components and weights (between brackets) used to estimate the possession suitability of argan parcels. All variables are standardized (between 0 and 1). Variables with * are inversely correlated (1—variable value) to its immediate goal. (FPA for Forest and Pasture Aptitude; EC for Exploitation Convenience; SAL for Social Attractiveness Level; SDL for Stand Degradation Level; TLM for Traditional Land Management).

4.5. Model Validation

For model validation, we compared the simulated possession maps for BAU and PPP with the observed acquired parcels (n = 76 parcels) hereafter called "reference parcels". The reference parcels are those acquired during the period from 2006–2009 and used to calculate the transition matrices, as well as those that were acquired before 2006.

5. Results

5.1. Validation of the Possession Dynamics

The average proportions of the acquired parcels are respectively 5.33% (±0.24) and 47.78% (±0.53) for *BAU* and *PPP*. The possession dynamic is nearly 10 times higher in *PPP* than in the *BAU*.

Table 1 gives the overall simulation success, *i.e.*, the percentage of successful matching between the reference and possessed parcels, for selected distances to the actual center of the reference parcels and various possession occurrences scores (minimum number of times one parcel is possessed among the 30 repetitions).

Table 1. Successful matching (%) between the reference parcels and the simulated possession as a function of the distance to the actual center of the reference parcels (lines) and to the possession occurrence (PO, rows), for the *BAU* and *PPP* scenarios. PO is the number of times one parcel is possessed among the 30 repetitions.

	***BAU* Scenario**		
	PO ≥ 10	**PO ≥ 5**	**PO ≥ 1**
50 m	1.32	23.68	78.95
150 m	2.63	32.89	90.79
250 m	3.95	46.05	96.05
	***PPP* Scenario**		
	PO ≥ 20	**PO ≥ 10**	**PO ≥ 5**
Exact location	42.10	73.68	76.32
50 m	69.74	89.47	92.10
150 m	89.47	96.05	98.68
250 m	97.37	100	100

BAU successfully (>78.9%) predicts the location of the acquired parcels in the overall area although no exact match could be obtained with the reference parcels. The main reason could be that the offer of suitable parcels is much larger than the historical demand. Despite the difficulty to match the exact location of the reference-parcels, the model predicts rather successfully the immediate vicinity of such areas or the potentially favorable areas for possession.

In *PPP*, the predictions more successfully agree with the observed possession dynamic. Almost 70% of the reference-parcels match the simulated parcels acquired at least 20 times out of the 30 model repetitions. This rate can reach up to 97% with a buffer area of 250 m around the acquired parcels. These results show that the model captures in a realistic manner the decision-making process of parcel selection.

Figure 5 confirms the previous results, showing the great number of reference plots located in the close neighborhood of the early acquired plots in both scenarios. For *BAU*, the possession dynamics look underestimated in some areas (circled in dotted line in Figure 5); this is due to the fact that the pace of possession in those STU is not constant in the period 2006–2009 (as assumed in the model) and the period for carrying the reference-parcels (acquired parcels including those before 2006).

The box plot in Figure 6 shows the suitability distributions of the reference-parcels as well as of the acquired and not acquired parcels for the *PPP* and *BAU* scenarios. For both scenarios, the suitability differs significantly ($p < 0.0001$) between the reference and the not possessed parcels. Comparing the suitability between the reference and acquired parcels, the difference is not significant for *PPP* ($p > 0.74$) while it is highly significant ($p < 0.002$) for *BAU*. The currently acquired parcels (reference parcels) are not always the most suitable; thus, the suitability function calculated based on the biophysical and socio-economic data does not reflect the total complexity of the choices made by households. The use of the Monte Carlo technique allows including the uncertainty that could not be included in the suitability map. The decision to acquire a parcel indeed depends on variables that are difficult to identify and take into account explicitly into the model. Uncertainty affects possession dynamics, forest degradation prediction and the utility of subsequent decision. The outputs' results (possession dynamic and SDL) are therefore formulated as a probabilistic distribution, rather than a single value.

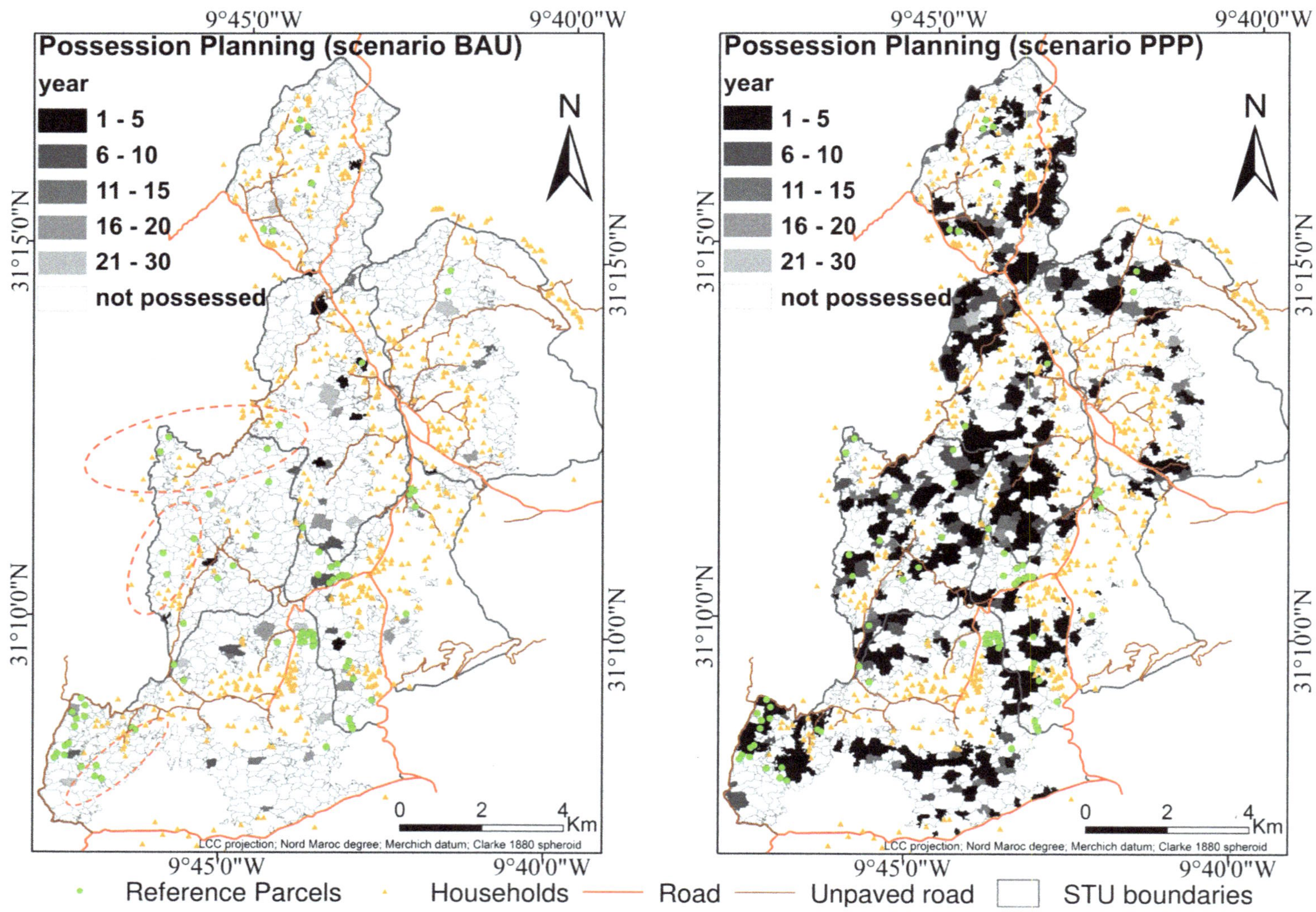

Figure 5. Simulated possession planning (schedule) dynamics for *BAU* (**left**) & *PPP* (**right**) *scenarios*. Dotted line circles delimits areas where the possession dynamics are under-estimated. The grey scale indicates the time steps for a parcel to change status tenure over the 30-year simulation period.

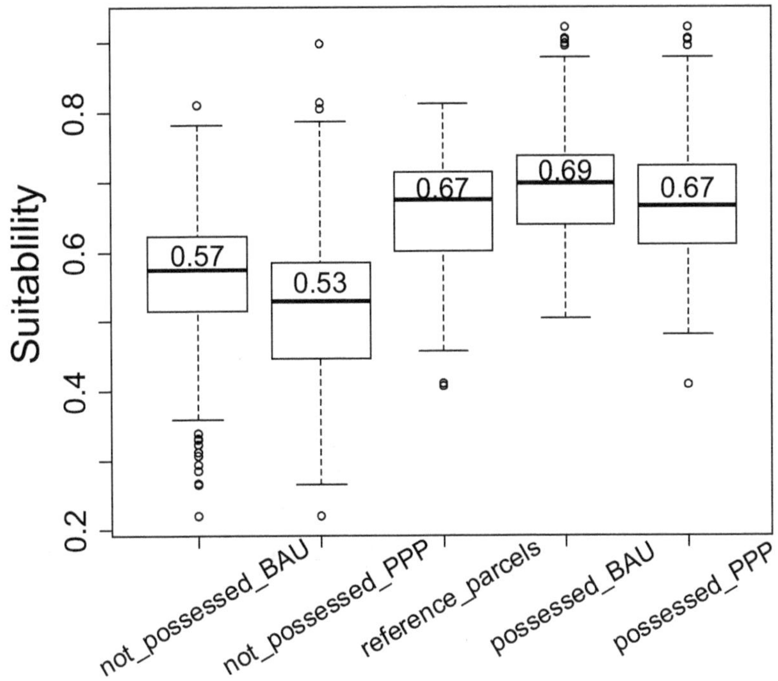

Figure 6. Suitability dispersion of reference-parcels, acquired parcels and not acquired parcels for the scenarios BAU and PPP.

5.2. Possession Transition Quantities and Spatial Maps

Figure 7 shows the number of parcels acquired each year. As imposed by the transition matrix, the quantity of parcels being acquired is stable (1.9 ± 0.4) over time for *BAU*. Following the actual possession dynamics, less than two parcels change their tenure status each year in the whole study area. The estimated confidence interval (with $\alpha = 0.05$ for $n = 30$) is relatively important showing an inter-annual variation of these dynamics related to the number of acquired parcels should be rounded to the nearest integer value.

The majority of new possessions (*PPP*) occurs at the beginning of the simulation period. In total, 75.37% of the acquired parcels are completed after the first five years. The yearly acquiring decreases quickly and becomes less than one parcel per year after the 24th year. Indeed, the suitable parcels become increasingly scarce and the constraint imposed by planting success is increasingly hard to meet (Figure 7).

This gradual decrease in the amount of newly acquired parcels adequately depicts the marginal utility of these parcels compared to the energy and investments that should be mobilized to restore a newly acquired parcel. The farmer already possessing a parcel would become less motivated to claim other parcels because of the constraints cited before.

The spatial distribution of the most suitable parcels (Figure 8) is mainly driven by the ecological potential gradient and is located in the most populated areas surrounding the farmlands.

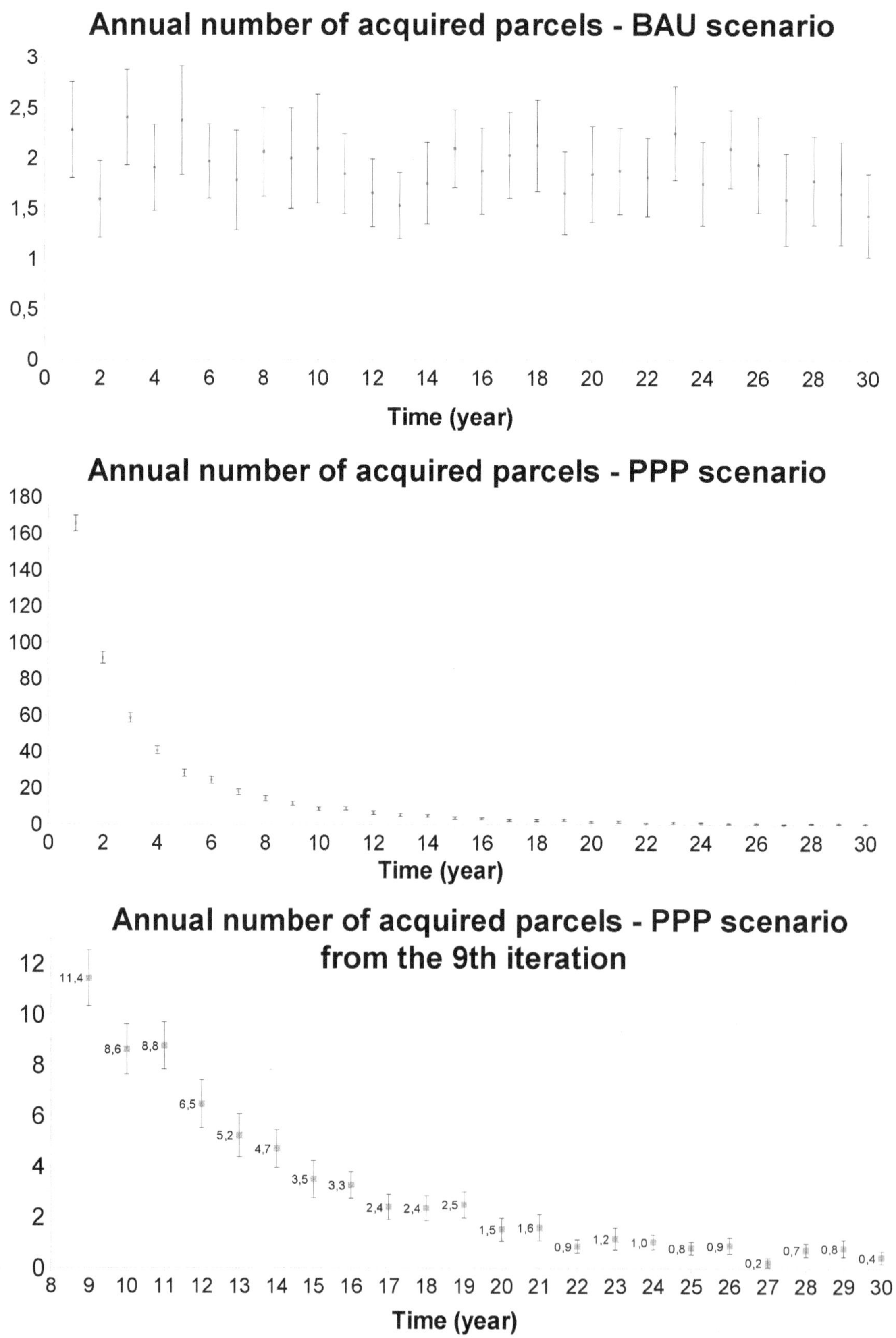

Figure 7. Annual number of acquired parcels and its confidence interval simulated over a 30-year period for the scenarios *BAU* and *PPP* (the sub-graph at the bottom shows the annual number of plots acquired from the ninth iteration for the PPP scenario).

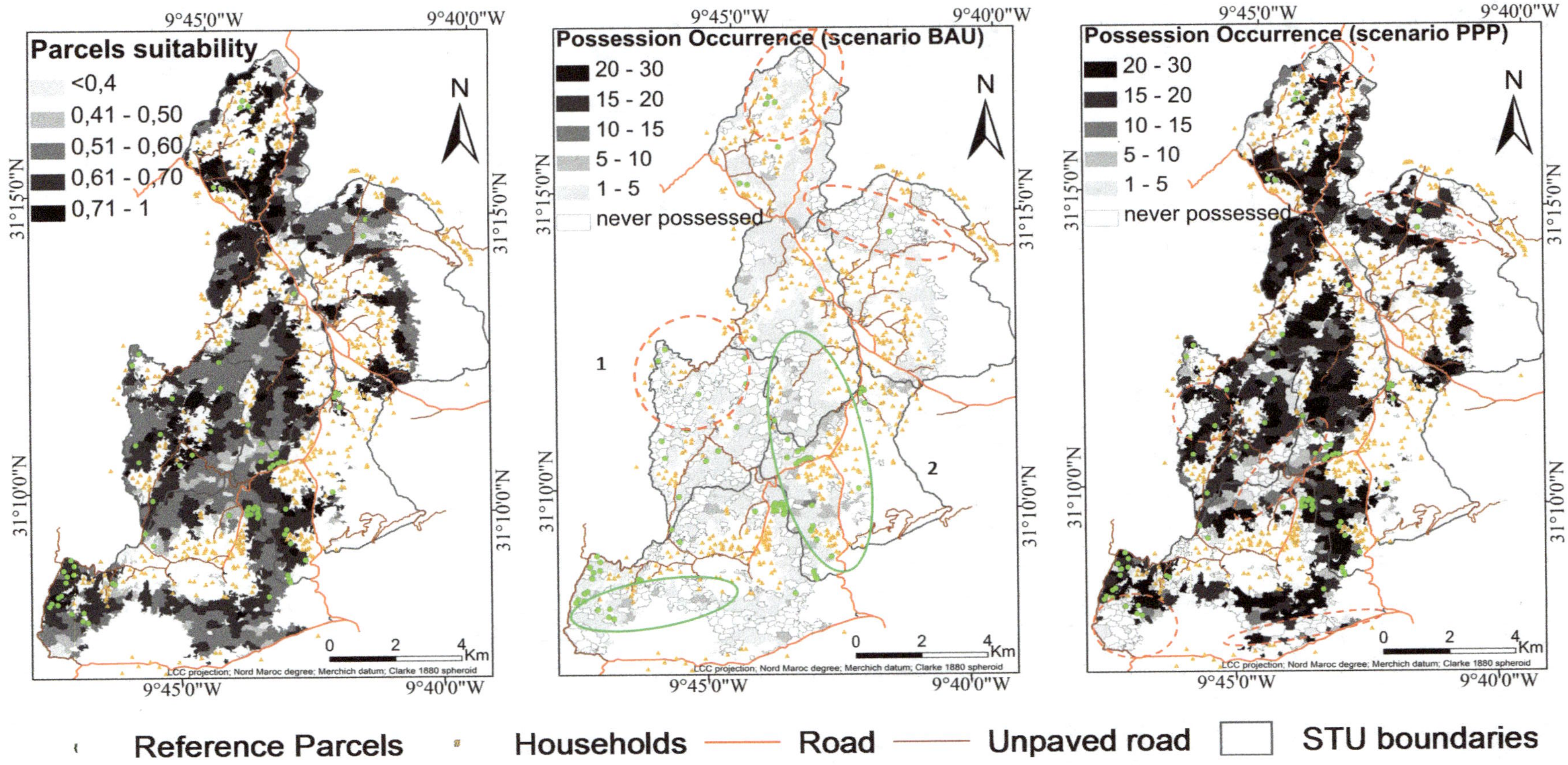

Figure 8. Spatial distribution of the parcels suitability and of the possession dynamics (BAU & PPP scenarios). Parcel suitability is quantified on a 0–1 scale according to Figure 4. Number of occurrences: number of times a parcel is acquired among 30 repetitions. Plain green circles: areas with a high demand of argan parcels; red dotted line circles: areas with a low demand of argan parcels. 1, 2 (*BAU* scenario) corresponds to the Socio-Territorial Units with the lowest (0.0014) and the highest (0.005) transition probabilities, respectively.

For the *BAU* scenario, we can distinguish between the areas with high demand of argan parcels (surrounded by a plain green line) and those with low demand (surrounded by a red dotted line). The possession dynamics occur in heterogeneous rhythm according to the transition probability estimated for each STU from 2006 to 2009. The highlighted circled areas (1, 2) correspond respectively to STU with the lowest (0.0014) and the highest (0.005) transition probabilities. These contrasted situations are driven by the territorial limits of each STU and their ethnic influence. In the *PPP* scenario, the possession dynamics are more important and more spread out in space. Indeed, the transition probabilities depend on the population density and on the available argan parcels. The non-acquired parcels in *PPP* scenario are the least accessible. In both scenarios, the possession dynamics occur by groups of adjacent parcels as observed in real cases. This pattern in contiguous groups is induced by using dynamic thresholds of scarcity level (*scenario BAU*) and the requirements for successful planting (*scenario PPP*), which also depend, among other factors, on the level of scarcity in a buffer zone of 1000 m.

5.3. Stand Degradation Level and Tree Density Changes

The frequency distribution (Figure 9) as well as the spatial pattern of SDL (Figure 10) show a strong convergence between the outcomes of *BAU* and *TDP*. The proportion of "*quite degraded*" parcels rises from 41.4% (current state) to 65.5% (±0.15) and 67.8% in *BAU* and *TDP*, respectively. Moreover, the proportion of "highly degraded" areas would be doubled in both simulated situations.

Conversely, in *PPP*, the total of "slightly degraded" and "restored" areas could reach up to 63.7% (±0.004). This change is mainly gained from the "quite degraded" areas 27.5% (±0.004). The "highly degraded" areas increased slightly from 8.4%–8.7% (±0.001). This overall improvement in stand health status is associated with an increased tree density (Figure 11). Although SDL transition matrices were the same for all three scenarios, the simulation results show a clear advantage of *PPP*.

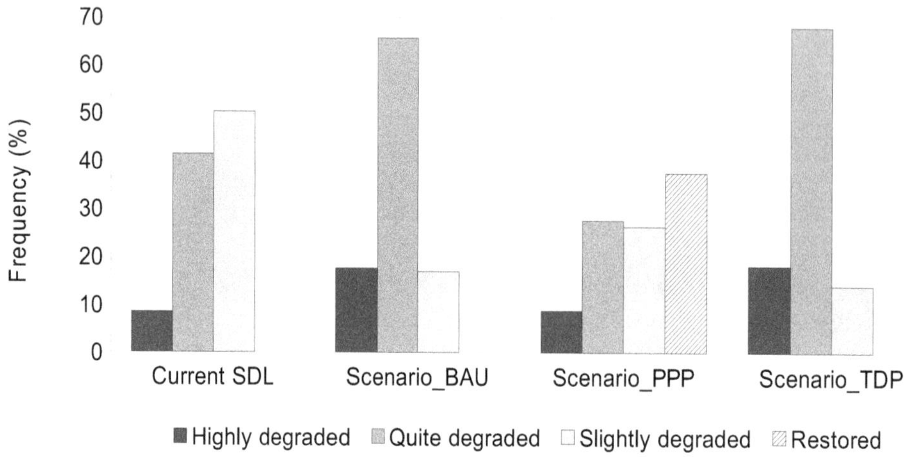

Figure 9. Simulated *versus* current SDL distribution frequencies according to the different scenario after 30 years.

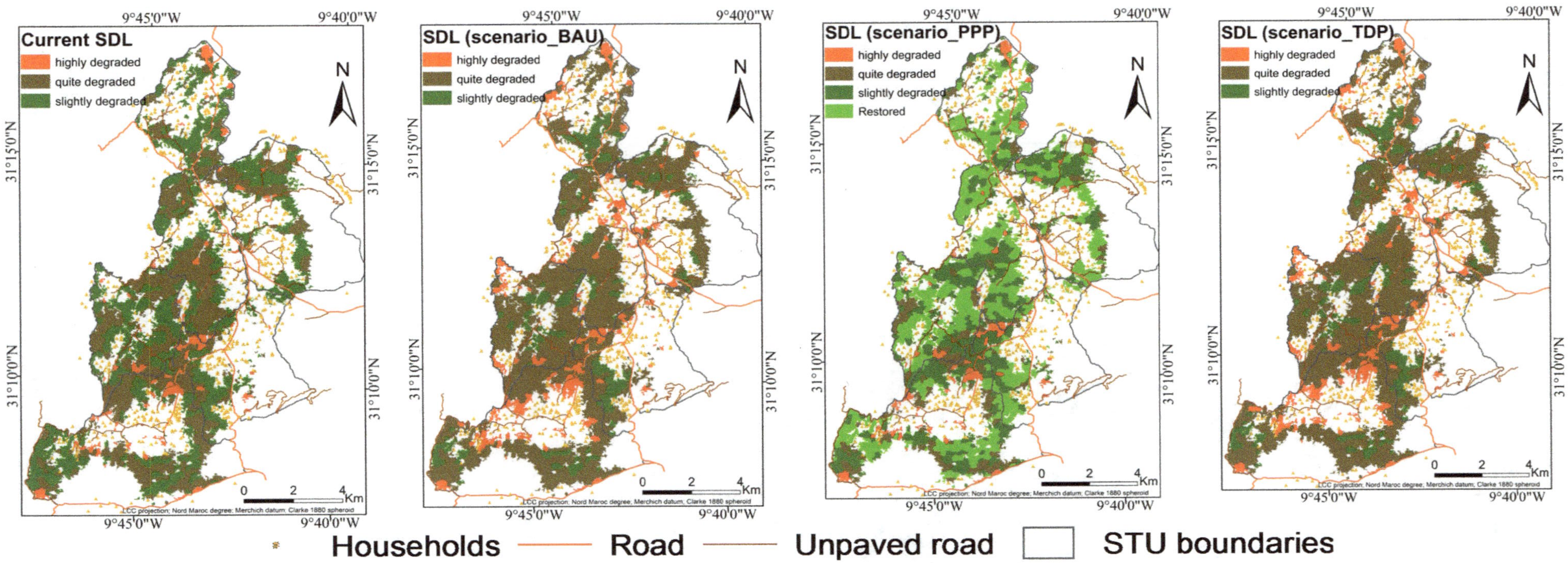

Figure 10. Current stand degradation level and simulated stand degradation level after a 30-year simulation period for scenarios *BAU*, *PPP* and *TDP* (the degradation level of the plots corresponds to the degradation level that has the maximum probability).

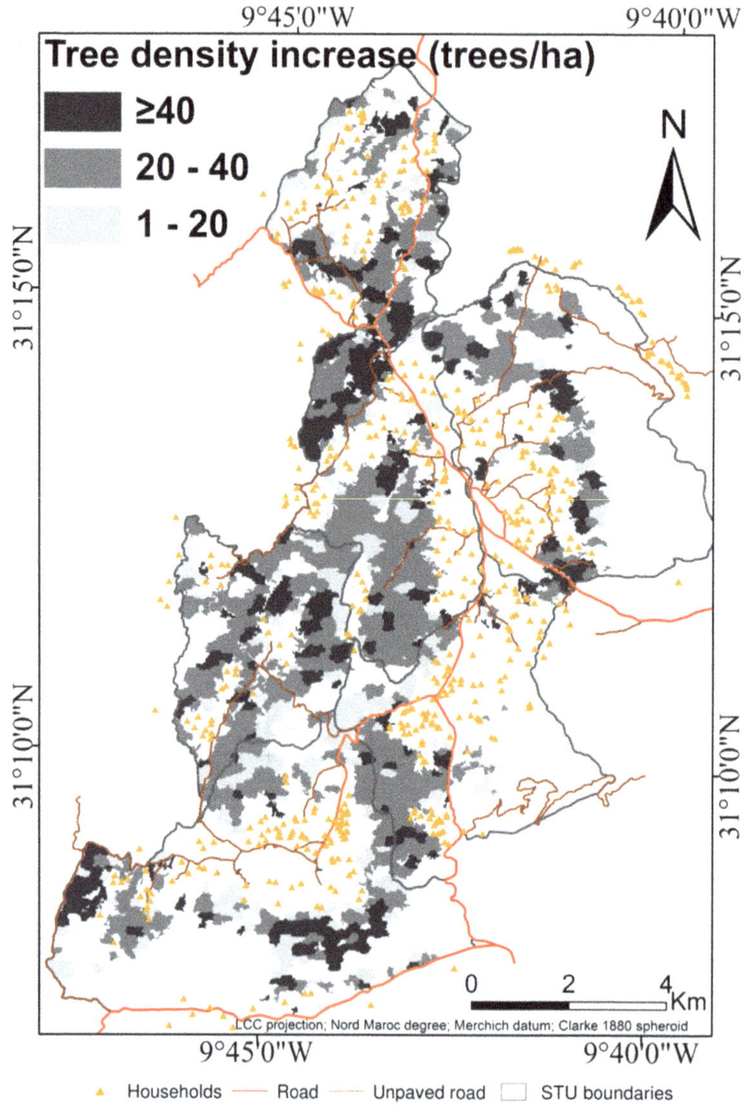

Figure 11. Simulated map of tree density change for *scenario PPP* after 30 years.

6. Discussion

The integrated overall approach allowed contrasting observed changes to simulated dynamics under hypothetical scenarios. It successfully combined multisource data related to biophysical and socio-economic data, as well as cultural and ethnic behaviors, in order to compare the scenarios.

6.1. Methodological Contributions

Our model successfully accounts for the behaviour of the two main agents: farmers and forest officers. The farmer, as a local stakeholder (or group of stakeholders), acts in a delimited zone, depicted here as STU, to practice his rights of uses. He chooses to acquire (or not) an argan parcel; he goes through a selection process and determines at which moment it will be implemented. The decision making process is affected by the scarcity level of argan parcel implemented as a communication tool between farmers. The second agent is the forest officer who is in charge of applying state policies to manage the argan forest. This agent is incorporated into the model by taking into consideration three land tenure options and negotiating the application of the specified requirements for the use and conservation of argan parcels.

Incorporating locals' behavior in a realistic way is very challenging for many approaches used in land use modelling. The combination of a suitability function and a stochastic process for argan parcel selection more closely mimics the actual stakeholder's decision than a purely non-probabilistic approach.

6.2. Assumptions and Limitations

This modelling work is based on three assumptions derived from field observations. The first considers that, once a parcel is acquired, it is not abandoned later on. This abandonment seems unlikely, in particular for *PPP* which provides financial incentives (PES) to support the restoration of these parcels. Indeed, field based observation indicates an opposite trend of new parcel acquiring. Similarly, it was assumed that qualitative degradation stops once a parcel is acquired. From field observation, the argan stands located in enclosed parcels are more preserved. Indeed, closing reduces land use pressures typically supported by the argan parcel, being used exclusively by one household. The third assumption is related to the unlimited availability of the argan seedlings and the high planting success. This is realistic because the forest nurseries annually produce many thousands of argan seedlings which are made freely available to farmers. The seedlings settle and grow if they are protected from livestock. In this respect, farmers master traditional tools for protecting seedlings as is the case for other fruit trees planted in their private parcels.

The modelling approach also relies on methods/parameters which are defined by expert judgments. The impacts of this subjectivity and its variability according to experts could be assessed from simulation results. This concerns (i) the multicriteria evaluation method (AHP); (ii) the thresholds of required suitability (S_i') (*BAU*); and (iii) the required rate of planting success (T_i) (*PPP*).

The AHP provides a comprehensive and rational framework for structuring a decision problem, quantifying stakeholders' preferences and correlating these measured criteria to overall goals. It has one of the strongest theoretical foundations over other ranking or rating methods and the ability to easily incorporate mixed elements (social and economic data) in the decision process [51]. Therefore, it is widely used for calculating suitability functions [12,24]. To limit the sensitivity of suitability function, Garfi, *et al.* [51] recommends relying on variables mixing external experts and local people or stakeholders. These provide universally relevant knowledge and place specificities to decision process.

The second type of arbitrary parameters used in the modelling approach is the dynamic thresholds, *i.e.*, S_i' and T_i. The former threshold doesn't have any effect on the simulation results because it was implemented upstream of the parcel selection process. Being a variable threshold according to the parcels scarcity, it favours the contiguity of the acquired parcels as observed in the field. The T_i threshold retrieves among the available parcels, those showing very low ecological potentialities and/or high risk of overgrazing. It serves as a cultural rule necessary for the establishment and success of argan plants and allows contiguity of the possession process. This threshold arbitrarily calculated from the argan parcels scarcity could have an effect on the model simulation. Indeed, the stochastic component included in the model is responsible for variable results for the same parameter value, making it difficult to assess the model sensitivity to the T_i parameter, as long as it remains in the typical range of the T_i values.

6.3. Forest Impacts and Social Benefits

Our analysis shows that a successful *PPP* might serve to increase tree density and to better control the qualitative degradation. The possession dynamics could involve 48% of the argan area and make argan tree stands denser by an average of 25 trees/ha after 30 years. Two afforestation types could be done in this scenario namely "tree planting" by users in acquired parcels and "perimeter planting" as routinely implemented by the forest administration. The planting perimeters could involve the remaining available argan parcels, which amounts to 52% of the total argan parcels in the present case. Furthermore, this scenario would provide greater management flexibility.

With the *BAU*, the current dynamics of qualitative forest degradation continues and the proportions of "*quite degraded*" and "*highly degraded*" stand rise by 58% and 108%, respectively, compared to the present state. *BAU* appears to be quite similar to the pessimistic scenario (*TDP*), which could be due to the low rate of illegal acquiring. It is however surprising that the current practices observed in a context of argan oil boom lead to the same qualitative degradation than a scenario without any interest in argan forest.

In addition to the argan forest ecological and environmental roles, the increasing tree capital (25 trees/ha in average over 30 years for *PPP*) should contribute to the survival of several hundreds of domestic livestock and improve household income. The annual harvest of nuts would increase by an amount equivalent to 750 kg/ha (30 kg of nuts per tree with 25 trees/ha). It could contribute to the household budget with an additional financial input of about 200–370 US $ per year. Such an increase will undoubtedly have a significant effect on local poverty mitigation.

Beyond the quantitative results comparing these three scenarios, the proposed model leads to the establishment of a decision support system. The proposed system is an operational framework to further investigate forest management scenarios by assessing their expected impact in the context of a real landscape. In this study, this system provides different kind of answers:

(i) "if we give more latitude rights to locals and economic incentive for argan tree planting", what would be the impact on land possession and stand degradation changes of argan forest. It could be used to identify the order of magnitude of the on-going process and the possible alternatives of the future landscape scenarios (*how much*);
(ii) Quantifying household preferences and perceptions of their territory for *adverse possession* pretentions and anticipating possible land competition induced by the new policy (*feedback*);
(iii) Pinpointing the high priority areas with high potential for land conservation and degradation hot-spots (*where*).

The public/private partnership as a framework for co-management of the argan forest could restore trust between the forest staff and the local communities. Redefining this relationship by giving more local communities more opportunity to take initiative with full recognition of the rights of use and by providing interesting economic incentives to support sustainable forest management as a payment for ecosystem services could increase their involvement in managing and restoring this ecosystem. Additional incentive mechanisms already implemented in the region (cooperatives, IGP, oil marketing, *etc.*) should be maintained in synergy with those explored by the *PPP* scenario to sustain forest conservation and rural development. The system of incentive mechanisms could fit in existing traditional system of land management to determine beneficiaries and delineate land management units (*agdal* parcels). The compliance with existing rights when implementing incentives would help in preventing conflicts between local stakeholders.

To improve predictions in the mid- to long-term, the modelling approach should include the following three drivers: (1) the demography which slightly decreases in the studied area; (2) the climatic change that seems to induce an increasing aridity; and (3) the exponential impact of aging forest stands on tree dieback and tree density decrease.

Restoring the trust between forest officers and local stakeholders and providing incentive to the population for taking over the management of their own forests might play in favor of a better conservation of the argan forest. Nevertheless, it is also interesting to note that the risk of degradation becomes higher for parcels that remain not acquired according to the theory of tragedy of the commons. Such zones could be used to install reforestation perimeters, without any local conflict.

7. Conclusions

This study successfully developed a decision support system to investigate the impacts of contrasting management/policy options on argan forest dynamics. This entity-based approach combined a Markov chain analysis with a suitability function derived from natural and socioeconomic data to simulate three scenarios of land possession dynamics and argan forest degradation changes over a 30-year period.

The modelling approach provided insight into policy assessment. In particular, it highlighted the positive impacts of forest policies combining economic incentives and recognition rights on argan forest conservation.

Acknowledgments

The study was conducted within the framework of the PIC-07-argan tree project (Projet Inter-Universitaire Ciblé) funded by the University Commission for Development (CUD) as a part of Ph.D. research. High spatial resolution imagery was provided from the ISIS program of CNED providing Sport imagery. The authors would like to thank the Forest Research Centre (CRF) for logistical support during field campaigns.

Author Contributions

Farid El Wahidi collected field data, conducted data analysis and led the preparation of the manuscript with conceptual inputs from Julien Radoux and Pierre Defourny. Pierre Defourny and Quentin Ponette also helped in structuring and editing the manuscript.

Conflicts of Interest

The authors declare no conflict of interest.

Appendix A

Transition matrices between qualitative degradation levels.

Table A1. Transition matrices between qualitative degradation levels for parcels managed as *agdal* or *mouchaa*. Degradation is faster in *mouchaa* than in *agdal*.

	Agdal			***Mouchaa***		
	Slightly	**Quite**	**Highly Degraded**	**Slightly**	**Quite**	**Highly Degraded**
Slightly	0.996	0.0036	0	0.995	0.005	0
Quite	0	0.999	0.0009	0	0.999	0.00047
Highly degraded	0	0	1	0	0	1

Appendix B

Uniform randomization of the selection probability.

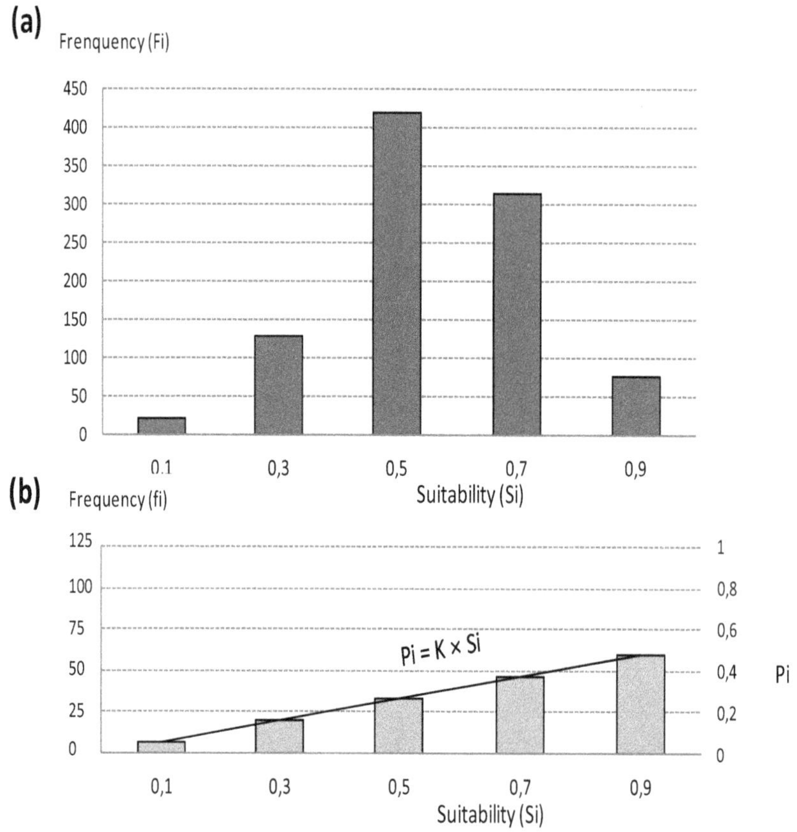

Figure A1. Frequency distribution of all the parcels (F_i) (**a**); and of the selected parcels (f_i) (**b**) resulting from uniform randomisation using the suitability index (S_i).

$$n = \sum_{i=1}^{l} F_i \tag{A1}$$

where n is the total available parcels, l is the number of suitability classes (i) and F_i is the number of parcels belonging to each suitability class.

$$f_m^p = \frac{\sum_1^l f_i}{n} = \frac{\sum (p_i \times F_i)}{n} \tag{A2}$$

f_i is the number of selected parcels by suitability class (i).

p_i is the proportion of selected parcels by suitability class (i).

As the proportion to be chosen by class of suitability (i) is proportionally correlated to its suitability score (S_i)

$$p_i = k \times S_i \ (\mathrm{k} \in \mathbb{R}^+) \quad \text{(A3)}$$

The Equation (A2) becomes:

$$f_m^p = \frac{\sum k \times S_i \times F_i}{n} = \frac{k \times \sum S_i \times F_i}{n} \quad \text{(A4)}$$

Then

$$k = \frac{n \times f_m^p}{\sum S_i \times F_i} \quad \text{(A5)}$$

Knowing the value of k, we can calculate the probability of possession (p_{ij}) for each parcel "j" in class "i" with a suitability S_i. If f_{ij} is larger than the outcome of a uniform random variable, the parcel is selected.

References

1. M'Hirit, O.; Benzyane, M.; Benchekroun, F.; El Yousfi, M.; Bendaanoun, M. *L'arganier: Une Espèce Fruitière-Forestière à Usages Multiples*; Mardaga: Sprimont, Belgique, 1998; p. 145.
2. Naggar, M.; M'hirit, O. L'arganeraie: Un parcours typique des zones arides et semi-arides marocaines. *Sécheresse* **2006**, *17*, 314–317.
3. Aspinall, R. Modelling land use change with generalized linear models—A multi-model analysis of change between 1860 and 2000 in Gallatin Valley, Montana. *J. Environ. Manag.* **2004**, *72*, 91–103.
4. Guan, D.; Li, H.; Inohae, T.; Su, W.; Nagaie, T.; Hokao, K. Modeling urban land use change by the integration of cellular automaton and Markov model. *Ecol. Model.* **2011**, *222*, 3761–3772.
5. Kamusoko, C.; Aniya, M.; Adi, B.; Manjoro, M. Rural sustainability under threat in Zimbabwe—Simulation of future land use/cover changes in the Bindura district based on the Markov-cellular automata model. *Appl. Geogr.* **2009**, *29*, 435–447.
6. Myint, S.W.; Wang, L. Multicriteria decision approach for land use land cover change using Markov chain analysis and a cellular automata approach. *Can. J. Remote Sens.* **2006**, *32*, 390–404.
7. Young, O.R.; Lambin, E.F.; Alcock, F.; Haberl, H.; Karlsson, S.I.; McConnell, W.J.; Myint, T.; Pahl-Wostl, C.; Polsky, C.; Ramakrishnan, P.S.; *et al.* A portfolio approach to analyzing complex human-environment interactions: Institutions and land change. *Ecol. Soc.* **2006**, *11*, 31.
8. Zhang, J.; Wang, K.; Chen, X.; Zhu, W. Combining a fuzzy matter-element model with a geographic information system in eco-environmental sensitivity and distribution of land use planning. *Int. J. Environ. Res. Public Health* **2011**, *8*, 1206–1221.
9. Verburg, P.H.; Overmars, K.P. Combining top-down and bottom-up dynamics in land use modeling: Exploring the future of abandoned farmlands in Europe with the Dyna-CLUE model. *Landsc. Ecol.* **2009**, *24*, 1167–1181.
10. Aurbacher, J.; Dabbert, S. Generating crop sequences in land-use models using maximum entropy and Markov chains. *Agric. Syst.* **2011**, *104*, 470–479.
11. Gong, J.Z.; Liu, Y.S.; Chen, W.L. Land suitability evaluation for development using a matter-element model: A case study in Zengcheng, Guangzhou, China. *Land Use Policy* **2012**, *29*, 464–472.

12. Strager, M.P.; Rosenberger, R.S. Incorporating stakeholder preferences for land conservation: Weights and measures in spatial MCA. *Ecol. Econ.* **2006**, *58*, 79–92.
13. Bousquet, F.; Le Page, C. Multi-agent simulations and ecosystem management: A review. *Ecol. Model.* **2004**, *176*, 313–332.
14. Hare, M.; Deadman, P.; Lim, K. Further towards a taxonomy of agent-based simulation models in environmental management. *Math. Comput. Simul.* **2004**, *64*, 25–40.
15. Couclelis, H. "Where has the future gone?" Rethinking the role of integrated land-use models in spatial planning. *Environ. Plan. A* **2005**, *37*, 1353–1371.
16. Verburg, P.H. Simulating feedbacks in land use and land cover change models. *Landsc. Ecol.* **2006**, *21*, 1171–1183.
17. Parker, D.C.; Manson, S.M.; Janssen, M. a; Hoffmann, M.J.; Deadman, P. Multi-agent systems for the simulation of land-use and land-cover change: A review. *Ann. Assoc. Am. Geogr.* **2003**, *93*, 314–337.
18. Kaimowitz, D.; Angelsen, A. Economic Models of tropical deforestation: A review. *Environ. Dev. Econ.* **1998**, *6*, 139.
19. Baskent, E.Z.; Jordan, G.A. Forest landscape management modeling using simulated annealing. *For. Ecol. Manag.* **2002**, *165*, 29–45.
20. Sklar, F.H.; Costanza, R. The development of dynamic spatial models for landscape ecology: A review and prognosis. *Quant. Methods Landsc. Ecol.***1991**, 239–288.
21. Simmonds, D.; Feldman, O. Alternative approaches to spatial modelling. *Res. Transp. Econ.* **2011**, *31*, 2–11.
22. Ludeke, A.K.; Maggio, R.C.; Reid, L.M. An analysis of anthropogenic deforestation using logistic regression and GIS. *J. Environ. Manag.* **1990**, *31*, 247–259.
23. Mertens, B.; Lambin, E.F. Spatial modelling of deforestation in southern Cameroon. *Appl. Geogr.* **1997**, *17*, 143–162.
24. Eastman, J.R. *IDRISI Kilimanjaro Guide to GIS and Image Processing*; Clark Labs, Clark University: Worcester, MA, USA, 2003; p. 306.
25. Moreno, N.; Wang, F.; Marceau, D.J. Implementation of a dynamic neighborhood in a land-use vector-based cellular automata model. *Comput. Environ. Urban Syst.* **2009**, *33*, 44–54.
26. Castella, J.C.; Verburg, P.H. Combination of process-oriented and pattern-oriented models of land-use change in a mountain area of Vietnam. *Ecol. Model.* **2007**, *202*, 410–420.
27. Jenerette, G.D.; Wu, J.G. Analysis and simulation of land-use change in the central Arizona-Phoenix region, USA. *Landsc. Ecol.* **2001**, *16*, 611–626.
28. Kocabas, V.; Dragicevic, S. Assessing cellular automata model behaviour using a sensitivity analysis approach. *Comput. Environ. Urban Syst.* **2006**, *30*, 921–953.
29. Ménard, A.; Marceau, D.J. Exploration of spatial scale sensitivity in geographic cellular automata. *Environ. Plan. B* **2005**, *32*, 693–714.
30. Flache, A.; Hegselmann, R. Do Irregular grids make a difference? Relaxing the spatial regularity assumption in cellular models of social dynamics. *J. Artif. Soc. Soc. Simul.* **2001**, *4*, 27.
31. Semboloni, F. The growth of an urban cluster into a dynamic self-modifying spatial pattern. *Environ. Plan. B Plan. Des.* **2000**, *27*, 549–564.

32. Liang, H.; Falta, R.W.; Henderson, J.K.; Shoemaker, S. Probabilistic simulation of remediation at a site contaminated by Trichloroethylene. *Gr. Water Monit. Remediat.* **2012**, *32*, 131–141.
33. Lybbert, T.J.; Barrett, C.B.; Narjisse, H. Market-based conservation and local benefits: The case of argan oil in Morocco. *Ecol. Econ.* **2002**, *41*, 125–144.
34. El Aich, A.; Bourbouze, A.; Morand-Fehr, P. *La Chèvre dans l'Arganeraie*; Actes Editions; IAV Hassan II: Rabat, Maroc, 2005; p. 123.
35. USAID. *Morocco Biodiversity and Tropical Forestry Assessment*; Morocco 118/119 Assessment Report; United States Agency for International Development (USAID): Washington, DC, USA, 2008; p. 92.
36. Adlouni, A.; Christon, R. L'arganier et l'huile d'argane au Maroc: Vers la mutation d'un système agroforestier traditionnel? Enjeux et contribution de la recherche. *Cah. Agric.* **2005**, *14*, 351–356.
37. Romagny, B. Des souks aux marchés internationaux. La valorisation économique de l'huile d'argan marocaine: Un cas d'école des contradictions du développement durable. In *Colloque International "Localiser les Produits"*; UNESCO: Paris, France, 2009; pp. 1–16.
38. Bejbouji, J.; Mougenot, C.; Mormont, M.; Qarro, M.; Aziz, L. Appropriation progressive de l'arganeraie: Exploration de l'histoire depuis la genèse des droits jusqu'à leur refonte. In *Forets et Foresterie. Mutations et Décloisonnements*; Farcy, C., Peyron, J.-L., Poss, Y., Eds.; L'Harmattan: Paris, France, 2013; pp. 99–121.
39. Lhafi, A. Les espaces boisés méditerranéens: Institutions et Gouvernance. In Procedings of 3ème Conférence Scientifique Méditerranéenne du GID—Espace Méditerranéen de la Science: "Richesse et Diversité Méditerranéennes; Biologie et Culture", Alexandrie, Egypt, 22–24 June 2010; p. 13.
40. UNCCD. *Incentive and Market Based Mechanisms to Promote Sustainable Land Management Practises by Land Users*; United Nations Convention to Combat Desertification (UNCCD): Bonn, Germany, 2012.
41. Jack, B.K.; Kousky, C.; Sims, K.R.E. Designing payments for ecosystem services: Lessons from previous experience with incentive-based mechanisms. *Proc. Natl. Acad. Sci. USA* **2008**, *105*, 9465–9470.
42. Le Polain de Waroux, Y. *Social and Ecological Change in the Argan Woodlands, Morocco*; Université catholique de Louvain: Louvain la Neuve, Belgium, 2012.
43. Sarin, M. Cogestion des forêts en Inde: Réalisations et défis. *Unasylva* **1995**, *46*, 30–36.
44. Lefhaili, A. *FAO Forest Resources Assessment: Morocco Country Report*; FAO: Rome, Italy, 2010.
45. Benomar, I. *Diagnostic des Systèmes D'utilisation des Terres et Identification des Unités Socio-Territoriales Dans les Zones Forestières et Péri-Forestières du Massif D'amsitten (cas de Trois Communes Rurales: Smimou, Imin'Tlit et Ida ou Azza)*; Ecole Forestière des Ingénieurs Thesis: Salé, Marocco, 2008.
46. El Wahidi, F.; Mounir, F.; Sabir, M.; Defourny, P.; Ponette, Q. Modélisation logistique et analyse multivariée pour l'élaboration des indicateurs de suivi de la dynamique de dégradation qualitative. In Proceedings of the First International Congress of the Argan Tree, Agadir, Morocco, 15–17 December 2011; pp. 237–251.

47. El Wahidi, F.; Mounir, F.; Defourny, P. GDEM ASTER et SRTM: Qualité et Exactitude dans les plateaux des Haha (Essaouira, Maroc). *GÉO Obs.* **2011**, *19*, 55–78.
48. El Wahidi, F.; Bellefontaine, R.; Ponette, Q.; Defourny, P. Dynamique de changement de l'arganeraie entre sur-usage et mutations sociales: Une opportunité d'équilibre socio-écologique? *J. Agric. Environ. Int. Dev.* **2014**, *108*, 109–133.
49. Saaty, T.L.*The Analytic Hierarchy Process*; McGraw-Hill Education (MHE): New York, NY, USA, 1980.
50. Alados, C.L.; El Aich, A. Stress assessment of argan (*Argania spinosa* (L.) Skeels) in response to land uses across an aridity gradient: Translational asymmetry and branch fractal dimension. *J. Arid Environ.* **2008**, *72*, 338–349.
51. Garfi, M.; Ferrer-Marti, L.; Bonoli, A.; Tondelli, S. Multi-criteria analysis for improving strategic environmental assessment of water programmes. A case study in semi-arid region of Brazil. *J. Environ. Manag.* **2011**, *92*, 665–675.

11

Patterns of Wastewater Infrastructure along a Gradient of Coastal Urbanization: A Study of the Puget Sound Region

Daniele Spirandelli

Department of Urban & Regional Planning, University of Hawai`i-Manoa, 2424 Maile Way, Saunders 107, Honolulu, HI 96822, USA; E-Mail: danieles@hawaii.edu

Academic Editor: Andrew Millington

Abstract: The aim of this paper is to explore patterns of wastewater infrastructures (sewers *vs.* septic tanks) in urbanizing watersheds across a coastal metropolitan region. This research combines an urban-rural gradient with spatial metrics at the patch and watershed scale (proportion of parcels on a treatment system, septic density, lot size and percent imperviousness) to analyze wastewater patterns in the Puget Sound, WA, USA. Results show that most urban residential parcels are hooked up to a sewer, although there remain urban residences on a septic tank with small lots. I find a complex arrangement of wastewater treatment in suburban watersheds representing a patchwork of parcels on sewers and septic tanks. Sewers dominate in total numbers, while the density of septic tanks is highest in this portion of the urban gradient. Lot size decreases from rural to urban; however, it varies depending on the type of wastewater treatment system. In urban watersheds, lots on septic tanks are significantly smaller than lots in suburban and rural watersheds and of a similar size compared to lots on sewers. I also find a significant difference in the amount of impervious surfaces in watersheds dominated by sewers *vs.* septic tanks. In the urban portion of the gradient, the amount of paved surfaces in parcels with septic tanks is also similar in level as parcels with sewers. I discuss how these patterns emerge from the interplay of biophysical, socio-economic and technological factors and how different regulatory regimes for septic tanks and sewers may further induce these patterns.

Keywords: urban patterns; wastewater infrastructure; septic tanks; sewers; coastal urbanization

---

1. Introduction

Coastal areas in the United States and in the world are experiencing some of the highest rates of population growth and land use change [1,2]. Coastal urbanization introduces new sources of pollution [3] and transforms natural landscapes [4–6], which effects the structure and function of coastal ecosystems [7,8]. As a result, urbanization and the human use of shoreline environments are associated with increased incidence of water-borne diseases from contact with polluted water and the consumption of contaminated seafood [9,10].

The treatment of sewage is critical for urban development, the protection of coastal water quality and ecosystem health. As shorelines urbanize, centralized and decentralized wastewater infrastructures are built to control the release of pathogens, viruses and other potentially harmful pollutants. Centralized wastewater infrastructures collect sewage from multiple sources and transport it through a network of sewers and pumps to be treated by a central plant before discharging effluent to surface and/or groundwater. Decentralized wastewater systems, also known as onsite wastewater treatment systems (OWTS), collect and treat wastewater from individual homes, clusters of homes and subdivisions at or near the point of waste generation [11]. Most OWTS consist of a septic tank and soil absorption field, although a range of alternative treatment technologies exist to process wastewater on site [12]. Since the passage of the Clean Water Act (1972) and subsequent environmental policies, sewers and wastewater treatment plants have improved coastal water conditions [13]. Despite this improvement, sewage and non-point sources of pollution continue to be leading causes of coastal contamination [9,14,15]. Studies identify pollutants from failing septic tanks, leaky sewer pipes, wastewater treatment plants and domestic and wild animal feces in stormwater run-off [16–18]. Efforts to formulate prevention measures are complicated by disagreements over leading stressors at the watershed scale. For example, there is evidence that septic density in watersheds with low amounts of urban development lead to contamination in coastal waters [19,20]. In suburban areas, urban land cover, impervious surfaces and septic density are also found to positively correlate with fecal pollution [21]. Other research has documented higher sewage contamination in sewered watersheds with urban development compared to unsewered watersheds with septic tanks and similar high levels of development [22]. Some suggest the effect of sewers and impervious surfaces can offset the gains achieved in controlling pollution from septic tanks [17,23]. These studies highlight the need to better understand how wastewater treatment systems (for example, sewers *vs.* septic tanks) relate to urban development patterns and coastal ecosystem function. A first step in understanding this relationship is to examine how wastewater infrastructures relate to the built environment and surrounding landscape patterns across a range of urbanizing environments. This paper addresses this first step by evaluating patterns of wastewater infrastructures and discusses the interacting socio-economic and bio-physical factors that influence their geographic location, patch attributes and the urban landscape patterns associated with different systems across an urbanizing coastal region.

To assess the spatial patterns of wastewater infrastructures, quantitative spatial analysis methods are needed. Among several approaches, an urban-rural gradient along with landscape analysis is useful for studying changes in patterns across urbanizing landscapes [24,25]. Researchers have used urban-rural gradients to represent a continuum of urban intensity and to study the ecological effects of urbanization [25–27]. Urbanization gradients also serve as a framework for assessing interactions among landscape patterns, built infrastructure and ecosystem processes [28]. In this study, I employ a coastal urban gradient combined with landscape pattern metrics to characterize spatial patterns of alternative wastewater infrastructures across watersheds in the Puget Sound, WA, USA. I measure spatial patterns at the patch scale by extracting parcel-level characteristics regarding sanitation system, lot size and estimating parcel density and impervious surfaces for watersheds dominated by different wastewater systems. I investigate two wastewater treatment systems, sewers and septic systems, and address three research questions related to their location, patch attributes and associated landscape pattern: (1) How does the proportion and density of parcels on wastewater treatment systems vary across an urban gradient? (2) How does the lot size of parcels on different treatment systems change across the gradient? (3) How does the wastewater treatment relate to patterns of imperviousness? Identifying spatial patterns of wastewater infrastructures is an important step to understand how alternative development patterns relate to wastewater treatment. Understanding how wastewater infrastructures are distributed and relate to surrounding development across a coastal region is also critical to provide planners and decision makers with the necessary information for managing land use and making wastewater choices in line with regional water quality goals.

2. Patterns of Wastewater Infrastructure

Scholars of ecology and urban planning have called on researchers to disentangle the role of wastewater infrastructure in the link between spatial patterns of urbanization and ecological processes [28,29]. Cities are highly heterogeneous landscapes made up of a range of development types and a mosaic of built-up and natural patches [26,30]. Alternative patterns of development contain different land uses with diverse arrangements of vegetation, impermeable surfaces and other land covers, with parcels that vary in size, shape and density [25,31]. Infrastructure is a defining feature of the urban landscape. Roads, drainage networks, sewers and septic systems introduce new geometries and land cover surfaces into the urban spatial structure [26,32,33] with consequences for hydrologic flows, nutrient cycling and aquatic ecosystems [16,17,29,34]. While it is clear that the degree of human impact on the landscape depends on the level and pattern of urban development, it is not clear what role wastewater infrastructure plays in mediating the impacts of urbanization on ecosystem functioning.

Few studies have specifically addressed development patterns associated with different wastewater infrastructures across an urban gradient. Some studies have examined the geographic distribution of different wastewater treatment systems in particular areas [32,35]. Harrison *et al.* (2012) [36] recorded the location and timing of residential development on sewers *vs.* septic tanks in the counties of the Baltimore region. LaGro (1998) [32] mapped the shape and size of residential parcels on septic tanks in rural Ozaukee County, Wisconsin. Hatt *et al.* (2004) [17] plotted the density of septic systems in suburban watersheds with a range of development using impervious surfaces and found a non-linear relationship. Sowah *et al.* (2014) [21] recorded the amount of impervious surfaces, forest, developed and agricultural land in suburban watersheds with high *vs.* low septic density. No studies, to the author's knowledge, have compared patterns associated with sewers and septic tanks across a range of urbanizing environments. In this study, I posit that spatial patterns of parcels with different wastewater treatment systems vary across a gradient of urbanization in complex ways. For instance, the proportion of sewered parcels may increase linearly with urbanization, while the opposite may not hold true for parcels on septic tanks. I expect the variation in lot size across an urban-rural gradient to display unique patterns depending on the wastewater treatment system. I hypothesize that parcels on sewers are primarily in urban areas, on small lots and with high amounts of impervious surfaces. I expect septic tanks, on the other hand, to be found in rural and suburban areas with low amounts of impervious surfaces, varying lot sizes and at varying densities.

2.1. Bio-Physical and Socio-Economic Factors

Wastewater infrastructure patterns emerge from the interplay of environmental, economic, technical, socio-cultural and political dynamics at multiple scales that evolve across and within an urban area [37,38]. Bio-physical conditions are a key consideration in the installation of wastewater systems [39]. At the same time, infrastructure decisions are also influenced by economics, sanitary infrastructure policies and public health regulations [37]. I summarize from existing literature key factors focusing primarily on bio-physical constraints, cost considerations and U.S. policy that either limit or induce the installation of sewers and septic tanks at the patch or parcel scale. I link how these factors influence their locations on an urban gradient and influence spatial attributes at the patch scale, namely lot size, density and impervious surfaces. I also discuss the role of technology and social acceptability in the adoption of alternative wastewater treatment systems.

2.1.1. Bio-Physical Factors

Geology, topography and the water table are key constraints to the location, siting and installation of sanitary systems [11]. Local environmental conditions control the location and suitability of septic tanks. Most important are soil permeability and depth to water table [40]. Sites characterized by a high water table, shallow bedrock, too slow or too porous soils are considered unsuitable for septic systems. In areas with marginal soil conditions, larger lots are necessary for adequate drainage and the treatment of nutrients and pathogens. Topography, water availability and the presence of waterways also influences the siting and physical form of both septic tanks and sewer networks.

2.1.2. Sanitary and Infrastructure Policies

Sanitation policies regulate all onsite wastewater systems, including septic tanks, with the primary objective of protecting water quality and public health. Health departments encode siting requirements that include prescriptive rules based on soil permeability, lot size and distance from water sources. Health codes generally require enough space for a drain-field and effective effluent treatment. In the U.S., some have argued that these sanitary codes act as a *de facto* growth control by communities that either exclude entire areas as developable lands or by using them for large-lot zoning [41–43]. As a result of these health codes and underlying environmental conditions, septic tanks are commonly perceived to primarily serve scattered rural or low-density areas. Alternative technologies with more advanced treatment capabilities may facilitate the installation of septic tanks on more marginally-suitable lands and at higher densities [35].

Sanitary infrastructure policies aim to direct development and the extension of sewers. Different states and regions in the United States adopt policies that function as implementation and phasing mechanisms that aim to promote development in designated areas and to avoid haphazard sewer extensions [44]. These policies either mandate it through the use of urban growth boundaries or incentivize it by targeting public spending for sewers within designated urban areas. For example, local jurisdictions in the United States have employed adequate public facilities ordinances (APFO) designed to link urban development to the extension of a public sewer [44,45]. At the state level, Washington adopted the Growth Management Act (GMA) in 1990, which requires county governments to work with cities to direct urban growth by requiring adequate infrastructure [45,46]. Similarly, Maryland approved its Priority Funding Area (PFA) program in 1997 to encourage local governments to focus all new development within specific PFA areas. The effectiveness of these policies, however, is uncertain. In a recent study of five counties surrounding Baltimore, Maryland, Harrison *et al.* (2012) [36] found that since the passage of the law, a significant proportion of new development with septic tanks occurred inside PFAs.

2.1.3. Economic Factors

Cost is also a major constraint [47]. Cost constrains the location of sewers and the conversion of old septic systems to sewers. The planning literature indicates that sewer infrastructure costs are sensitive to development patterns [48–51]. Sewer systems are more economically efficient and achieve higher economies of scale in compact urban areas with small lots. Dispersed patterns with larger lots require longer sewer mains and incur higher costs to local governments and residents [51]. Cost also influences the ability for households to convert older septic systems to sewer. This constraint might explain why many older neighborhoods in the U.S. still contain legacy septic systems that were built decades ago or even earlier during the postwar era, when suburban developments proliferated [43]. In the city of Olympia, in Thurston County, WA, for example, there are nearly 2000 septic systems in the city's sewer service area, many of which were installed before public sewers were constructed [52]. According to a study by the Olympia Wastewater Utility, the estimated cost to convert a home on a septic system to a public sewer can range from $4000 to $36,000 in construction expenses alone, depending on proximity to existing sewers [52].

While these factors are important for the selection of wastewater treatment, it is also vital to recognize the significant influence of the social, cultural and political context within which treatment decisions are made [53]. Some communities reject proposals to construct sewers, even as a solution to mitigate failing septic systems, based on the concern that sewers enable high density development and spiraling utility costs [35]. At the same time, household preferences for rural and suburban living and development pressures for additional housing creates a market demand for development beyond existing sewer systems. Figure 1 presents a conceptual framework that links the interplay of socio-economic, biophysical and technological factors to the location and spatial pattern of wastewater infrastructures. This framework identifies and organizes factors that determine the location, patch attributes and distribution of wastewater treatment systems across a region. The "patch attributes of wastewater systems" specifies parcel characteristics, such as a parcel's location on an urban-rural gradient, lot size, density and amount of impervious surfaces. "Landscape patterns" refer to the surrounding urban matrix of the built environment at a neighborhood or sub-watershed scale.

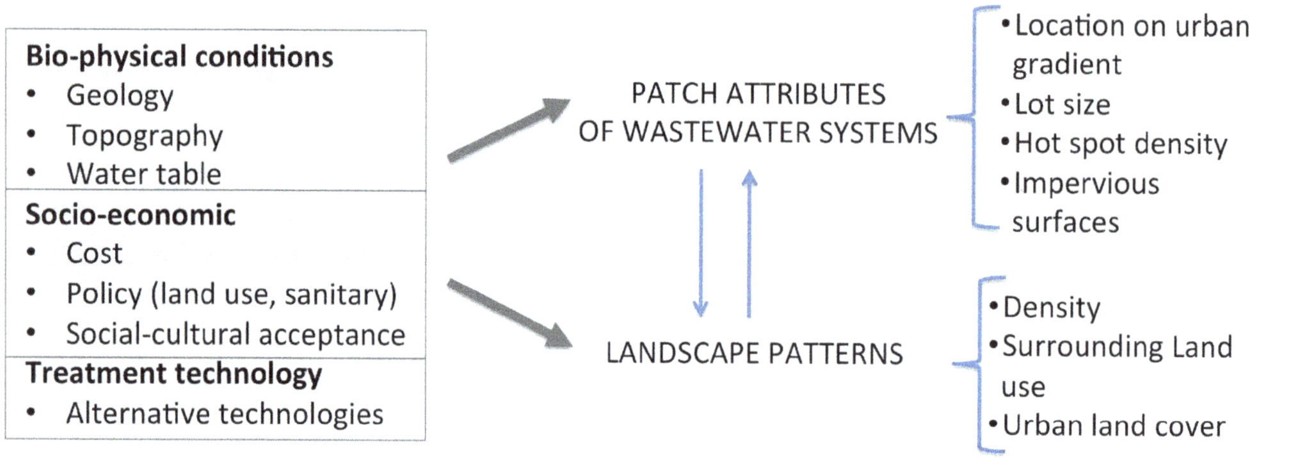

Figure 1. Conceptual framework of the regional-scale pattern of development associated with wastewater infrastructure, adapted from Spirandelli (2014) [54].

3. Materials and Methods

The approach I describe here builds on urban gradient studies [24] and previous research quantifying urban landscape patterns [25,26,55]. A commonly-used technique to characterize urban gradients has been the transect approach, which defines a linear distance from the urban core to the rural outskirt [25,56]. This approach assumes a monocentric model of the region, where a central urban core is surrounded by concentric rings of development. The Puget Sound coastal region is not made up of one major urban center, but instead of several urban centers and ports, including Seattle, Everett, Tacoma and Olympia (Figure 2). In contrast to the transect approach, I build off an approach developed by Alberti (2008), to construct a more complex urban gradient that explicitly takes into account social and natural processes driving coastal urban development. To quantify spatial patterns of wastewater infrastructures across an urban gradient, I estimate and summarize select parcel-scale spatial attributes most relevant to alternative wastewater infrastructures. Namely, I quantify the number of parcels on a wastewater system, lot size, "hot spot" septic density and percent imperviousness, and summarize these at a sub-watershed scale (Table 1). I use a coastal urban gradient to examine differences and change in patterns across a metropolitan coastal region and in relation to urban development.

3.1. Study Area and Data Description

3.1.1. Study Area

The Puget Sound metropolitan region is located in Western Washington and comprises 12 counties in total, of which this study focuses on seven with varying land uses and population density (Figure 2). The underlying geology and soil types in the Puget Sound region are mainly the result of the last glacial period, roughly 15,000 years ago [57]. The majority of watersheds in the region are dominated by poorly-drained glacial till and moderately well-drained glacial outwash soil types.

In 2010, there were just under 4 million people residing in the region, and it remains one of the fastest growing coastal area in the United States [1]. The state is ranked fourth in leading U.S. coastal population growth, with most people settling around the shores of the Puget Sound. Rapid urbanization in the region is the leading driver of land cover change. To accommodate growing populations, coastal areas have rapidly transformed into suburban and urban environments, with land-based pollution impacting the health and ecosystem function of the estuarine environment. The rate and scale of urban development and associated wastewater needs has resulted in a patchwork of wastewater treatment, which presents a unique opportunity to examine their spatial configuration across an urban-rural gradient.

In 1990, Washington State established the Growth Management Act (GMA) that directs all new development within urban growth boundaries. The GMA requires fast growing counties and their cities to develop comprehensive plans and to designate urban growth areas (UGA), with all new developments inside those boundaries to be connected to a public sewer system. However, counties can approve the installation of septic systems on either new developments or subdivisions within urban areas subject to the evaluation by the local health department [58]. Local health officers establish, for example, a minimum lot size based on soil conditions and water supply (private well *vs.* public).

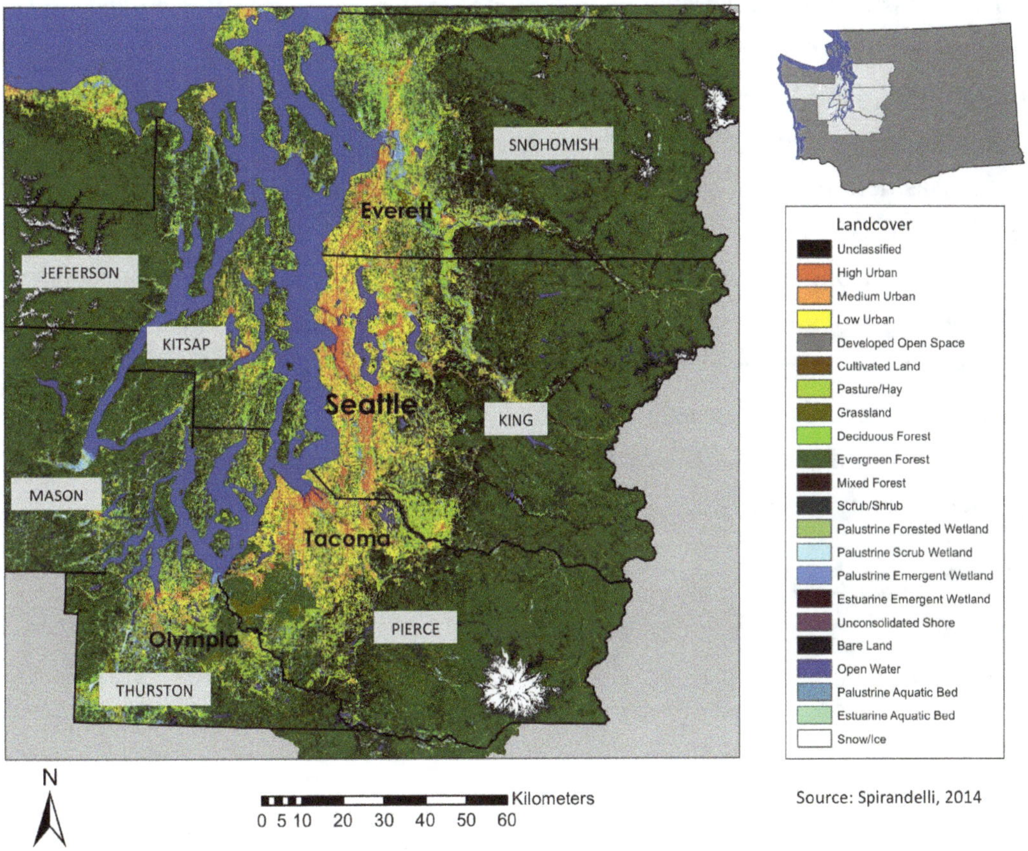

Figure 2. The study area includes seven counties of the Puget Sound, WA, estuary.

3.1.2. Sampling of Coastal Watersheds across an Urban Gradient

Coastal watersheds were selected from a spatial database developed by the Puget Sound Nearshore Ecosystem Restoration Project (PSNERP). This database contains small watersheds draining upland areas directly into the Puget Sound. The data were developed to support research on human-induced changes and ecosystem decline along Puget Sound's shoreline [59].

I stratified the watersheds by "urban", "suburban", "rural/exurban" and "wildlands" using a constructed urban-rural gradient (Figure 3). Adapting Alberti's (2008) methodology, I develop a coastal gradient that explicitly takes into account urban centers and port cities, population density and slope. I use publically-available GIS data on population from the 2000 U.S. Census, slopes derived from a digital elevation model (DEM) and location of ports and urban centers from the Puget Sound Regional Council. I randomly extracted 10,000 data points from each layer and log transformed each layer. Using principle components analysis, I re-expressed the layers as a linear combination and into an urban-rural gradient index. Eigenvectors from the first component (PC1) were used to construct the index, which retained 74% of the variation of the original data. The final gradient index was mapped and expressed as a range from 10 representing areas that are most urban to −2 representing wildland areas. The continuous gradient was subsequently partitioned into four discrete classes, "urban", "suburban", "rural" and "wildlands", using the distribution of 2006 C-CAP Washington State land cover data from NOAA's Coastal Change Analysis Program (C-CAP) [60]. I selected cut-off points for each gradient class by visually observing the distribution of land cover across the gradient index values (Figure 3). Break points were determined between urban and suburban classes based on the peak distribution of light urban land cover. I field checked land uses, such as major office complexes, with lower density populations and classified these as suburban. The break between the suburban and rural/exurban class was determined based on the distribution of pastureland and cultivated land across the two classes. The break was identified between the two classes at just below the peak of the pastureland and near the trough of the cultivated land. The break between rural and wildlands was determined based on the transition between mixed-forest cover and coniferous forest where wildlands represent a dominant alpine coniferous forest. Identifying appropriate break points between classes was an iterative process visually comparing class break locations (e.g., division between "urban" and "suburban") on the gradient map with areas of known land uses using high-resolution digital ortho-photos (1-meter resolution) [61].

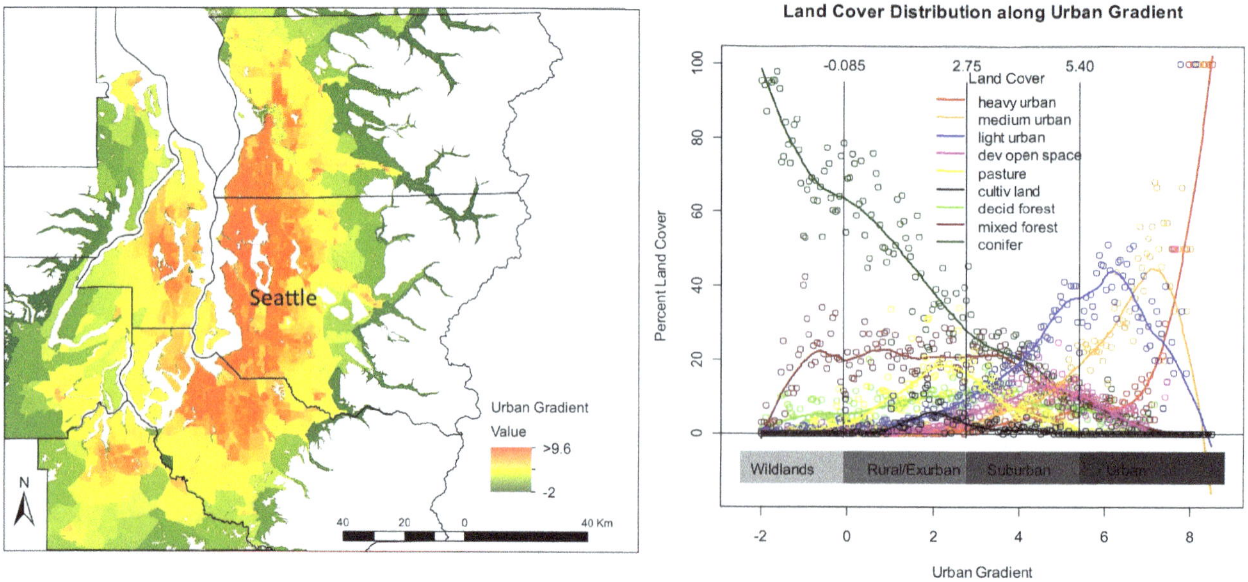

Figure 3. Map of the urban gradient with land cover distribution. Gradient index ranges from 9.6 (most urban) to −2 (wildlands). Source of map and graph: Spirandelli, 2014 [54].

All PSNERP watersheds were classified as either "urban," "suburban" or "rural" using the gradient map. Watersheds in the "wildlands" portion of the gradient were excluded from the study, since the objective of this project was to identify parcels on wastewater treatment systems in urbanizing watersheds. A choropleth map was created (Figure 4) by intersecting the PSNERP watersheds with the gradient map and classifying each watershed based on the amount of area dominated by one of the three classes. If fifty percent or more of the watershed's area was dominated by one of the gradient classes, it was labeled as such. For example, if a watershed contained 60% "urban", 25% "suburban" and 15% "rural," it was classified as an urban watershed. A stratified random sampling routine was employed to select 30 urban, 30 suburban and 30 rural watersheds [62].

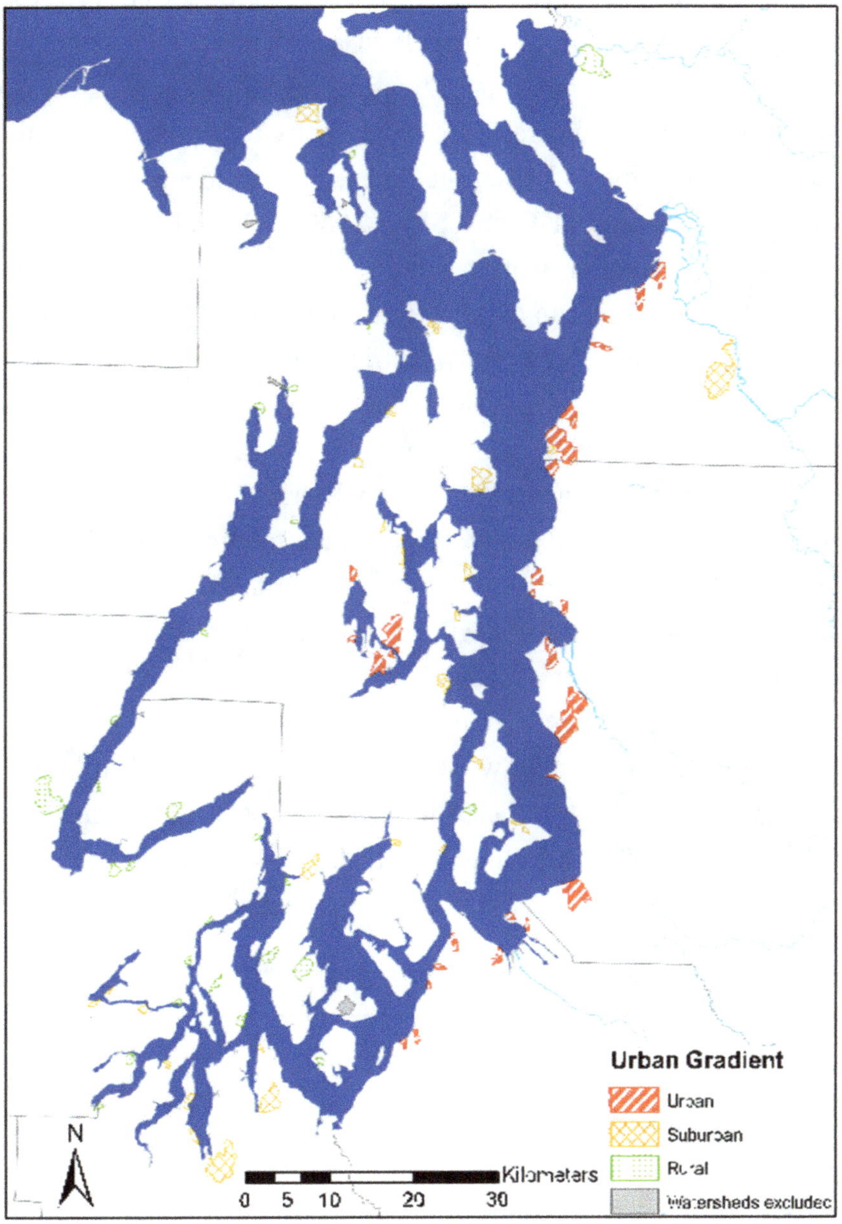

Figure 4. Location of 90 sampled coastal watersheds across the study area. Each watershed was classified as urban (30 watersheds), suburban (30 watersheds) and rural (30 watersheds). Five watersheds were removed from the sample because they lacked wastewater treatment.

3.2. Watershed Analysis

Spatial Metrics

Data records with information about parcels on a septic or hooked up to a sewer system was collected from local public agencies, including county departments of health, tax assessors, public works, utilities and city service departments. Details on data sources, location information and year collected are provided in the Supplementary Material. Wastewater treatment location data were spatially joined with the Washington State Parcel Database [63]. I calculated five spatial metrics at the parcel scale and summarized them at the sub-watershed scale (Table 1). Four measures are patch attributes of parcels on a wastewater treatment system, either septic or sewer. The first two measures are the number of parcels and the proportion of parcels in a watershed on a sewer or septic tank to determine the relative dominance of residences on a treatment system. Parcel lot was extracted from the Washington State Parcel Database. I calculated lot size because it may be a constraining factor with regard to the choice in treatment type [40,50]. The fourth metric identifies "hot spots" of septic intensity using a kernel density estimate (KDE) in ESRI's ArcGIS Spatial Analyst 9.3. The KDE uses a quadratic function to produce a smooth density surface of points over space by computing point intensity as density estimation [64]. I calculated clustered "hot spots" using a radius of 1 kilometer around points of known septic tanks. The fifth metric, percent impervious, is a measure of urbanization. The percent area covered by impervious surfaces is strongly correlated with population density and is a well-documented environmental indicator [5,6,65]. Previous studies have shown that impervious surfaces increase run-off volumes, enhance the transport of pollutants and alter bio-geochemical processes that cycle nutrients [66–68]. As such, the metric is a useful urban development pattern. I derived information on imperviousness for parcels on a wastewater treatment type using classified Landsat TM data from NOAA's C-CAP 2006 land cover data [60]. The mean percent impervious was extracted for parcel location with either a septic system or sewer.

I use several non-parametric tests to assess differences across groups. The Kruskal–Wallis rank sum test allows the assessment of group differences without assuming a normal distribution. In addition, I use ANOVA and scatterplots to examine differences in watershed patterns and check for relationships with measures of urbanization. The mean count of parcels on wastewater and the mean percent imperviousness were log transformed to account for normality assumptions.

Table 1. Spatial metrics.

Metric	Description
Mean count of sewers and septic tanks	The sum of all parcels on either a septic system, sewer system or no system
Proportion of parcels on wastewater system	Total number of parcels on a septic or sewer system divided by the sum of parcels on a wastewater type
Mean parcel-lot	Lot size of parcels on sewers and septic tanks averaged across watershed
Kernel density of parcels on a septic system	Proxy for density of septic systems, which may be a stressor on the nearshore
% Impervious	Measure of urbanization

4. Results

4.1. Coastal Watershed Patterns

Coastal watersheds in the sample are small, roughly 1.7 km^2 (420 acres), on average. Some are as small as 0.3 km^2 (74 acres) and as large as 10.8 km^2 (2668 acres). Watershed area does not vary significantly by gradient classes ($\chi^2 = 5.8$, df = 2, $p < 0.06$), meaning rural watersheds are not larger than urban or suburban, and urban watersheds are not significantly smaller than their suburban or rural counterparts. The watersheds contain a range of land uses (Figure 5). While the Puget Sound nearshore is dominated by single-family homes, the coastal shoreline also contains many areas that are undeveloped. Among the parcels identified as being on a wastewater treatment type, a more uneven distribution of land use is evident. The bar chart in Figure 5 shows that most parcels on a wastewater treatment system (87%) are residential properties. These residential lands include single-family and multi-family residences. The remaining parcels with wastewater services are on lands classified as "undeveloped", "services" (including government and educational), "trade" (including retail), "transport/utilities" and "cultural/recreation". Because the majority of parcels on a wastewater system are dominated by residential development, I extracted these parcels for the subsequent analysis.

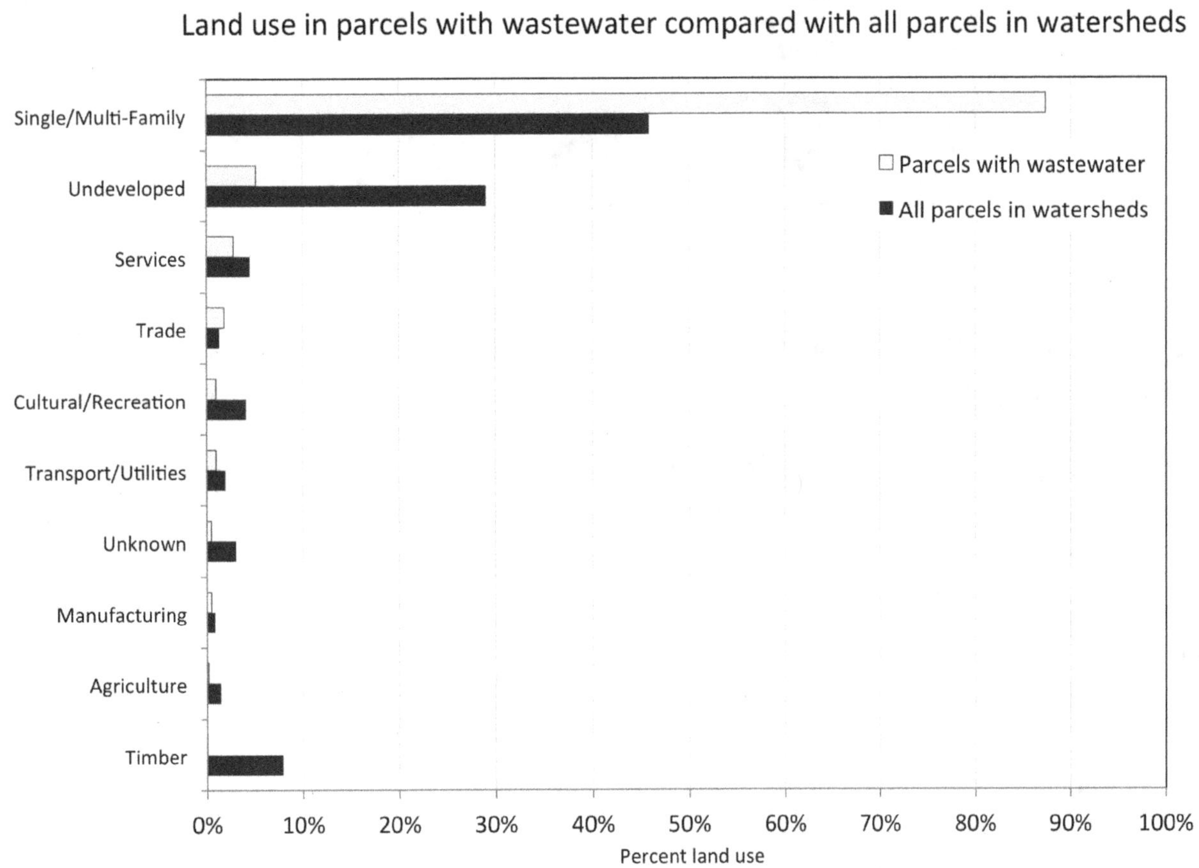

Figure 5. Proportion of land uses in parcels with wastewater treatment (septic or sewer) and the proportion of land uses in all parcels in sample watersheds. Over 87% of parcels with wastewater are single-family or multi-family residential.

4.2. Wastewater Infrastructure Patterns

Across the 85 sampled watersheds, there are a total of 44,629 residential properties either hooked up to a sewer or on a septic system (Figure 6). Most of these residences and the population (86%) are on a sewer system. Figure 6 and Table 2 show the distribution of residential properties on a wastewater treatment system across the gradient. An overwhelmingly majority of urban residences (>95%) are on a sewer system. In the suburban coastal areas, there is a 40:60 ratio of residences on a septic *vs.* sewer, respectfully. Nearly all rural properties use a septic system.

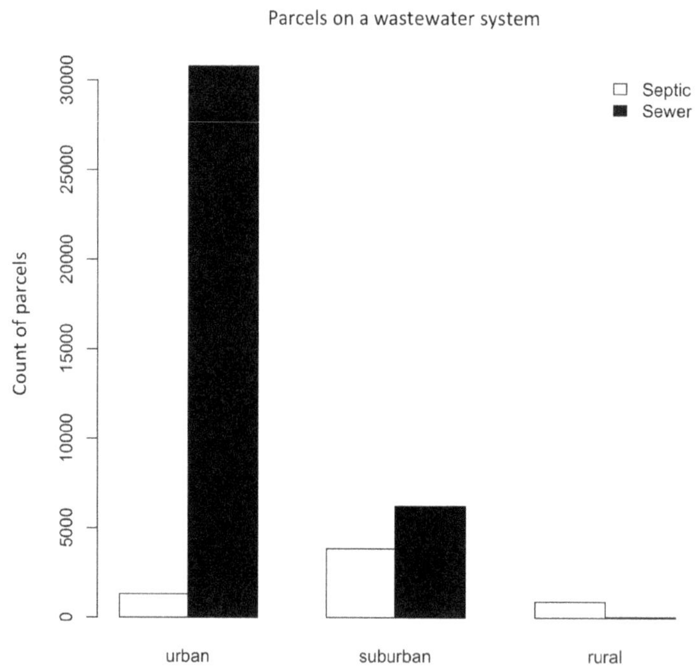

Figure 6. Number of residential parcels on a sewer system or septic system distributed across the urban gradient.

Consistent with the decline in the number of households, the total number of residential parcels on either wastewater system type decreases with the urban-rural gradient. However, when the data are partitioned between properties with a septic tank and properties on a sewer, the proportion of parcels on a sewer decreases exponentially across the gradient, where 85% of sewered parcels are urban, 15% are suburban and less than 1% are rural. By contrast, the proportion of residences relying on a septic tank is greatest in the suburban portion of the gradient (63%) followed by urban (21%) and rural (14%).

The density of parcels on a septic system displays a similar trend across the urban-rural gradient. The septic density is higher, on average, in suburban watersheds (90 ± 103 units/km^2) compared to urban (26 ± 60 units/km^2) and rural watersheds (18 ± 20.7 units/km^2). Results from a one-way ANOVA revealed that the density of septic tanks differed significantly among urban, suburban and rural watersheds, $F(2, 76) = 15.7, p = <0.001$.

The average lot size of all residential parcels increases with the urban gradient (Table 2). However, this lot size varies by wastewater treatment type. The average lot of sewered properties is 0.3 acres across all urban, suburban and rural watersheds. There is some variation in lot size among sewered urban watersheds (0.3 ± 1.21 acres) and suburban watersheds (0.3 ± 2.10 acres), but very little in rural

(0.3 ± 0.2 acres). The average lot size of properties on a septic system is smallest on average in urban basins (0.5 ± 2.8 acres) compared to suburban (1.37 ± 3.07 acres) and rural basins (2.8 ± 5.3 acres).

Table 2. Summary results of residential wastewater infrastructure patterns across a coastal urban gradient. Data are from Spirandelli, 2014 [54].

Count of parcels	**Urban**	**Suburban**	**Rural**	**Total (*n*)**
Septic (*n*)	1323	3845	889	6057
Percent of total	(22%)	(63%)	(14%)	(14%)
Sewer (*n*)	32,305	6220	47	38,572
Percent of total	(85%)	(15%)	(0.05%)	(86%)
All parcels on wastewater (*n*)	33,628	10,065	936	44,629
Population count	Urban [1]	Suburban [2]	Rural [2]	Total
Population on septic	3056	9497	2195	14,748
	(21%)	(64%)	(15%)	(14%)
Population on sewer	74,624	15,363	116	90,103
	(83%)	(17%)	(.001%)	(86%)
Population using wastewater system	77,680	24,860	2311	104,851
Lot size (mean acre)	**Urban**	**Suburban**	**Rural**	
Septic	0.5 ± 2.8	1.37 ± 3.07	2.8 ± 5.3	
Sewer	0.3 ± 1.21	0.32 ± 2.10	0.3 ± 0.2	
All residential parcels	0.32 ± 1.5	0.94 ± 4.22	4.07 ± 9.7	
Septic density (units/km^2)	**Urban**	**Suburban**	**Rural**	
Mean density	26 ± 60	90 ± 103	18 ± 20	
Mean impervious (%)	**Urban**	**Suburban**	**Rural**	**Mean %**
Septic dominated				8 ± 14
Sewer dominated				34.5 ± 16
All watersheds	38 ± 16	13 ± 16	5 ± 13	

[1] Population data were estimated using households and persons per household data collected by the U.S. Census American Community Survey, 2013. An average of 2.31 people per household was estimated from the four largest cities in the Puget Sound (Seattle, Olympia, Tacoma and Everett) and used as a multiplier for the urban parcels. [2] An average of 2.47 people per household was estimated from the seven counties in the study area and used as a multiplier for suburban and rural parcels.

A wide range of impervious surfaces (0%–70%) was observed across all watersheds (Figure 7a). Urban watersheds contain, on average, just over 38% (±16%) of their area covered by impervious surfaces. Suburban watersheds are more moderately developed with an average of 13% (±16%). Rural watersheds contain minimal imperviousness, ranging mostly in the single digits. At the parcel scale, sewered parcels contain an average of over 20% impervious surface coverage regardless if they are in urban, suburban or rural watersheds (Figure 7b). Septic parcels in urban watersheds also contain high amounts of imperviousness, in the range of 23%–45% (Figure 7c). By contrast, septic parcels in suburban watersheds contain, on average, 10% imperviousness. Mean percent imperviousness was averaged across watersheds dominated by septic tanks *vs.* sewers. Sewer-dominated watersheds contain, on average, 34.5% (±16%) and septic dominated watersheds contain 8% (±14%) impervious surfaces. Among suburban watersheds, sewer *vs.* septic dominated watersheds contained different amounts of imperviousness (21% ± 9 *vs.* 13% ± 14). A positive, although weak, relationship was observed between the log of percent impervious and the log of all parcels on a wastewater treatment type ($r^2 = 0.34$, $p < 0.001$). When the data are partitioned across the urban gradient, this relationship is stronger among suburban residential properties ($r^2 = 0.56$, $p < 0.001$). This suggests that in suburban watersheds, as the number of residential parcels on wastewater treatment (sewer or septic) increases, so does the amount of impervious surfaces. When the data are partitioned by septic *vs.* sewer dominated among suburban watersheds, a positive relationship was observed between the number of septic tanks and impervious surfaces ($r^2 = 0.41$, $p < 0.001$).

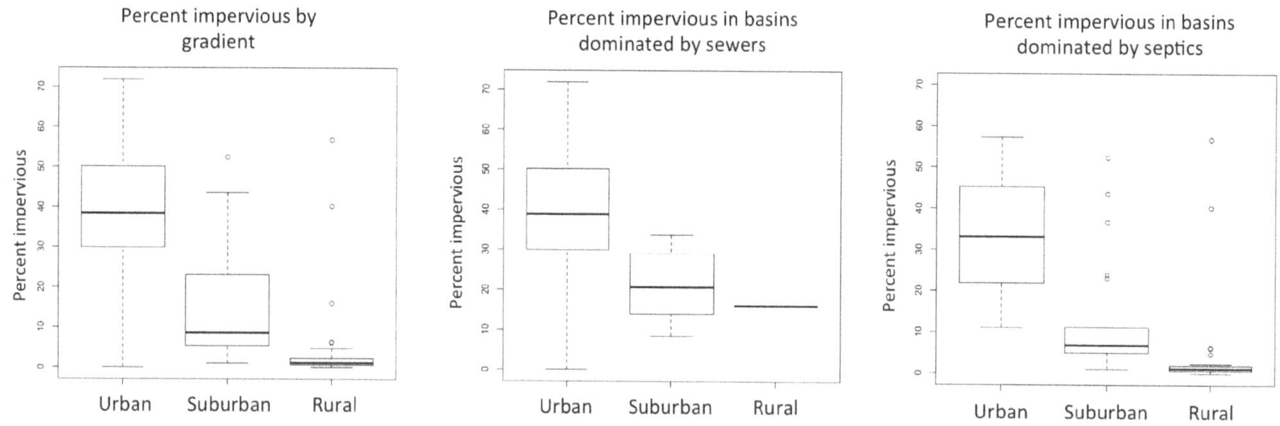

Figure 7. (**a**) Percent impervious across all watersheds; (**b**) percent impervious in sewered parcels; (**c**) percent impervious in septic parcels. Source of graphs: Spirandelli, 2014 [54].

5. Discussion

5.1. Proportion and Density of Wastewater Treatment across an Urban Gradient

It has been widely believed that sewers primarily serve urban areas and septic tanks are most prevalent in scattered rural developments [40]. Findings from this study find a more nuanced pattern that such generalizations do not necessarily hold true. I find most urban residences hooked up to a central sewer system; however, there are a remaining, albeit small, number of residential parcels on a septic system in the most heavily-urbanized watersheds. There might be two explanations for this finding. The first is that these are legacy septic tanks installed during a period before sewers were built. Converting these systems has been slow, partly due to the phenomenon of "path dependence": once a technology is chosen, it is very hard to change, even when it is out of date [37]. The second explanation is that these parcels were developed with septic tanks in urban areas despite existing sewer infrastructure.

The pattern of sewage treatment in suburban watersheds is more complex, representing a patchwork of parcels on sewers and septic tanks. Heimlich and Anderson (2001) state that suburban and rural development represent two fundamentally different types of development based on wastewater treatment: suburbia relies on access to sewer and comprises small residential lots; while rural development relies on septic systems with larger lots [69]. Findings from this study do not support this view. While sewers still dominate in total numbers, the relative number of septic tanks and their density is significantly higher in the suburban portion of the urban gradient. The suburban development pattern in these watersheds contains a mixture of sewers and septic tanks, but with local neighborhoods containing very high clusters of septic tanks, on the order of 6–8 septic tanks per acre.

The high density of septic tanks in suburban areas may be the result of rapid development and increased demand for housing, either in spite of or in response to urban containment policies, such as the Washington Growth Management Act (GMA). Between the period of 1990–2002, the rate of urban development was significantly higher in areas outside of urban growth boundaries than inside across six of the same counties in this study [70]. Whether septic tanks or sewers facilitated this increased rate would require additional information on the year when these different systems were installed. Whatever the role of the Growth Management Act might be, septic tanks may also be a preferred option for developers inside urban growth areas. In Baltimore, for example, since the passage of Maryland's smart growth legislation, development in urban areas that use septic tanks has grown significantly and, in some cases, inside the priority sewer funding area [36]. Harrison *et al.* (2012) [36] explains this may be because the local county government cannot afford sewer expansion and state subsidies are lacking, and/or there is a failure in adequate infrastructure planning.

5.2. Lot Size of Parcels with Different Wastewater Treatment

Previous studies have observed lot size to decrease with urbanization, from the rural outskirts to the urban core [25,26]; however, lot size was not discriminated by wastewater treatment system. This study finds residential lots vary in size depending on the wastewater treatment system. A surprising result was the relatively small lot size of urban septic tanks, suggesting that there may be more available area to build with septic tanks than previously thought.

5.3. Wastewater Treatment and Patterns of Imperviousness

Previous landscape studies have shown that impervious surfaces and the amount of vegetation varies widely in residential patches and that urban density and the location on the urban gradient influences the distribution [25,26,71] This study finds a significant difference in the amount of impervious surfaces in watersheds dominated by different wastewater treatment systems. At the patch scale, urban lots on septic tanks contain high amounts of impervious surfaces (>35%), an amount that at the watershed scale can negatively impact coastal ecosystem functions [72,73]. This is of concern, since septic tanks require adequate unpaved area in the yard for a leach-field that can support the efficient processing of effluent. At the watershed level, the suburban watersheds display two distinct patterns. On the one hand, watersheds dominated by septic tanks contain very high densities of septic tanks, but with low amounts of impervious surfaces. By contrast, sewer-dominated watersheds contain high amounts of impervious surfaces.

6. Conclusions

The formulation of septic regulations is generally confined to the narrow task of site-scale controls based on bio-physical characteristics and concerns for human health and water quality. As such, these policies fail to address the wider impacts of development patterns associated with septic tanks within and outside areas served by centralized sewer systems. In contrast, planning departments and public agencies develop sewer infrastructure policies that serve as implementation mechanisms for land use planning [44]. Although these policies are designed to direct development and discourage haphazard development, they also fail to address the proliferation of residences with septic systems in suburbia at very high densities.

If an important challenge for ecological scholars and urban planners is to figure out how best to balance human needs with ecosystem function in urban ecosystems, then understanding the role of alternative wastewater infrastructures may be an important component. The link between wastewater treatment systems and urban patterns has reinvigorated the attention of recent researchers and policy makers who view septic systems as encouraging development on the urban fringe in the form of sprawl, particularly in the wake of growth management policies [70,74].

This study provides an important step in the examination of urban development patterns associated with sewers and septic tanks at the patch and watershed scale. Yet, much remains to be explored. For example, more research is needed to understand the role that land use policies such as the Growth Management Act play in the distribution and pattern of wastewater treatment across an urban gradient. Additional research is needed to explore the relationship between sewage treatment types and patterns of land use and land cover; for instance, whether the fragmentation of forest and other patterns of sprawl-like development are associated more with septic tanks *vs.* sewers. Furthermore, an explicit exploration of the relationship between patterns of wastewater infrastructure and watershed ecosystem function is necessary. The pattern of urban development and its associated wastewater infrastructures may be key to devise better strategies to ensure that ecological services support human needs and coastal ecosystems.

Acknowledgments

The research presented in this paper was partially funded by the United States National Oceanic and Atmospheric Administration (NOAA) and the Joint Institute for the Study of the Atmosphere and Ocean (JISAO).

Conflicts of Interest

The author declares no conflict of interest.

References and Notes

1. Crosset, K.; Culliton, T.; Wiley, P.; Goodspeed, T. *Population Trends along the Coastal United States: 1980–2008*; Coastal Trends Report Series; NOAA's National Ocean Service Management and Budget Office: Silver Spring, MD, USA, 2005.
2. Small, C.; Nicholls, R.J. A global analysis of human settlement in Coastal Zones. *J. Coast. Res.* **2003**, *19*, 584–599.
3. Islam, M.S.; Tanaka, M. Impacts of pollution on coastal and marine ecosystems including coastal and marine fisheries and approach for management: a review and synthesis. *Mar. Pollut. Bull.* **2004**, *48*, 624–649.
4. Burton, G.; Pitt, R. *Manual for Evaluating Stormwater Runoff Effects in Receiving Waters*; Lewis Publishers: Boca Raton, FL, USA, 2001.
5. Arnold, C.L.; Gibbons, C.J. Impervious surface coverage. *J. Am. Plan. Assoc.* **1996**, *62*, 243–258.

6. Leopold, L. *Hydrology for Urban Land Planning: A Guidebook on the Hydrologic Effects of Urban Land Use*; U.S. Geolgoical Survey: Washington, DC, USA, 1968.
7. Wear, S.L.; Thurber, R.V. Sewage pollution: Mitigation is key for coral reef stewardship. *Ann. N.Y. Acad. Sci.* **2015**, *1355*, 15–30.
8. Halpern, B.S.; Walbridge, S.; Selkoe, K.A.; Kappel, C.V.; Micheli, F.; D'Agrosa, C.; Bruno, J.F.; Casey, K.S.; Ebert, C.; Fox, H.E.; *et al.* A global map of human impact on marine ecosystems. *Science* **2008**, *319*, 948–952.
9. Stewart, J.; Gast, R.; Fujioka, R.; Solo-Gabriele, H.; Meschke, J.S.; Amaral-Zettler, L.; del Castillo, E.; Polz, M.; Collier, T.; Strom, M.; *et al.* The coastal environment and human health: Microbial indicators, pathogens, sentinels and reservoirs. *Environ. Health* **2008**, *7* (Suppl. 2), S3.
10. Shuval, H. Estimating the global burden of thalassogenic diseases: Human infectious diseases caused by wastewater pollution of the marine environment. *J. Water Health* **2003**, *1*, 53–64.
11. Tchobanoglous, G.; Burton, F.L.; Stensel, H.D. *Wastewater Engineering: Treatment and Reuse*, 3rd ed.; Metcalf and Eddy Inc.: Boston, MA, USA, 2003.
12. US EPA. *Onsite Wastewater Treatment Systems Manual*; EPA/625/R-00/008; Office of Water and Office of Research and Development: Washington, DC, USA, 2002.
13. National Research Council (N.R.C.). *Managing Wastewater in Coastal Urban Areas*; National Academy Press: Washington, DC, USA, 1993.
14. Mallin, M.A. Wading in waste. *Sci. Am.* **2006**, *294*, 52–59.
15. National Research Council. *Clean Coastal Waters: Understanding and Reducing the Effects of Nutrient Pollution*; National Academies Press: Washington, DC, USA, 2000.
16. Carey, R.O.; Hochmuth, G.J.; Martinez, C.J.; Boyer, T.H.; Dukes, M.D.; Toor, G.S.; Cisar, J.L. Evaluating nutrient impacts in urban watersheds: Challenges and research opportunities. *Environ. Pollut.* **2013**, *173*, 138–149.
17. Hatt, B.; Fletcher, T.; Walsh, C.; Taylor, S.L. The influence of urban density and drainage infrastructure on the concentrations and loads of pollutants in small streams. *Environ. Manag.* **2004**, *34*, 112–124.
18. Lipp, E.K.; Kurz, R.; Vincent, R.; Rodriguez-Palacios, C.; Farrah, S.R.; Rose, J.B. The effects of seasonal variability and weather on fecal pollution and enteric pathogens in a subtropical estuary. *Estuaries* **2001**, *24*, 266–276.
19. Cahoon, L.B.; Hales, J.C.; Carey, E.S.; Loucaides, S.; Rowland, K.R.; Nearhoof, J.E. Shellfishing closures in southwest Brunswick County, North Carolina: Septic tanks *vs.* storm-water runoff as fecal coliform sources. *J. Coast. Res.* **2006**, *222*, 319–327.
20. Kelsey, H.; Porter, D.E.; Scott, G.; Neet, M.; White, D. Using geographic information systems and regression analysis to evaluate relationships between land use and fecal coliform bacterial pollution. *J. Exp. Mar. Biol. Ecol.* **2004**, *298*, 197–209.
21. Sowah, R.; Zhang, H.; Radcliffe, D.; Bauske, E.; Habteselassie, M.Y. Evaluating the influence of septic systems and watershed characteristics on stream faecal pollution in suburban watersheds in Georgia, USA. *J. Appl. Microbiol.* **2014**, *117*, 1500–1512.
22. Young, K.; Thackston, E. Housing density and bacterial loading in urban streams. *J. Environ. Eng.* **1999**, *125*, 1177.

23. Walsh, C. The urban stream syndrome: Current knowledge and the search for a cure. *J. N. Am. Benthol. Soc.* **2005**, *24*, 706.
24. McDonnell, M.; Hahs, A. The use of gradient analysis studies in advancing our understanding of the ecology of urbanizing landscapes: current status and future directions. *Landsc. Ecol.* **2008**, *23*, 1143–1155.
25. Luck, M.; Wu, J. A gradient analysis of urban landscape pattern: a case study from the Phoenix metropolitan region, Arizona, USA. *Landsc. Ecol.* **2002**, *17*, 327–339.
26. Alberti, M. The effects of urban patterns on ecosystem function. *Int. Reg. Sci. Rev.* **2005**, *28*, 168–192.
27. McDonnell, M.J.; Pickett, S.T.A. Ecosystem structure and function along urban-rural gradients: an unexploited opportunity for ecology. *Ecology* **1990**, *71*, 1232.
28. Alberti, M. *Advances in Urban Ecology: Integrating Humans and Ecological Processes in Urban Ecosystems*; Springer: New York, NY, USA, 2008.
29. Pickett, S.; Cadenasso, M.; Grove, J.M.; Groffman, P.; Band, L.; Boone, C.; Burch, W.; Grimmond, S.; Hom, J.; Jenkins, J.; *et al.* Beyond urban legends: An emerging framework of urban ecology, as illustrated by the Baltimore Ecosystem Study. *BioScience* **2008**, *58*, 139.
30. Pickett, S.T.A.; Cadenasso, M.L.; Grove, J.M.; Nilon, C.H.; Pouyat, R.V.; Zipperer, W.C.; Costanza, R. Urban ecological systems: Linking terrestrial ecological, physical, and socioeconomic components of metropolitan areas. *Ann. Rev. Ecol. Syst.* **2001**, *32*, 127–157.
31. Grimm, N.B.; Foster, D.; Groffman, P.; Grove, J.M.; Hopkinson, C.S.; Nadelhoffer, K.J.; Pataki, D.E.; Peters, D.P. The changing landscape: Ecosystem responses to urbanization and pollution across climatic and societal gradients. *Front. Ecol. Environ.* **2008**, *6*, 264–272.
32. LaGro, J., Jr. Landscape context of rural residential development in southeastern Wisconsin (USA). *Landsc. Ecol.* **1998**, *13*, 65–77.
33. Walsh, C.J.; Papas, P.J.; Crowther, D.; Sim, P.T.; Yoo, J. Stormwater drainage pipes as a threat to a stream-dwelling amphipod of conservation significance, Austrogammarus australis, in southeastern Australia. *Biodivers. Conserv.* **2004**, *13*, 781–793.
34. Alberti, M.; Booth, D.; Hill, K.; Coburn, B.; Avolio, C.; Coe, S.; Spirandelli, D. The impact of urban patterns on aquatic ecosystems: An empirical analysis in Puget lowland sub-basins. *Landsc. Urban Plan.* **2007**, *80*, 345–361.
35. Curtis, K.J. Growth-liberating technology meets fragmented regulation: Decentralized wastewater systems in Wilson County, Tennessee. *Geogr. Bull.* **2008**, *49*, 3–17.
36. Harrison, M.; Stanwyck, E.; Beckingham, B.; Starry, O.; Hanlon, B.; Newcomer, J. Smart growth and the septic tank: Wastewater treatment and growth management in the Baltimore region. *Land Use Policy* **2012**, *29*, 483–492.
37. Melosi, M.V. *The Sanitary City: Urban Infrastructure in America from Colonial Times to the Present*; Johns Hopkins University Press: Baltimore, MD, USA, 2000.
38. Healey, P. *Urban Complexity and Spatial Strategies: Towards a Relational Planning for our Times*; Routledge: New York, NY, USA, 2007.
39. Marsh, W.M. *Landscape Planning: Environmental Applications*, 5th ed.; Wiley: Hoboken, NJ, USA, 2010.
40. Berke, P.R.; Godschalk, D.R.; Kaiser, E.J. *Urban Land Use Planning*; University of Illinois Press: Chicago, IL, USA, 2006.

41. Popper, F.J. Unshaping growth: The potential land-use implications of the sand filter and related alternative wastewater treatment technologies. *Am. Land Forum* **1980**, *2*, 8–14.
42. Jacobs, H.M.; Hanson, M.E. Land use impacts of private sewage systems in Wisconsin. *J. Soil Water Conserv.* **1989**, *44*, 149–152.
43. Rome, A. *The Bulldozer in the Countryside*; Cambridge University Press: New York, NY, USA, 2001.
44. Pendall, R.; Martin, J.; Fulton, W. *Holding the Line - Urban Containment in the United States*; Brookings Institute: Washington, DC, USA, 2002.
45. Bengston, D.N.; Fletcher, J.O.; Nelson, K.C. Public policies for managing urban growth and protecting open space: policy instruments and lessons learned in the United States. *Landsc. Urban Plan.* **2004**, *69*, 271–286.
46. Growth Management Act. *Revised Code of Washington (RCW)*, Title 36, Chapter 36.70A, 1990.
47. Miles, M.; Berens, G.; Eppli, M.; Weiss, M. *Real Estate Development : Principles and Process*, 4th ed.; Urban Land Institute: Washington, DC, USA, 2007.
48. Real Estate Research Corporation. *The Costs of Sprawl : Environmental and Economic Costs of Alternative Residential Development Patterns at the Urban Fringe : Prepared for the Council on Environmental Quality, The Office of Policy*; U.S. Govt. Print. Off.: Washington, DC, USA, 1974.
49. Windsor, D. A critique of the costs of sprawl. *J. Am. Plan. Assoc.* **1979**, *45*, 279–292.
50. Speir, C.; Stephenson, K. Does sprawl cost us all?: Isolating the effects of housing patterns on public water and sewer costs. *J. Am. Plan. Assoc.* **2002**, *68*, 56–70.
51. Burchell, R. *Sprawl Costs : Economic Impacts of Unchecked Development*; Island Press: Washington, DC, USA, 2005.
52. *Wastewater Management Plan*; City of Olympia: Olympia, WA, USA, 2010.
53. Balkema, A.J.; Preisig, H.A.; Otterpohl, R.; Lambert, F.J.D. Indicators for the sustainability assessment of wastewater treatment systems. *Urban Water* **2002**, *4*, 153–161.
54. Spirandelli, D. Examining the Effects of Wastewater Infrastructure on Puget Sound Near-Shore Water Quality across a Gradient of Urbanization. Ph.D. Thesis, University of Washington, Seattle, WA, USA, 2014.
55. Wu, J. Urban ecology and sustainability: The state-of-the-science and future directions. *Landsc. Urban Plan.* **2014**, *125*, 209–221.
56. McDonnell, M.; Pickett, S.; Groffman, P.; Bohlen, P.; Pouyat, R.; Zipperer, W.C.; Parmelee, R.; Carre, M. Ecosystem processes along an urban-to-rural gradient. *Urban Ecosyst.* **1997**, *1*, 21–36.
57. Kruckeberg, A. *The Natural History of Puget Sound Country*; Weyerhaeuser Environmental Books; University of Washington Press: Seattle, WA, USA, 1995.
58. Plats - Subdivisions - Dedications. *Revised Code of Washington (RCW)*, Title 58, Chapter 58.17, Section 010, 1974.
59. Puget Sound Nearshore Ecosystem Restoration Project. PSNERP Change Analysis Geodatabase, Version 3.0, 2009. Available online: http://www.pugetsoundnearshore.org/index.html (accessed on 9 February 2009).
60. National Oceanic and Atmospheric Administration. *The Coastal Change Analysis Program regional land cover (C-CAP)*; National Oceanic and Atmospheric Administration: Silver Springs, MD, USA, 2006. Available online: https://coast.noaa.gov/dataregistry/search/collection/info/ccapregional (accessed on 9 February 2009).

61. United States Department of Agriculture. National Agricultural Imagery Program, National Agriculture Statistics Service, 2006. Available online: http://rocky2.ess.washington.edu/data/raster/naip2006/index.html (accessed on 10 January 2010).
62 Cochran, W. *Sampling Techniques*, 3rd ed.; John Wiley & Sons: New York, NY, USA, 1977.
63. Rogers, L.; Cooke, A.; McLaughlin, M. The Washington State Parcel Database, Version 2009n, Edition 9.2, Release 1.1, 2010.
64. Silverman, B.W. *Density Estimation for Statistics and Data Analysis*, Monographs on Statistics and Applied Probability, Chapman and Hall: London, UK, 1986.
65. Brabec, E.; Schulte, S.; Richards, P. Impervious surfaces and water quality: A review of current literature and its implications for watershed planning. *J. Plan. Lit.* **2002**, *16*, 499–514.
66. Wollheim, W.; Pellerin, B.; Vorosmarty, C.; Hopkins, C. N Retention in urbanizing headwater catchments. *Ecosystems* **2005**, *8*, 871–884.
67. Groffman, P.M.; Law, N.L.; Belt, K.T.; Band, L.E.; Fisher, G.T. Nitrogen fluxes and retention in urban watershed ecosystems. *Ecosystems* **2004**, *7*, 393–403.
68. Paul, M.; Meyer, J.L. Streams in the urban landscape. *Ann. Rev. Ecol. Syst.* **2001**, *32*, 333.
69. Heimlich, R.; Anderson, W. *Development at the Urban Fringe and Beyond: Impacts on Agriculture and Rural Land*; Agricultural Economic Report 803; Economic Research Service, U.S. Department of Agriculture: Washington, DC, USA, 2001.
70. Hepinstall-Cymerman, J.; Coe, S.; Hutyra, L.R. Urban growth patterns and growth management boundaries in the Central Puget Sound, Washington, 1986–2007. *Urban Ecosyst.* **2013**, *16*, 109–129.
71. Kim, J.; Zhou, X. Landscape structure, zoning ordinance, and topography in hillside residential neighborhoods: A case study of Morgantown, WV. *Landsc. Urban Plan.* **2012**, *108*, 28–38.
72. Mallin, M.; Johnson, V.; Ensign, S. Comparative impacts of stormwater runoff on water quality of an urban, a suburban, and a rural stream. *Environ. Monit. Assess.* **2008**, *159*, 475–491.
73. Holland, A.F.; Sanger, D.M.; Gawle, C.P.; Lerberg, S.B.; Santiago, M.S.; Riekerk, G.H.; Zimmerman, L.E.; Scott, G.I. Linkages between tidal creek ecosystems and the landscape and demographic attributes of their watersheds. *J. Exp. Mar. Biol. Ecol.* **2004**, *298*, 151–178.
74. Newburn, D.; Berck, P. Exurban development. *J. Environ. Econ. Manag.* **2011**, *62*, 323–336.

12

"Nothing Is Like It Was Before": The Dynamics between Land-Use and Land-Cover, and Livelihood Strategies in the Northern Vietnam Borderlands

Sarah Turner [1,*] and Thi-Thanh-Hiên Pham [2]

[1] Department of Geography, McGill University, 805 rue Sherbrooke Ouest, Montréal, QC H3A 0B9, Canada

[2] Département D'études Urbaines et Touristiques, Université du Québec à Montréal, 315 rue Sainte-Catherine Est, Montréal, QC H2X 3X2, Canada; E-Mail: pham.thi_thanh_hien@uqam.ca

* Author to whom correspondence should be addressed; E-Mail: sarah.turner@mcgill.ca

Academic Editors: Claudia A. Radel and Jacqueline M. Vadjunec

---

Abstract: Land uses are changing rapidly in Vietnam's upland northern borderlands. Regional development platforms such as the Greater Mekong Subregion, state-propelled market integration and reforestation programs, and lowland entrepreneurs and migrants are all impacting this frontier landscape. Drawing on a mixed methods approach using remote sensing data from 2000 to 2009 and ethnographic fieldwork, we examine how land-use and land-cover change (LULCC) has occurred across three borderland provinces—Lai Châu, Lào Cai and Hà Giang—with high proportions of ethnic minority semi-subsistence farmers. After a broad examination of regional land-use changes, we select three case studies to further analyze the underlying relationships between specific LULCC and local livelihood diversification strategies. These include specific patterns of urban growth due to a range of political decisions in Lai Châu and Lào Cai Provinces; reforestation due to non-timber forest (NTFP) product cultivation in the west of Lào Cai Province; and a stable landscape that restricts government attempts at refashioning upland livelihoods in the east of Hà Giang. Our findings point to the difficulties of completing LULCC maps for this highly heterogeneous region and the complexity of LULCC and livelihood interactions and relationships examined on the ground.

Keywords: livelihood diversification; ethnic minorities; northern Vietnam; land-use and land-cover change; reforestation; cardamom; urbanization; mixed methods

1. Introduction

Two hundred million people, more than half of whom are ethnic minorities, reside in the Southeast Asian Massif, a broad expanse of uplands extending southeast from the Himalayan Plateau and shared today among 10 countries. Dwelling on the physical, economic, and often cultural fringes of their respective states and predominantly undertaking semi-subsistence rural agriculture, residents of these frontiers face numerous land-use and livelihood challenges. While studies of the causal mechanisms of land-use and land-cover change (LULCC) in the Southeast Asian Massif are growing [1–5], there remains an urgent need to refine our understandings of how and why potential factors—be they socio-economic, political, or ecological—underpin land-use changes in these uplands. Not only do we need to better comprehend how specific land policies and market-led changes, for instance, drive LULCC, but also how depleted soil nutrients, population growth, or household agricultural knowledge can trigger land-use change [6]. Furthermore, since rural household livelihoods often have direct relationships with land use, the interactions between land-use change and livelihoods merit greater attention [7–10].

The livelihoods of those residing in the Southeast Asian Massif are undergoing important transformations due to market integration, agrarian transformations, globalization, and the closing of land frontiers. Indeed, political scientist James C. Scott in his 2009 book, *The Art of Not Being Governed,* noted that since 1945, these uplands have become increasingly claimed and incorporated by modern states through processes labeled as "development, economic progress, literacy, and social integration" ([11]; p. 4). For most local residents, this has resulted in the replacement of communal property with private land-use rights, the establishment of cash cropping, and a drive to convert shifting cultivators into permanent farmers. One such regional development approach is the Greater Mekong Sub-Region (GMS), which aims to connect the uplands of Vietnam, Cambodia, Laos, Myanmar, Thailand, and Yunnan Province, China. Via the GMS project alone, the Asian Development Bank (ADB) and regional states have poured more than US$14 billion into infrastructure, including highways and railways, telecommunications, and energy and urban development [12,13]. With this region also increasingly targeted by state officials and private entrepreneurs for natural resource extraction and agrarian possibilities, livelihood approaches and land-use patterns are being impacted profoundly [14,15].

Within the Southeast Asian Massif, Vietnam's northern borderlands—with their complex political context on the Chinese border, ethnic diversity, and dynamic frontier location—provide an important case study with which to examine the mechanisms of LULCC through a livelihood lens. These borderlands are undergoing critical socio-economic transformations created by both trans-national and local forces. Two of the most notable implications for LULCC are increasing urbanization and changing forest dynamics. Urbanization, supported by the government's pro-urban policies since economic reforms were initiated in the mid-1980s (*Đổi Mới*) [16], is increasing market interactions and opening up possibilities of new livelihood diversification trajectories for local populations. Forest changes, including deforestation, afforestation (often for timber production), and reforestation (natural succession, as well as production policies) are in turn impacted by interwoven and oftentimes conflicting land-use and forest policies [14], creating different livelihood opportunities as well as restrictions.

In this paper we examine how livelihoods and LULCC are intertwined within three of Vietnam's northern borderland provinces, namely Lai Châu, Lào Cai, and Hà Giang. Each province is home to a high proportion of ethnic minority semi-subsistence farmers. We focus specifically on the 14 border districts of these three study provinces (Figure 1) and temporally, on changes that have occurred from 2000 to 2009.[1] After examining meso-level results across our study region, we select three case studies and examine the underlying relationships between specific LULCC and local livelihood diversification strategies by combining quantitative mapping with qualitative fieldwork in a mixed methods approach. Such an approach is increasingly recognized and promoted by scholars from different disciplinary backgrounds to provide insight into the causes of land-use change [17]. We find that combining ethnographic fieldwork with land-cover change mapping reveals hidden causes of LULCC in these three provinces. For instance, specific patterns of urban growth have been caused by pro-urban policies, cross-border trade, political boundary redrawing, and resettlement in Lai Châu and Lào Cai Provinces. In the west of Lào Cai Province, reforestation has occurred due to the increased cultivation of cardamom, a non-timber forest product (NTFP), while in the east of Hà Giang a particular landscape has restricted government attempts at refashioning upland livelihoods from those already built around local maize cultivation.

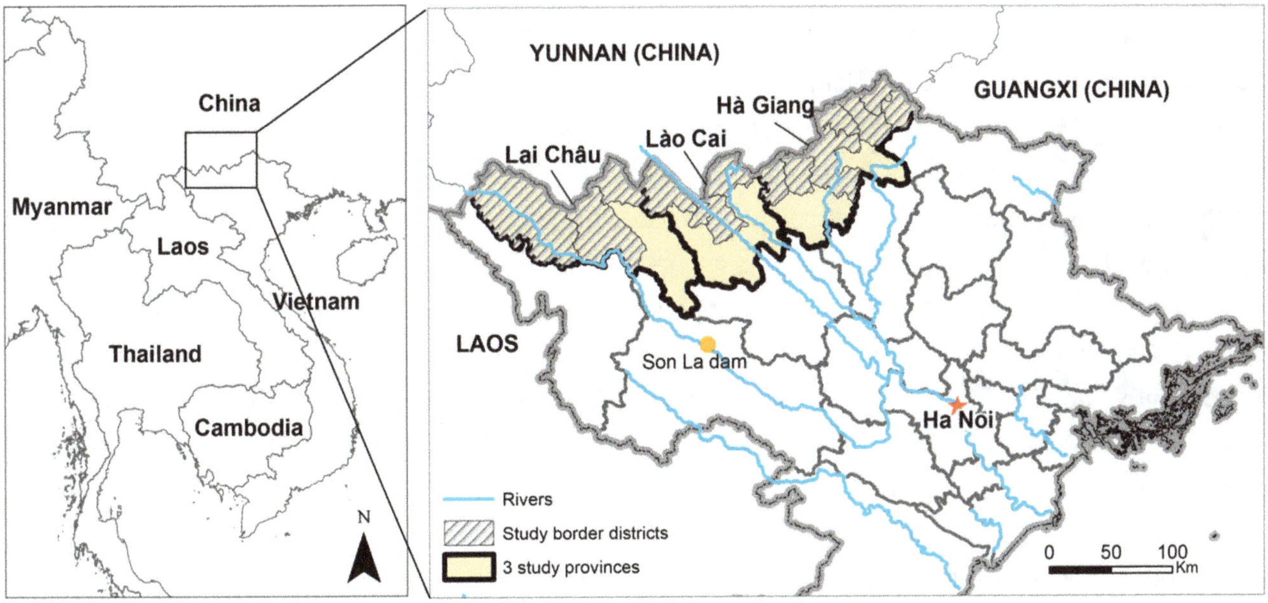

Figure 1. Location of the study border districts and provinces in northern Vietnam.

2. Conceptual Framework: Land Change Science, Frontiers and Livelihood Diversification

Conceptually, we draw from land change science (LCS), frontier studies, and livelihood diversification literature. LCS is an emerging interdisciplinary area of study that "focuses on material

[1] These years correspond with the last two national censuses completed for Vietnam (1999 and 2009), and the most unobstructed satellite images available. They also cover a dynamic time in the contemporary political and economic history of the uplands.

related to the nature of land use and land cover, their change over space and time, and processes that produce those patterns and changes"([9]; p. 3). We direct attention to three of the core objectives of LCS, namely the observation of land changes, understanding the causes of these changes, and assessing their impacts [9,18–20]. Of specific note here are causes of LULCC, which can be separated into predisposing environmental factors, underlying forces, and proximate causes [21]. Predisposing environmental factors include elevation, slope, soil conditions, and natural disasters. Underlying causes can include population, economic change, policy, ethno-cultural aspects, and technology [21]. Proximate causes are activities that directly alter land use. For example, one common proximate cause of deforestation is agricultural expansion, which could be the result of an underlying cause such as population growth or in-migration [22].

Frontier regions are both physical entities and social constructs [23,24]. Frontier studies are important to help analyze the degree to which the Sino-Vietnamese borderlands are experiencing natural resources extraction, agricultural intensification, in-migration, increasing cross-border exchanges, and state control [25,26]—all of which are altering land use and land cover in the region. Frontier regions offer numerous economic attractions such as access to natural resources, newfound business opportunities, and cross-border employment and trade prospects [13,27–32]. Agricultural expansion and natural resource exploitation are two particularly common vectors of frontier exploitation by both state and private entrepreneurs [33–36], often resulting in "spontaneous land opening, environmental degradation and the taming of ethnic minorities" ([37]; p. 14). Moreover, culturally speaking, frontier regions are often depicted as "backward" and in need of improvement, in contrast with the relative social and cultural "modernity" and dominance of the core [24,38], in this case the majority Vietnamese (Kinh or *Việt*) lowlands. Used to ease demographic pressure on core areas, state and private investment in frontier regions across the Southeast Asian Massif has also been argued to establish territorial domination by the state, further integrating ethnic minorities into the Nation [11,26,39].

For some, livelihoods in this region are changing substantially as one livelihood portfolio is replaced by another, such as when agricultural semi-subsistence is replaced by waged labor in a rural town. Yet more frequently, households undergo livelihood diversification, "the process by which rural families construct a diverse portfolio of activities and social support capabilities in their struggle for survival and in order to improve their standards of living" ([40]; p. 4). Livelihood diversification needs to be examined through a dynamic analysis incorporating the context-specific sociocultural (e.g., ethnic, gender, class), environmental, historical, and spatial dynamics of how people create and sustain viable and meaningful livings [41,42]. In turn, these responses are situated within broader spheres of institutional knowledge and power, with social actors supported or constrained by discrepancies in access to resources and information [43–45]. In rural locales in the Global South, livelihood diversification is often closely linked with changing land uses and, in turn, changes in land cover [10,46,47].

Together, these three bodies of literature inform each other and allow us to better interpret the LULCC that we find in Vietnam's northern uplands. This conceptual framing offers an integrated approach to examining the processes and results of changing land cover and land use as well as how they are interwoven with livelihood diversification decision making, power relations, access regimes, institutions, and state-society relations in this political and cultural frontier zone. Yet, combining these three conceptual approaches is challenging, as they are rooted in different paradigms, and, hence, use different research methodologies. Land change scientists usually use statistical models to identify environmental and socio-economic determinants of LULCC, but social scientists often criticize such methodological approaches as oversimplifying and generalizing social groups, and hence omitting intra-group livelihood diversity [17]. Combining such conceptual approaches thus needs to be done with caution, as explored below.

3. Context: Vietnam's Northern Uplands

The Socialist Republic of Vietnam recognizes 54 ethnic groups: the majority lowland Vietnamese and 53 "national minorities" (*các dân tộc thiểu số*). Though the latter comprise 14.8 percent of Vietnam's population [48], in the three upland provinces of Lai Châu, Lào Cai, and Hà Giang (Figure 1), ethnic minorities are in fact in the majority. They comprise 87 percent of the total population of 724,000 in Hà Giang Province, 85 percent in Lai Châu (total pop. 370,000), and 66 percent in Lào Cai (total pop. 614,000) [48]. Many of these ethnic minorities, such as the Hmong and Yao (Dao), belong to geographically-dispersed, politically-fragmented, lineage-based minority populations [49–51]. These populations frequently undertake diversified livelihoods through a composite agricultural system. A combination of maize and rice fields are commonly mixed with home gardens, small (officially banned) swidden plots, the gathering of firewood, honey, and herbs from forests, as well as small-scale commercial exchanges of cardamom, livestock, textiles, or homemade alcohol [15,52].

Land access and quality play central roles for these livelihoods, while it must be remembered that all land in Vietnam is officially owned by the state. Land-use rights have been assigned to households since the 1993 Land Law, following the dismantlement of socialist collectives in 1988, initiated by Resolution 10 [53,54]. Forests are also important for many upland livelihoods, providing fuel wood and non-timber forest products. The northern Democratic Republic of Vietnam (DRV) designated all hills and mountains with slopes over 25° as forestland in 1954 and founded a series of State Forest Enterprises (SFEs) [55,56]. These SFEs have had rather ambiguous functions, logging forests in some areas, while driving reforestation endeavors in others. Nonetheless, forest cover for the whole country dropped to 30 percent by 1985 due to the actions of these SFEs and individual farmers [57]. After the initiation of the *Đổi Mới* economic reforms of the mid-1980s that shifted Vietnam to a more open economy from a stricter socialist one [58], reforestation and forest protection became a priority in state policy (e.g., the 1991 Act on Forest Protection and Development and the 2004 Forest Protection and Development Law). A new management system was introduced with three classifications of forests: production forests (for timber extraction); protection forests (especially around watersheds); and special-use forests (focusing on strict preservation for biodiversity and cultural values) [59]. These classifications were integrated into the 1993 Land Law, which initiated a distribution of land tenure certificates providing production forest or protection forest allocations to upland households [60]. Through the late 1990s, a forest transition is thought to have occurred due not only to this classification of forests, but also due to a ban on opium and logging implemented in 1992/93, further reforestation programs including the "Greening the Barren Hills Program" (also named Program 327) from 1992, a similar program labeled Program 556, and the "Five Million Hectare Reforestation Program" from 1998. Nonetheless, the outcomes of these reforestation programs have been rather ambiguous, as case studies suggest that forest density and quality are diminishing [56,61–63].

The agricultural systems of upland ethnic minority groups have likewise undergone rapid transformations, including an expansion of wet rice farming due to migration of lowland Vietnamese from the Red River Delta since the 1960s and the *Đổi Mới* economic reforms. A move from swidden farming to fixed crops has been encouraged by the granting of long-term land-use rights to households as well as the introduction of high-yield hybrid rice and maize seeds since 1999 [26]. A transition from subsistence crops to cash crops in some regions has also been driven by the expansion of the market economy [2,52]. In addition, since 2000, new exogenous factors such as increasing weather extremes, and the 2008 food price crisis have further impacted livelihoods in these communities.

LULCC studies in upland Vietnam have started to analyze such transformations since the country's independence in 1954 [1,2,61,62,64–66]. However, all these studies have been conducted elsewhere in the country, either in the Central Highlands or in northern provinces at lower altitudes and more closely connected to the capital, Hanoi, than the borderlands. In addition, apart from our own prior research [14,67] including a systematic literature review, no study has addressed the relationships between LULCC and livelihoods along the northern borderlands with their ethnically diverse, quantitatively "poor" populations compared to the rest of the country [57] (but see [68] with one case study in a borderland district and [5] with a study neighboring, but not in, a borderland district). Consequently, LULCC dynamics in the northern border uplands—a political, resource, and cultural frontier—remain poorly understood.

4. Methods

Taking a mixed methods approach, this study draws on both quantitative and qualitative methods, namely land-cover change mapping and ethnographic interviews. More specifically, we take an "interactive design" to mixed methods, in that our approach emphasizes quantitative and qualitative methods equally. We do not, however, aim for data transformation (e.g., we do not take our qualitative data and transform them into quantifiable data for statistical analysis). Instead, we aim for "data importation", namely that data from one approach was reflected upon mid-stream in our analysis, and fed into the analysis of the other data set, and *vice versa* in an iterative manner [69]. We began by mapping land-cover change via remote sensing. We then chose cases representing the most important dynamics between LULCC and livelihoods, and drew on previous interviews completed since 1999 regarding livelihoods as well as additional interviews and observations focusing on the case sites during the summers of 2012, 2013, and 2014 to guide our interpretations. An important element in this interactive process was our long-term knowledge of the region, which allowed us to be confident about our choice of cases.

4.1. LULC and LULCC Mapping

Given the lack of accurate data regarding land use in rural Vietnam, especially historical data [70], we had to compromise somewhat methodologically, opting for an analysis of a mixed system of land-use and land-cover types. LULC maps were derived from Landsat thematic mapper (TM5) and enhanced thematic mapper (ETM+) images obtained from the United States Geological Survey (USGS). All images were taken during the winter season (Table 1) when sparse vegetable crops were growing, with no maize or wet rice. To identify land-cover types from our Landsat images, we drew on the Level I land-cover types proposed by Anderson in 1976 [71] and previous LULCC studies of Vietnam using Landsat images [52,61].

Table 1. Dates of images.

Province	Dates
Lai Châu	2009 November 03
	2000 November 02
Lào Cai	2009 November 12
	1999 December 27
Hà Giang	2009 November 05
	2000 November 04

Initially, we identified five key land-*cover* types across the study region from satellite images (shrubs, bare soil, open canopy trees, closed canopy trees, and water). To provide further detail, we then added one land-*use* type, namely built-up areas, working from the five original land-*cover* types. It should be noted that there is rarely any grassland in this mountainous area; when small patches of grass are mixed up with shrubs, we identify them only as "*shrubs*". Cropped land is composed of subsistence crops (lowland and upland rice, corn, cassava) and cash crops (banana, pineapples, tobacco, among others). According to our interviews and local crop calendars, areas with subsistence crops are often covered by *bare soil* in the winter, when our images were acquired. This is also observed in other upland areas of Vietnam [4,52,61]. Pineapples are small and fairly dispersed plants, hence pineapple areas are spectrally similar to *bare soil* in the images. Areas having bananas remain vegetated and identified as *shrubs* in the Landsat images. However, bananas were not very common in the region in 2009 (interviews and observations), and were usually found next to houses or mixed with shrubs. Built-up areas are also identified as *bare soil* in the images, although dense urban areas have stronger spectral reflectance. To separate built-up from cropped land, we used information regarding road systems (see below). *Open canopy trees* in this region tend to be bamboo, plantations for timber production, or natural succession since the 1990s (interviews). *Closed canopy trees* are old and nearly intact forests that are often protected by the local government, such as special-use forests (often labeled national parks). This class has sometimes been called "mature and evergreen forest" in Southeast Asian forest studies [72]. Separating forests into two categories—open and closed canopy—is usual in Southeast Asian upland studies [4], although some authors tend to group them into one category [61,73]. Interpretation of classification results was done with these details in mind. Illustrations of these LULC types are presented in Figure 2, while we detail our classification method below.

The images were pre-processed to remove distortions caused by sensor errors, atmospheric interference, and surface irregularities. Cloud and cloud shadow masking was conducted on the 2000 image of Lai Châu and the 2009 image of Lào Cai (more specific technical details are provided in [14]). We used an object-based approach [74] to identify the five land-cover types listed above plus clouds and shadow. Although associated most often with very high-resolution images, this approach has proven accurate to produce rural land-cover types from middle-resolution images such as Landsat [75,76]. Cultivation systems in upland Southeast Asia are complex, mixing young forest, shrubs, and different types of crops [52,72]. In this context, an object-based approach is very helpful in creating segments (objects) and incorporating textural information of segments into classification. This approach has also allowed us to work at two spatial levels: segmentation of large-size patches (to separate large-scale forest and crops) and of small-size patches with relatively homogenous texture and signals (to identify smaller plots).

Water (Red River, near Lào Cai City) — Shrubs — Closed canopy trees — Rice (bare soil in winter) — Open canopy trees — Corn (bare soil in winter) — Built-up areas — Pineapples

Figure 2. Illustration of the six LULC classes. (Photo credit: Lê Mạnh An, Thi-Thanh-Hiên Pham).

Segmentation parameters in eCognition (bands, scales, color/shape ratio, and compactness/smoothness ratio) were tested at different values. Segmentations and rule-based classifications were undertaken at different scales using the same band composition (bands 1–5, 7), color/shape ratio (0.2/0.8), and compactness/smoothness ratio (0.3/0.7). Segmentation values of 5, 10, 20, 50, 100, and 200 were tested, aiming at creating segments of different sizes. Segments were visually examined to determine the visibility of the main land-cover classes. We chose two values that produced the most homogenous segments in terms of spectral values and texture. The first segmentation was conducted at a scale of 50 to obtain large size objects. Rules were then used to classify those objects into *clouds and shadow*, *water,* and *bare soil* classes. The second segmentation was conducted at a scale of 10, then a second set of rules was used to classify segments into three vegetation classes of *shrubs*, *open canopy trees,* and *closed canopy trees*.

Bare soil was then separated into *bare soil* and *built-up*. Built-up pixels were assigned by evaluating road density. The road network was separated into solid (gravel or concrete) and non-solid (compacted soil susceptible to flooding during the rainy season) roads. A density map was developed by assigning solid roads an importance value of 3 and non-solid roads a value of 1. Pixels having high road density ($3m/km^2$ in 1999 and $4m/km^2$ in 2009) were visually compared to aerial images on Google Earth and those that coincided with more urban areas (built-up and having a street network) were renamed as *built-up*. The remaining *bare soil* pixels remained *bare soil.*

We did not use a set of training points *per se* to define rules of classification. Rather, we chose roughly 200 objects (segments) that corresponded to land-cover types that we knew on the ground from observations. Then we created rules composed of textural and spectral indicators for each class from the chosen objects. Refining and adjusting rules were based on interactive "trial and error". The classification process included initial field observations in summer 2013, and field verification in summer 2014 by the second author.

4.2. Assessing the LULC Mapping

A ground truth assessment was conducted for the 2009 classification by using ground control points (GCPs). There are no historical air photos for the region covering 1999/2000. There are a few photos on Google Earth from 2009/2010 but they are dispersed and cover only 15 percent of the study area. We hence opted for GCPs. We collected 365 GCPs: 142 points in Lào Cai Province (September 2012), 101 points in Lai Châu, and 122 points in Hà Giang (September 2013). Given the difficult access to many locations in the region, points were sampled along roads, focusing on typical land-cover types in the region. There were fewer points in Lai Châu given the lack of roads and access difficulties caused by landslides.

Each point was registered in GPS (precision of 5m) and photographed in the four cardinal directions to capture potential mixtures of land cover. Descriptions of land use and land cover were included at each point and for the four directions. Land-cover types were assigned to each point based on photos and descriptions. Unfortunately, we do not have ground points for the 1999 images, but since we used the same image processing procedure for the 1999 and 2009 images, we believe that the accuracies of the 1999 images are similar to those of the 2009 images.

In order to evaluate the accuracy of the 2009 classification, we created a confusion matrix (Table 2) using the 365 GCPs. The overall accuracy is 71 percent, mostly due to confusions between *open canopy* and *closed canopy* classes. Accuracies varied from 46 percent (*open canopy*) to 96.88 percent (*closed canopy*). In this mountainous area undergoing complex forest transitions, *open canopy trees* are usually mixed with *shrubs*, making it difficult to separate them from Landsat images. The confusion between *open canopy* and *closed canopy* is most likely due to the fact that open forests in the three provinces are highly heterogeneous. What is defined on the ground as *open canopy* in areas with plantations (since the mid-1990s) could be similar spectrally to *closed canopy* in other areas. When grouping these two categories into "forest" (as done in several Southeast Asian LULCC studies [61,73]), we obtained a "user accuracy" of 79.30 percent and a "producer accuracy" of 72.22 percent, raising overall accuracy to 73.70 percent.

Other important confusions related to *bare soil*, our proxy for rice or corn fields. In these mountainous areas, dispersed houses and remote roads are often located close to rice or corn fields, creating a spectral mixture in the image. In Đồng Văn District, Hà Giang Province, this gets complicated further by limestone outcrops within fields and near houses (Figure 3a). Lastly, the confusion between *shrubs* and *bare soil* is explained by the fact that upland dry rice and maize are commonly planted in areas close to shrubs (Figure 3b).

Table 2. Confusion matrix and producer's and user's accuracy for accuracy assessment of the 2009 classification (based on ground control points).

		Ground Reference						Total	Producer Acc. (%)
		Water	**Closed Canopy**	**Open Canopy**	**Shrubs**	**Bare Soil**	**Built-Up**		
Classification	Water	**16**	0	0	0	1	2	19	84.21
	Closed canopy	3	**31**	10	7	3	0	54	57.41
	Open canopy	0	1	**23**	8	2	2	36	63.89
	Shrubs	0	0	13	**42**	7	2	64	65.63
	Bare soil	3	0	3	10	**109**	11	136	80.15
	Built-up	0	0	1	3	15	**37**	56	66.07
Total		22	32	50	70	137	54	365	
User Acc. (%)		72.73	96.88	46.00	60.00	79.56	68.52		

(a)

(b)

Figure 3. (**a**) Typical karst landscape and (**b**) mixture of maize and shrubs in Đồng Văn District, Hà Giang Province. (Photo credit: Sarah Turner and Lê Mạnh An).

4.3. Computing LULCC

To measure LULCC, we computed the percent change and the annual rate of change for each class, where *A1* is the cover of arable land at an initial time (t_1) and *A2* is the cover of arable land at a later time (t_2), and where $t=t_2-t_1$ (Equations (1) and (2)) [77].

$$R = \frac{\text{A2} - \text{A1}}{\text{A1}} \quad (1)$$

$$Rate = \ln(A2/A1)/t \quad (2)$$

We also computed the absolute values of changes for each LULCC type per district. In other words, we computed the magnitude of change (R in Equation (1)) without taking into consideration the direction of change (decreasing or increasing over time) because we wanted to capture the total amount of relative change. Finally, in order to adjust the changes by size of district, we computed the ratio of absolute value of change over the district area (without including areas covered by *clouds and shadow*). This allowed us to identify the most dynamic and the most stable districts in terms of LULCC. The higher the ratio, the more dynamic the district with regards to LULCC.

4.4. Qualitative Fieldwork

The qualitative fieldwork that supports this study was completed by the first author during repeated research visits to Lào Cai Province since 1999, Lai Châu Province since 2004, and Hà Giang Province since 2009, and by the second author in Lào Cai Province since 2012. This includes over 100 in-depth unstructured (conversational) interviews with ethnic minority farmers (Tày, Hmong, Yao, Nùng) in Lào Cai Province, and 50 each in Lai Châu and Hà Giang Provinces. Approximately 75 percent of farmer interviewees were women, as they were more likely to be in the house during the day, yet had a wide knowledge of land uses and changes. Farmer ages ranged from 25 to 80 years old. Interviews ranged in duration from 20 minutes to over an hour, focusing on local livelihood diversification, agricultural practices, land-use changes, and state-society relations. Twenty-eight semi-structured interviews were also completed with provincial officials (mostly Kinh) working in government departments linked to agriculture, planning, labor, and natural resources. The core themes of these interviews were livelihoods, market integration, land-cover change, and the impacts of state policies on each of these. Interviews were completed with the aid of local ethnic minority interpreters for farmers of the same ethnicity, or with Kinh interpreters or alone for Kinh farmers, urban dwellers, and state officials. All interviews were transcribed and coded using a mix of constant comparative, axial, and thematic qualitative coding approaches. Concurrently, observations of LULCC have been completed and noted annually.

5. Results

Our land-cover mapping shows that the most important changes in terms of area (Table 3) include an increase in *closed canopy forest* (2437 km^2), followed by a decrease in *open canopy forest* (roughly 1694 km^2), a decrease in *bare soil* (537 km^2), and a decrease in *shrubs* (324 km^2). Examining the 2000 and 2009 maps (Figure 4), spatial patterns for these changes are visible. The increase in *closed canopy forest* is most notable in all the border districts of Lào Cai Province (except Si Ma Cai and Lào Cai City), in Mường Tè District (Lai Châu Province), and in the southwestern parts of Hà Giang Province (also see Table 4). In turn, a decrease in *open canopy forest* has occurred in almost all these same locales where *closed canopy forest* increased.

Table 3. Land-cover change in the region, 2000–2009.

Land Cover	Area in 2000 (km^2)	Area in 2009 (km^2)	2009–2000 Area Change (km^2)	2009–2000 Percent Change (%)	Annual Rate of Change (%/year)
Closed canopy	1723.89	4161.03	2437.14	141.37	8.81
Open canopy	3281.12	1587.14	−1693.98	−51.63	−7.26
Shrubs	3357.08	3033.28	−323.80	−9.65	−1.01
Bare soil	3668.91	3141.54	−527.37	−14.37	−1.55
Built−up	50.58	108.29	57.70	114.08	7.61
Cloud shadow	1068.04	918.51	−149.53	−14.00	−1.51
Water	40.23	59.29	19.06	47.37	3.88
Total	13 189.86	13 009.07*			

* The Landsat scenes for the two years were slightly different, hence different total land-cover area.

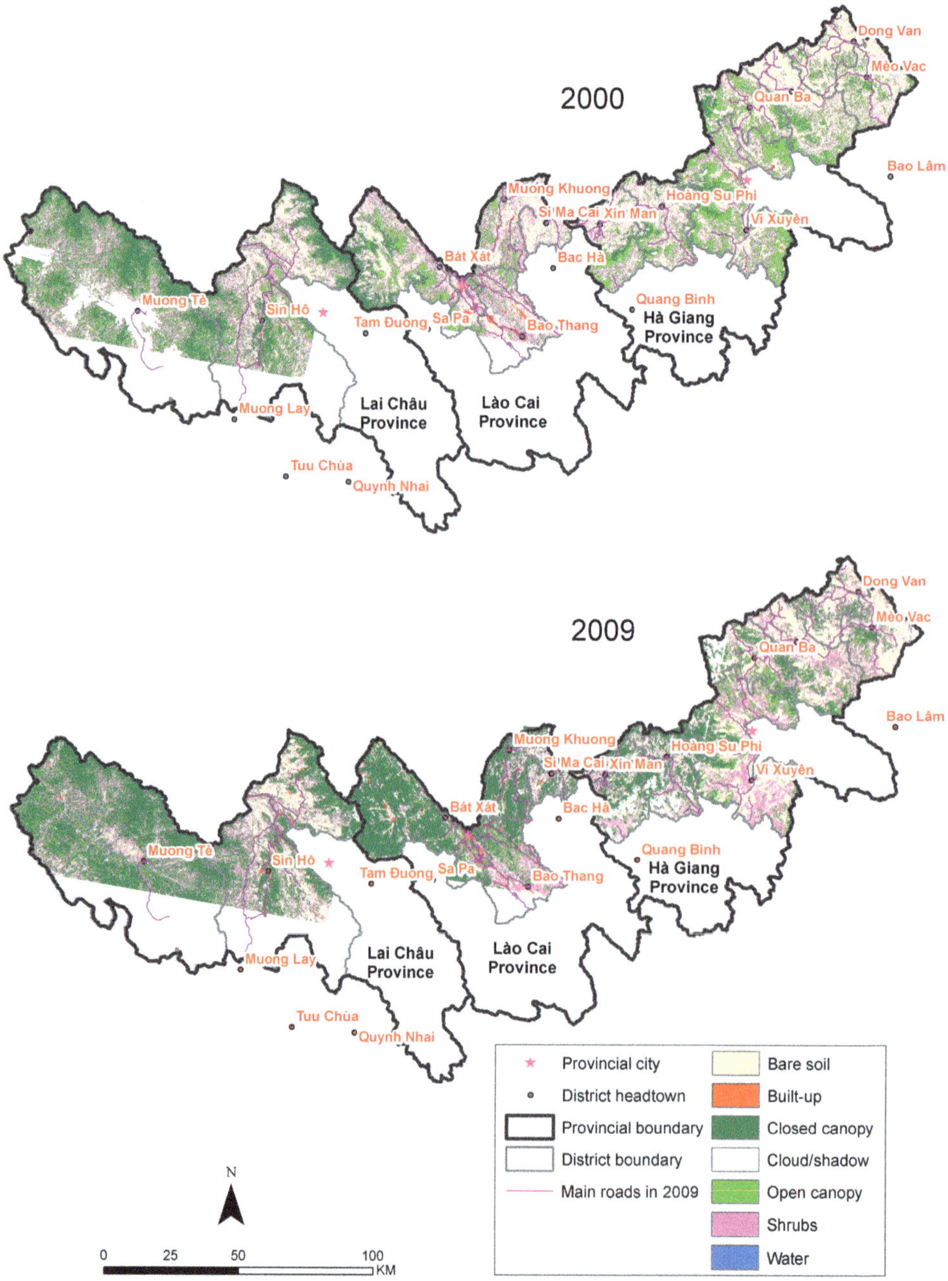

Figure 4. Land cover maps for the region, 2000 and 2009.

Table 4. Land-cover change in the region by district (km^2), 2000–2009.

Province/District	Closed Canopy	Open Canopy	Shrubs	Bare Soil	Built-up	Cloud Shadow	Water	Absolute Change (km^2)*	Absolute Change/ Total Area**
Lai Châu Province									
Mường Tè	*503.14*	**−231.83**	−30.54	−75.52	3.90	−243.95	7.22	852.15	0.35
Sìn Hồ	93.47	**−175.11**	14.07	32.50	*17.05*	−71.93	−2.71	334.90	0.27
Phong Thổ	69.50	−77.95	−43.30	28.69	*11.45*	8.59	2.64	233.53	0.28
Lào Cai Province									
Bát Xát	*571.35*	**−218.24**	**−234.72**	**−149.47**	*13.27*	13.07	4.61	1191.66	1.13
Lào Cai City	71.92	−18.50***	−27.74	−47.02	17.76	0.65	2.94	185.89	0.87
Bảo Thắng	*105.41*	22.83	−32.25	−88.88	−7.94	0.85	−0.31	257.63	0.49
Mường Khương	*313.35*	−89.74	**−121.43**	**−112.09**	6.69	2.91	0.87	644.18	1.16
Si Ma Cai	91.42	−6.96	−18.21	−73.69	5.74	0.26	1.25	197.27	0.85
Hà Giang Province									
Xín Mần	*161.50*	**−120.03**	−51.27	−41.19	0.00	48.63	2.20	376.20	0.66
Hoàng Su Phì	*188.44*	**−116.39**	−62.59	−52.84	−0.99	43.73	1.12	422.37	0.67
Vị Xuyên	*131.31*	**−420.03**	*117.64*	*170.33*	−7.97	10.74	−1.66	848.94	0.58
Quản Bạ	74.81	**−119.75**	−3.97	11.60	−0.35	36.86	1.28	211.76	0.39
Yên Minh	22.00	−68.72	94.37	−47.10	−0.69	0.00	0.11	232.98	0.30
Đồng Văn	28.76	−11.20	37.74	−61.65	−0.22	−0.01	−0.48	140.05	0.31
Mèo Vạc	10.75	−42.36	38.41	−21.04	0.00	0.05	−0.03	112.59	0.20

Note: Blue italics and red bold values: important increases and decreases, respectively. * Absolute change is the sum of absolute values of each LULCC, excluding areas covered by *cloud/shadow*. ** Total areas do not include *cloud/shadows*. *** The change in *open canopy* class in Lào Cai city is numerically small, but was 37.83 percent of the total *open canopy* area in 2000.

Coverage by *shrubs* declined across our study region, with the exception of the eastern districts of Hà Giang (Vị Xuyên, Yên Minh, Đồng Văn, and Mèo Vạc). The most noticeable loss was in Bát Xát District, Lào Cai Province, and a close look at the transition map (Figure 5) shows this area changed mostly to closed canopy forest. Two main transitions to *shrubs* occurred; first, when *open canopy forest* converted to *shrubs* (in Vi Xuyên District, Hà Giang Province) and second, and more pronounced, when *bare soils* changed to *shrubs* (mostly in Yên Minh, Đồng Văn, and Mèo Vạc). It should be noted that this so-called transition from *bare soils* to *shrubs* probably reflects the confusion noted above in Section 4.2, with observations and interviews (2010) supporting the contention that very little change in areas of crop cover have occurred here. *Bare soils* were reduced in most districts. They changed to *closed canopy forest* in Mường Tè (Lai Châu Province), in west Bát Xát and Mường Khương (Lào Cai Province), and in Xin Mần and Hoàng Su Phì (Hà Giang Province). However there were some gains of *bare soils* in southern Hoàng Su Phì, adjacent to Vị Xuyên. Another noticeable change, although not as large in square kilometers as the aforementioned changes, was the expansion of *built-up areas* corresponding to urban growth in the region. This expansion occurred mostly in Sìn Hồ, Phong Thổ (Lai Châu Province), and in Bảo Thắng and Lào Cai City (Lào Cai Province) (Table 4). The most important transitions in the region are shown in Table 5 and illustrated in Figure 5.

Examining the magnitude of change *versus* stability in each district (Table 4, final two columns) we note some important tendencies. Districts with the highest amounts of change were Mường Tè, Bát

Xát, Mường Khương, and Vị Xuyên, hence high rates of change are spread across the study region. When taking into account district size, Bát Xát, Lào Cai City, and Si Ma Cai were the most dynamic, all in Lào Cai Province. Inversely, the most stable districts were Sìn Hồ and Phong Thổ in Lai Châu Province, and Yên Minh, Mèo Vạc, and Đồng Văn in Hà Giang Province. Further analyses of the links between these LULCC patterns and livelihoods are explored next.

Table 5. Transitions of land-cover types in the region, 2000 to 2009 (km^2).

		In 1999							
		Cloud Shadow	Water	Closed Canopy	Open Canopy	Shrubs	Bare Soil	Built-up	Total
	Cloud shadow	506.81	1.52	110.40	142.63	93.13	61.92	0.87	917.28
	Water	8.51	15.45	1.68	3.86	8.41	19.70	1.82	59.45
	Closed canopy	139.74	0.97	1 285.01	**1 344.63**	**871.10**	**515.50**	1.66	4 158.61
In	**Open canopy**	98.04	1.73	115.34	650.42	472.50	245.52	3.21	1 586.76
2009	**Shrubs**	103.44	4.30	87.95	**662.15**	1 155.14	**1 005.88**	13.39	3 032.25
	Bare soil	66.78	14.72	114.72	**456.69**	**723.57**	1 744.98	18.33	3 139.79
	Built-up	2.58	1.61	0.98	6.08	**24.66**	**61.06**	11.29	108.27
	Total	925.90	40.31	1 716.09	3 266.45	3 348.51	3 654.56	50.58	13 002.40

Note: Grey cells indicate amount of land-cover type that has not changed, bold numbers indicate important changes.

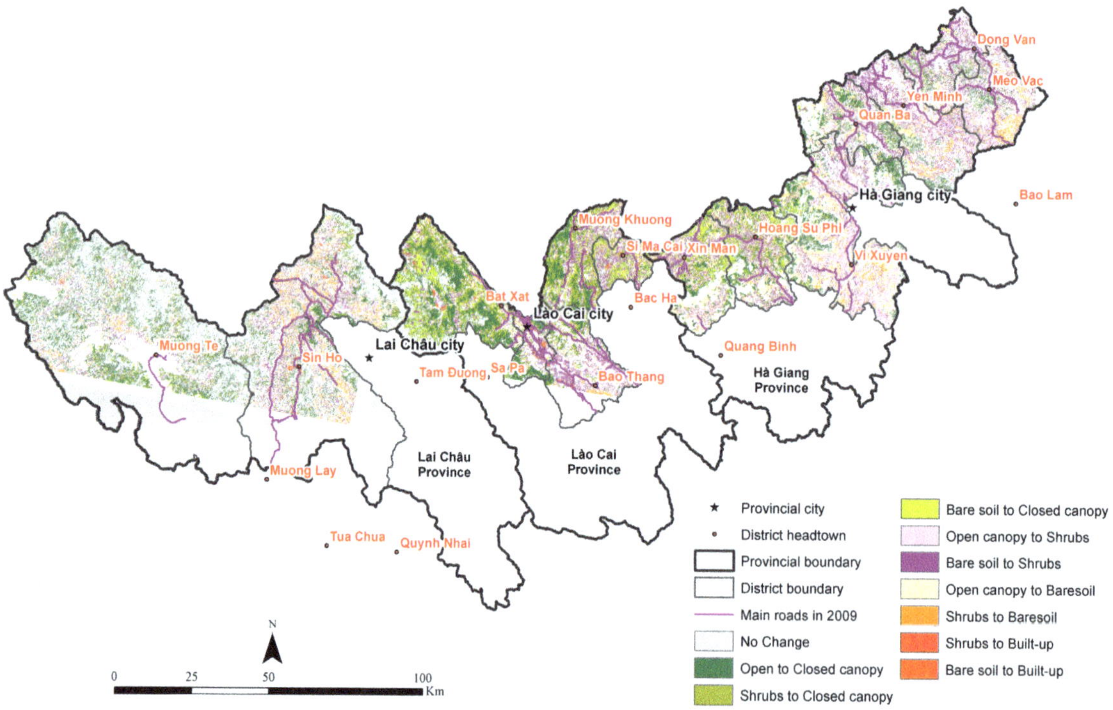

Figure 5. Transitions of land-cover types in the region between 2000 and 2009.

5.1. Examining the Underlying Relationships between LULCC and Livelihood Diversification

Drawing on our quantitative findings above, in this section we focus on three case studies, one from each of our study provinces. Specific case studies cannot represent the entire spectrum of change across these uplands, and our cases were chosen instead to highlight the breadth of LULCC types underway, the diversity of causes, some important impacts on local livelihoods, livelihood diversification responses, and some surprise findings.

5.1.1. Urbanization in the Uplands and the Case of Lai Châu's Changing Capitals

Across our study site there was a 114 percent increase in *built-up areas*, notably in Sìn Hồ and Phong Thổ (Lai Châu Province), as well as in Bát Xát, Lào Cai City, and Bảo Thắng (Lào Cai Province). Over half of this expansion transpired in the Red River valley, within the city municipality of Lào Cai, and to its immediate west in Bát Xát District. This increase is concurrent with national strategies to encourage cross-border trade and to stimulate the Greater Mekong Subregion North-South Economic corridor, begun in 1998. The Vietnamese government is investing heavily in Lào Cai City and its environs, supporting trade and improving road infrastructure. This has included the creation or upgrading of border crossing points for heavy vehicles carrying merchandise from both Bát Xát District and Lào Cai City to and from Yunnan, China. At the provincial level, the *2020 Lào Cai Economic Development Plan* continues to prioritize construction, renovation, and improvement of facilities at the Lào Cai–Hekou (Yunnan) border crossing, as well as the construction of new, smaller border crossings [78].

Livelihood opportunities have undergone notable diversification in Lào Cai City. In 1999 this was a sleepy, dusty frontier town with most activity focused on the border crossing with Hekou Town, Yunnan. By 2009, the metropolises on both sides of the border had expanded considerably, with Lào Cai City boasting a new six-story border gate administrative center and a brand new multi-story shopping center. The city's main marketplace was also renovated and expanded, selling numerous electronics and plastic goods imported from China. As one Kinh trader noted, sweeping her hands over her stall's plastic toys: "All of this, I get it all from China" (interview, August 2013). While these expanding trades and services have provided new livelihood opportunities, it is important to note that overwhelmingly, it is ethnic Kinh (Vietnamese lowlanders) who have taken up employment opportunities here, rather than ethnic minorities from the surrounding countryside. Without the formal education skills, social capital, and financial reserves necessary to enter urban livelihoods, upland urbanization has remained strongly ethnically divided. When asked why they do not move to Lào Cai City for work opportunities, a group of Hmong ethnic minority young women responded, "Why would we go there? It's scary—close to the border we might get kidnapped, and we don't want to work in the city anyhow. Our home is here" (interviews, January 2009; July 2014).[2] As such, young ethnic minority men and women tend to remain based in rural villages—either that of their parents or, for women after marriage, that of their husband—and maintain agricultural livelihoods, plus small-scale trade for some. While there is migration of ethnic minority individuals over the border for agricultural or mining labor, this remains strongly gender-divided for men only and is limited in size. This urban-rural ethnic divide is also apparent elsewhere in our study region, in Lai Châu and Hà Giang cities.

Growth in *built-up areas* in these borderlands is also due to the establishment of new administrative districts, and thus the need for new head towns. This transformation occurred in August 2000 for Si

[2] An increasing number of young ethnic minority women have been trafficked to China to become wives for rural farmers there. A common kidnapping approach is for a young man—often from the same ethnicity and posing as a friend of a friend—to lure a young woman to Lào Cai City on his motorbike 'to have fun'; she is then abducted and taken over the border (interviews with Hmong and Yao youth, 2009, 2014).

Ma Cai Town, which became the head town of a new district with the same name in the east of Lào Cai Province [79]. This resulted in a new suite of roads, public services, and administrative buildings [80], plus all the officials who go with such a bureaucracy. Locals whom we talked to in Si Ma Cai, who had lost their land to these new infrastructure developments, were none too pleased with the limited compensation (interviews, March 2009).

The cause of *built-up areas* in Lai Châu Province, the farthest west of our study provinces, is perhaps even more controversial. Here, urban "upgrading" has also occurred, but this time largely due to the creation of a large reservoir for a dam project further south in Sơn La Province (see Figure 1 for the dam's location). Lai Châu Province consists of Lai Châu City and seven districts (with the most recent, Nậm Nhùn, created in 2012) with a combined total of 98 communes [81]. The province's total land area is 9,112.3 square kilometers [82]. The province is considerably smaller than it used to be, however, as a new province called Điện Biên was created to the west of Lai Châu province in 2004, carving off Lai Châu Province's more industrialized and financially prosperous southern and western parts [48,83]. To the south of Lai Châu Province, construction began on the Sơn La dam in 2005; when it was completed in 2012, it was the largest hydropower plant in Southeast Asia, with a 224 km^2 reservoir [84]. The project required the displacement of over 91,000 ethnic minority individuals across three provinces, including Lai Châu Province, beginning in 2005 [85]. In Lai Châu, the resettlement of certain villages was delayed because of conflicts over land prices and an absence of infrastructure at resettlement sites [84]. These sites are mostly in rural areas and with resettlement being fairly recent, their impacts were not clear on our LULCC maps, although resettlement villages are now observable in the high-resolution imagery used by Google Earth.

Just south of our study area, it has been intriguing to observe the increase in urban land use around the newly relocated provincial capital, also called Lai Châu, inaugurated with its new name in 2004 [86]. Formerly a sleepy town called Phong Thổ, the "new Lai Châu" was given a substantial makeover to become the new provincial capital because the former Lai Châu Town was lost to the dam's reservoir.[3] Extensive new infrastructure including large government offices and wide boulevards were already complete by 2006—when we observed the roads being used for motorbike racing and not much else—and "new Lai Châu" was formally recognized as having gained "city status" in December 2013 [87]. By 2014, the total urban area of this new city was more than 7,000 ha with 52,500 residents and seven wards [87].

Back within our study site's border districts in Lai Châu Province, the most important growth of *built-up areas*—an increase of 17 km^2—has occurred in a linear fashion along Road 12, which crosses the district of Sìn Hồ from north to south (Figure 6). This increase is on par with the urban growth around Lào Cai City, in Lào Cai Province. While most interviewees suggested this growth was due to local household livelihoods diversifying into new, small-scale trade possibilities along the road as traffic (and tourism) increases with improved roads, some of this urban growth might also be due to resettlement from the dam. Only two interviewees of the 10 with whom we talked along Road 12 had moved to this area due to dam resettlement, but it would not surprise us if others had also done so.

[3] In 2004–2005, when former Lai Châu Province was split into Lai Châu and Điện Biên, former Lai Châu Town was renamed Mường Lay Town, now part of the new province of Điện Biên. Most of this town is now under the reservoir for the Sơn La dam.

These changes require further investigation as uncertainty also surrounds a new hydroelectric project, the Lai Châu dam, this time in the province itself (Nậm Hang Commune, Mường Tè District), which will be the third largest in the country after the Hoà Bình and Sơn La projects, also on the Đà River. The Lai Châu dam was initiated in 2011, was 80 percent completed in 2014, and is expected to be operational by 2017. This project alone requires the relocation of approximately 800 rural households [88] and may bring important LULCC to Mường Tè District and, in turn, changes in possible livelihoods.

Figure 6. Sìn Hồ Town on Road 12. (Photo credit: Lê Mạnh An)

5.1.2. The Impacts of Increased Cash Needs on Closed Canopy Forest: The Case of Bát Xát District

In terms of LULCC in rural areas of our study region, there was a total increase of 531 km^2 in *closed canopy forest* (over 300 percent). This is in line with a national study that also concluded that the extent of the forest network increased during this time period (though it did not note density, diversity, or quality) [70]. Our findings show that most forest change happened in Bát Xát and Mường Khương Districts (Lào Cai Province), Mường Tè (Lai Châu Province), and Hoàng Su Phì and Xín Mần (Hà Giang Province).

Lào Cai Province lies in the center of our three study provinces. It is divided into eight administrative districts with a combined total of 164 communes, plus Lào Cai City. The province's total land area is 6,383.9 square kilometers [82]. In the west of Lào Cai Province, Bát Xát District is characterized by high elevations and steep slopes, a sparse road network, and relatively few marketplaces. With a population of approximately 70,000, the dominant ethnic group in the district is Hmong (approx. 26 percent) [48,89]. In 1999, this region was principally classified as upland fields, identified as bare soil in our maps, with dry rice and maize production. From our interviews with Hmong and Yao uplanders here, we found that a transition towards state-sponsored high-yield hybrid rice had taken place alongside a decrease in upland fields/*bare soils*; the total decrease in agricultural land being nearly 150 km^2, or 54 percent of the district. Interviewees explained that many farmers had chosen to begin using intensive rice farming techniques as the state discouraged swidden agriculture with dry rice and maize. As one Hmong interviewee in Mường Hum, a market town within Bát Xát District, put it, "The corn and the dry rice, we had to stop" (August 2013). While reducing the land used for agriculture, this switch has had important repercussions for livelihood strategies, as farmers are now obliged to purchase (infertile) hybrid rice and maize seeds annually as well as chemical pesticides and fertilizers. Another Hmong farmer explained, "We need to buy the seeds every year now, rather than saving some, because the Chinese [hybrid] seeds are not so good to save" (March 2007). As a result, semi-subsistence households are far more dependent on the cash economy than in the past (interviews 2004, 2007, 2013, see also [26]).

Interviews with minority farmers in the district suggest that the increase in forest cover in Bát Xát District (one of the most dynamic districts across our fieldwork sites; see Table 4) could be the result of a declining agricultural footprint, reforestation policies, and an increase in cardamom cultivation to meet the increasing cash needs of local households (interviews 2013). Highly sought after in China, cardamom grows in the shade of mature trees and requires relatively little labor to cultivate [90] (Figure 7). In Bát Xát District, ethnic minority farmers are increasing the amount of cardamom they cultivate to earn cash income, with some farmers recently reporting an annual harvest of 500 kg from their cardamom plots if the weather is good and noting, "There's far, far more [cardamom] than ten years ago" (interview, August 2013). From Bát Xát, the cardamom is transported across the border at nearby local level crossings by the farmers themselves or by Kinh intermediaries who purchase it at the "farm gate" and transport it to Lào Cai City, where most often it is collected by Chinese wholesalers. The cardamom then makes its way to wholesale markets and processing plants in Kunming, the capital of Yunnan Province, or in neighboring Guangxi Province to satisfy the sizeable demand across China (interviews 2013, 2015).

(**a**) (**b**)

Figure 7. (**a**) Forest containing cardamom fields above a Hmong hamlet and (**b**) cardamom growing in the shade of mature trees Bát Xát District, Lao Cai Province. (Photo credit: Sarah Turner).

As the Vietnamese state attempts to homogenize agriculture in the uplands using hybrid varieties, and as demand for cash income among local farmers consequently rises, ethnic minority households are reacting by supporting reforestation through the protection and restoration of old growth forests for cardamom cultivation. Surprisingly, therefore, market integration and agrarian change have led to forest conservation and expansion in a rather roundabout way. Nonetheless, it is not clear how long this situation will continue. Despite the cash income that cardamom can bring, the increasing frequency of extreme weather events—especially extreme cold and hail storms—is resulting in unpredictable cardamom yields and unstable incomes for farmers, with one noting fluctuations of 60 to 400 kilograms a year (interviews, August 2013, June 2014). Additionally, prices fluctuate substantially from one year to the next, in part due to varying demand in China and in part due to volatile relationships between individual intermediary traders and cultivators, who tend to be of different ethnicities (interviews 2013) [15].

5.1.3. When Geomorphology Tests Livelihood Limits: Đồng Văn District, Hà Giang Province

The three districts with the least LULCC are the karst hill and mountain-peaked districts of Sìn Hồ and Phong Thổ (Lai Châu Province), and Đồng Văn and Mèo Vạc (Hà Giang Province). Here, the ratio of change is the lowest, between 0.20 and 0.30 (Table 4). To explain this lack of LULCC, we focus on the case of Đồng Văn and Mèo Vạc in Hà Giang Province. Hà Giang, the farthest east of our study provinces, is divided into 10 administrative districts with a combined total of 195 communes, plus Hà Giang Town; its total land area is 7945.8 square kilometers [82]. The province spans three officially recognized distinct agro-ecological zones. The northern borderland districts of Mèo Vạc, Đồng Văn, and Quản Bạ, within our case study, are in Zone 1. This is a reasonably homogenous high plateau agro-ecological zone. About 90 percent of the surface area is limestone, reflecting the region's karst geography. This zone supports the cultivation of maize, often on steep slopes, and the raising of livestock, mostly cattle, horses, goats, and poultry. Zone 1 also includes Yên Minh District, which is somewhat different, having large areas of open land both with and without forest cover. Zone 2 covers the mountainous western borderland districts of Hoàng Su Phì and Xín Mần, also in our case study. This zone has an average elevation of 1,600 meters, many steep slopes, and poor soil. Agriculture includes rice and maize, cash crops such as tea, and livestock rearing. Zone 3 includes Hà Giang Town as well as the districts of Bắc Mê, Bắc Quang and Vị Xuyên, which fall outside this study. These are less mountainous (500 to 1,000 meters), with old forests and valleys alternating with rivers and large streams (interviews and observations 2009, 2010) [91].

On our LULCC maps, the northernmost districts of Đồng Văn and Mèo Vạc recorded important areas of *bare soil* (234.5 km^2 and 285 km^2 in 2009, or roughly 50 percent of the total area in both districts). Particularly interesting in these two districts are the ways by which local ethnic minority farmers maintain livelihoods, as these *bare soils* on the maps represent a specific karst landscape with extremely small pockets of usable soil just a few meters square between large numbers of rocky outcrops. In these small pockets, minority farmers have diligently added soil and traditional fertilizers (a mix of household fire ash and cattle dung) to be able to grow traditional local maize. Running up the stalks of the maize are beans, and other vegetables such as pumpkins are sometimes interspersed among these (Figure 8).

As noted earlier, across much of the upland northern provinces—and indeed elsewhere in Vietnam—farmers are being strongly encouraged by the government to switch to hybrid varieties of rice and maize seeds to increase outputs. A Yao farmer in Đồng Văn market explained, "We have to spend 3 million VND on fertilizer for the hybrid corn seeds. I get it in advance from the government and pay them back with some of my crop. Each year I have to buy new seeds" (June 2010). Our research and that of our former graduate students has shown that these crops can actually increase food insecurity in these uplands rather than improve it, due to the agro-ecological limits of the new seeds, difficulties with the appropriate quantity of seeds being supplied, and (un)timely distribution [26,92].

Notably, hybrid maize is *not* grown in the small pockets of land in Mèo Vạc and Đồng Văn's karst landscape because farmers insist it requires a more level growing surface. This points to the importance of landraces and traditional ecological knowledge for maintaining livelihoods here. Local maize diversity is maintained by traditional seed-saving within households, as well as exchange among households [92]. Hmong farmer interviewees noted that they far prefer traditional maize to hybrids due to its suitability in the rugged topography and the means by which farmers can intercrop it—this is more difficult or impossible with hybrid maize, which is planted closer together. Traditional maize also has superior long-term storage properties, being less susceptible to mold, and was declared by numerous interviewees to taste far better. As noted earlier, it should be remembered that in Table 4, the increase and decrease in *bare soil* and *shrubs* in these districts might counter-balance each other to some degree, given the results of the confusion matrix and interviews confirming very little change in land uses. Overall, when considering *bare soil* and *shrubs* together, these districts have remained extremely stable in land-cover and land-use types compared to other areas of our case study. We interpret this as being due to a harsh terrain that is not conducive to the diversification of local livelihoods away from traditional, carefully adapted land uses.

(a) (b)

Figure 8. (**a**) Typical small fields with numerous karst outcrops, and (**b**) local maize with runner beans, Đồng Văn District, Hà Giang Province. (Photo credit: Sarah Turner).

6. Discussion: Market Integration, State Policies, and Land-Use Change

In general, land uses in our frontier study region are undergoing substantial and often conflicting transformations. In one interview, a forest ranger noted, "They [the state] are razing the mountains," in relation to new hydropower dam projects and recently built urban areas in Lào Cai Province, both reflecting the modernization drive of state and private actors in this frontier zone. Yet elsewhere in the same province, another ranger said, "Look at the hills there. I planted and protected them, those hills and forest there. I worked six to seven years there. People understand, they dare not burn forest to make new fields," reflecting an awareness (corroborated by local farmers) of the importance of forests for local watershed protection. Not only our interviews but also our maps demonstrated a complex LULCC picture. In some areas, forests were expanding, while in others they were declining; in yet others they remained remarkably stable due to the agro-ecological limits of the local geology, but what are the underlying causes of LULCC in this region? How are livelihood strategies related to these highly variable dynamics of land use across this upland border region?

During the last 25 years, upland northern Vietnam has been considered a frontier region by the state, a region to be modernized and civilized as rapidly as possible [93]. The 2013 Vietnam Constitution promotes "cultural development" (*phát triển văn hóa*), and "social progressiveness" (*tiến bộ*) while focusing on the "modernization" (*hiện đại hóa*) of the country [94]. The northern uplands are the focus of numerous policies that have drawn on these principles and concurrently have substantively challenged or changed the livelihoods and livelihood diversification strategies of ethnic minorities [2,3,68,90,95]. Infrastructure to encourage upland integration increased considerably between 1999 and 2009, marked by an expanded road network that reduced distances to marketplaces newly renovated by the state and connected upland and lowland towns and cities. Such infrastructure has created additional livelihood opportunities for local households to sell agricultural surpluses, gain access to market knowledge or resources, and acquire non-agricultural income [96]. Yet other state-driven interventions and market-oriented economic reforms, including the introduction of and strong state support for hybrid rice and maize seeds, have rarely favored ethnic minorities' indigenous knowledge and historical land uses, but instead have frequently challenged their land-use strategies, often with negative impacts on food security [3,26,45,97].

Ongoing state-supported "development" projects in these borderlands trigger multidimensional responses from local populations that face challenges accessing livelihood resources or who are confronted with the state's vision of what their livelihoods should be (resettled, based on hybrid crops, cash-based, and so on). In our study locale, many rural upland residents have maintained composite agricultural systems as their primary livelihoods, rather than transitioning to plantation crops such as coffee or rubber that are often (but not always) grown on a larger scale [98]. Instead, livelihood diversification has occurred in these borderlands for a range of other complex reasons. The state's drive for farmers to use hybrid rice and maize seeds, for instance, means farmers must have access to more cash for inputs than ever before. In turn, some individuals with the human capital—appropriate skills and know how—and access, have chosen to cultivate cardamom under the shade of closed canopy trees. Providing cash income in this way, farmers have let go of some of their former hillside swidden agricultural lands to focus on hybrid seed crops; in turn, forest cover is increasing. While this livelihood diversification approach and chain of events is a surprising avenue for forest cover regeneration, resulting in a specific LULCC, less surprising is the urban development occurring along the Red River, as again the state intervenes in frontier "development". This time, pushing for greater market integration via inclusion in the Greater Mekong Subregion, the state has heavily promoted urbanization and industrialization adjacent to the Lào Cai–Hekou border crossing, with imports and exports rising steadily (trade data shows the value of goods traded across this border crossing rose from US$92 million to US$993 million from 2000 to 2011 [99]). Urban livelihood opportunities here have diversified dramatically for Kinh residents and traders as this frontier city expands, and an ever-increasing range of goods and services has become available. The complexity of this region is again highlighted with the contrasting tale of eastern Hà Giang Province, where ethnic minority farmers have worked within fairly constraining agro-ecological limits to maintain livelihoods that are diverse within themselves, but yet have remained extremely stable over time. This stability has been reflected in the near lack of LULCC in the most north-eastern districts.

7. Conclusions

State policies for this frontier region, market opportunities (both state-supported and private), and agro-ecological conditions have resulted in highly complex and heterogeneous land uses and land covers. Yet can LULCC mapping really help us to unravel these changes? While completing this project we hit many roadblocks. Mapping LULCC in the mountainous regions of Vietnam is challenging, not only due to the political and physical complexities of accessing the field, but also due to a lack of fine-scale spatial data. Landsat images are the most available data, but the resolution is not fine enough to distinguish bare soils from built-up areas, shrubs from planted crops, and various crops from each other. We found many interesting tales on the ground of livelihood diversification at a scale the maps could not register. For instance, Hmong farmers in Mường Khương District, Lào Cai Province have been experimenting with pineapple plantations, bringing inputs (pineapple plants) and crop-cultivation skills with them from former employment in plantations just over the border in China. Will such crop diversification change land cover in important ways in this area? Others in Mèo Vạc District, Hà Giang Province have been experimenting with honey production and sales to Hanoi distributors, with beehives kept in forests near specific trees to produce highly desired flavored honey. Could honey be the new cardamom and start to protect (the limited) forests there? Only time will tell. As such, the interpretation of LULCC maps must be done with caution and be combined with in-depth, on-the-ground knowledge.

Nonetheless, at the same time, the LULCC maps we developed allowed us to take a step back from our ethnographic results and place them within the bigger picture. We knew that cardamom was an increasingly important crop for many ethnic minority livelihoods, but we needed to see these maps to begin to realize that there were dramatic changes occurring to forest cover that seemed to be connected. This motivated us to return to local villages and ask the relevant questions to confirm these causal links. Likewise, we knew of livelihood constraints in Hà Giang Province, but not the degree to which land-use types are static at the district level over time, proving that the state really does reach limits in trying to push specific agricultural policies. The numerous urban growth patterns across the region are noticeable as one drives through these built-up areas, but the causes of their growth became easier to deconstruct while examining the bigger picture. Although not part of this paper, our findings regarding urban change have stimulated us to start a new project examining the growth of small cities and towns in these uplands with a focus on how LULCC is occurring in peri-urban zones, the quality of life that local residents enjoy, and rural-urban migration processes.

Our mixed methods approach, drawing on an interactive design, provided scope for raw data to be examined and re-examined in an iterative process, moving back and forth between quantitative and qualitative results so as to socialize the pixel as well as pixelize the social [100]. Moreover, a conceptual framework drawing on land change science, frontier studies, and livelihood diversification literature allowed for the integration and recognition of a number of important variables across multiple scales. These ranged from frontier transformation and resource exploitation projects (hydropower dams, hybrid seed programs, urban infrastructure), to culturally-rooted livelihood diversification decisions (specific knowledges of agro-ecological limits, watershed needs, non-timber forest product responses to extreme weather events, and so on). We also found that a key benefit of a mixed methods approach emerged from the justifications and explanations we had to provide to each other regarding our contributions. The first author was forced to find answers beyond the household and village levels regarding processes she had previously most often examined at the micro-scale, and then be able to defend her arguments. Likewise, the second author found that she needed to justify and explain meso-scale results to the first author, who required convincing of the percentage changes that had been calculated. We constantly went back to check our fieldwork notes, examine the data again, and tighten our arguments. While all researchers should do this, of course, our different methodological perspectives and assumptions had to be explained and "typical findings" defended. We believe that this cyclical verification added further strength to our results. Our study thus contributes to the emerging trend of using grounded, in-depth fieldwork to help explain regional land change [17,101], while also shedding light on the benefits meso-scale studies can bring to micro-scale ethnographies.

Our findings point to the importance of policy makers having access to complementary methods and an integrated conceptual framework for implementing appropriate, sustainable land-use and livelihood policies in the region. Yet, at the same time, many upland residents may not be that receptive to state officials asking them probing questions regarding their land uses and livelihoods: farmers are sometimes cultivating cardamom within national parks where doing so is illegal; those being strongly encouraged to plant hybrid rice and maize often prefer to covertly maintain a more diverse livelihood approach that they know will work within local agro-ecological limits; and urban growth in this frontier locale includes illegal (smuggling, prostitution, trafficking, *etc.*) as well as legal livelihood opportunities. Perhaps this points to an important bridging role for non-state researchers here [102]. Nonetheless, if officials are sympathetic to local resident concerns over livelihood opportunities, greater knowledge of LULCC and livelihoods in this region could also forge important opportunities. For instance, officials recognizing an increase in closed canopy forest on a LULCC map could encourage the regulated cultivation of cardamom with strict firewood rules (for drying the pods) and set harvest dates to help protect forests, as has been implemented with success just over the border in Yunnan (interviews 2015). Likewise, greater awareness of agro-ecological limits—noting the lack of change in certain areas in LULCC tables and maps—could mean that unsustainable agricultural policy options are rejected. Instead, the traditional ecological knowledge of upland farmers could be acknowledged and taken on board, as is beginning to occur in lowland regions regarding the limits of hybrid seeds [103].

The question then turns to how to incorporate such findings into decision-making processes. With an extremely hierarchical state apparatus, this is not a particularly easy task in Vietnam [102,104]. We recommend a diverse approach, making the most of opportunities to collaborate with local academics such as members of the North-West Research Program at Vietnam National University (Hanoi), discussing findings at public conferences in the country and regionally, and contributing to meetings such as those held by the Ethnic Minority Working Group supported by the Non-Governmental Organization (NGO) Resource Centre in Hanoi. Through numerous discussions and collaborations it is hoped that greater awareness among policy makers and NGOs of these layers of LULCC complexity can effectively advance relevant research and culturally aware policies for appropriate household strategies and livelihood diversification across the uplands.

Acknowledgments

We would like to thank Lê Mạnh An for fieldwork assistance, Melody Lynch, Thomas Kettig, and Kate Trincsi for research assistance, and special issue editors, Claudia Radel and Jackie Vadjunec, for feedback on an earlier draft. We acknowledge funding from the Social Sciences and Humanities Research Council, Canada.

Author Contributions

Pham completed the remote sensing data analysis and initial interpretations; Turner completed the ethnographic data analysis and interpretations. Turner drafted the context and conceptual framework, with input from Pham. Using a mixed methods approach, explained in the paper, both authors worked together to explore emerging interpretations and themes from the combined data. Both authors contributed equally to the final paper in an iterative process.

Conflicts of Interest

The authors declare no conflict of interest.

References

1. Fox, J.; Rindfuss, R.R.; Walsh, S.J.; Mishra, V. *People and the Environment: Approaches for Linking Household and Community Surveys to Remote Sensing and GIS*; Kluwer Academic: Boston, MA, USA, 2003.
2. Castella, J.-C.; Manh, P.H.; Kam,S.P.; Villano, L.; Tronche, N.R. Analysis of village accessibility and its impact on land use dynamics in a mountainous province of northern Vietnam. *Appl. Geogr.* **2005**, *25*, 308–326.
3. Jakobsen, J.; Rasmussen, K.; Leisz, S.; Folving, R.; Quang, N.V. The effects of land tenure policy on rural livelihoods and food sufficiency in the upland village of Que, North Central Vietnam. *Agric. Syst.* **2007**, *94*, 309–319.
4. Chi, V.K.; van Rompaey, A.; Govers, G.; Vanacker, V.; Schmook, B.; Hieu, N. Land transitions in northwest Vietnam: An integrated analysis of biophysical and socio-cultural factors. *Hum. Ecol.* **2013**, *41*, 37–50.
5. Jadin, I.; Vanacker, V.; Hoang, H.T.T. Drivers of forest cover dynamics in smallholder farming systems: The case of northwestern Vietnam. *Ambio* **2013**, *42*, 344–356.
6. Lambin, E.F.; Meyfroidt, P. Land use transitions: Socio-ecological feedback *versus* socio-economic change. *Land Use Policy* **2010**, *27*, 108–118.
7. Chhabra, A.; Geist, H.; Houghton, R.A.; Haberl, H.; Braimoh, A.K.; Vlek, P.L.G.; Patz, J.; Xu, J.; Ramankutty, N.; Coomes, O.; Lambin, E.F. Multiple impacts of Land use/cover change. In *Land Use and Land Cover Change: Local Processes and Global Impacts*; Geist, H., Lambin, E., Eds.; Springer: Berlin, Germany, 2006; pp. 71–116.
8. Xu, J.; Fox, J.; Melick, D.; Fujita, Y.; Jintrawet, A.; Jie, Q.; Thomas, D.; Weyerhaeuser, H. Land use transition, livelihoods, and environmental services in Montane Mainland Southeast Asia. *Mt. Res. Dev.* **2006**, *26*, 278–284.

9. Aspinall, R.J.; Hill, M.J. *Land Use Change: Science, Policy and Management*; CRC Press: Boca Raton, FL, USA, 2008.
10. Makalle, A.M.P.; Obando, J.; Bamutaze, Y. Effects of land use practices on livelihoods in the transboundary sub-catchments of the Lake Victoria Basin. *Afr. J. Environ. Sci. Technol.* **2008**, *2*, 309–317.
11. Scott, J. *The Art of Not Being Governed: An Anarchist History of Upland Southeast Asia*; Yale University Press: New Haven, CT, USA, 2009.
12. Glassman, J. *Bounding the Mekong: The Asian Development Bank, China, and Thailand*; University of Hawai'i Press: Honolulu, HI, USA, 2010.
13. Turner, S. Under the state's gaze: Upland trading-scapes on the Sino-Vietnamese border. *Singap. J. Trop. Geogr.* **2013**, *34*, 9–24.
14. Trincsi, K.; Hien, P.T.T.; Turner, S. Between state-led policies and mountains: Land use land cover change of ethnic minorities in Vietnam's borderlands. *Land Use Policy* **2014**, *41*, 484–497.
15. Turner, S.; Bonnin, C.; Michaud J. *Frontier Livelihoods. Hmong in the Sino-Vietnamese Borderlands*; University of Washington Press: Seattle, WA, USA, 2015.
16. Douglass, M.; DiGregorio, M.; Bunjamin, W.; Pichaya, V.; Taniguchi, K. Poverty: From basic needs to entitlements in Vietnam during urban transition. In *New Challenges for Sustainable Development in Millennia*; Liu, C.H., Yeh, J.-R., Huang, C.-H., Eds.; CIER Press: Taipei, Taiwan, 2003; pp. 127–174.
17. Brannstrom, C.; Vadjunec, J.M. Notes for avoiding a missed opportunity in sustainability science: Integrating land change science and political ecology. In *Land Change Science, Political Ecology, and Sustainability*; Brannstrom, C., Vadjunec, J.M., Eds.; Routledge: London, UK, 2013; pp. 1–23.
18. Gutman, G.; Janetos, A.C.; Justice, C.O.; Moran, E.F.; Mustard, J.F.; Rindfuss, R.R.; Skole, D.; Turner, B.L., II; Cochrane, M.A. *Land Change Science Observing, Monitoring and Understanding Trajectories of Change on the Earth's Surface*; Kluwer Academic: Dordrecht, The Netherlands, 2004.
19. Rindfuss, R.R.; Walsh, S.J.; Turner, B.L., II; Fox, J.; Mishra, V. Developing a science of land change: Challenges and methodological issues. *Proc. Natl. Acad. Sci. USA* **2004**, *101*, 13976–13981.
20. Turner, B.L., II; Lambin, E.F.; Reenberg, A. The emergence of land change science for global environmental change and sustainability. *Proc. Natl. Acad. Sci. USA* **2007**, *104*, 20666–20671.
21. Geist, H.J.; Lambin, E.F. Proximate causes and underlying driving forces of tropical deforestation: Tropical forests are disappearing as the result of many pressures, both local and regional, acting in various combinations in different geographical locations. *BioScience* **2002**, *52*, 143–150.
22. López-Carr, D.; Davis, J.; Jankowska, M.M.; Grant, L.; López-Carr, A.C.; Clark, M. Space *versus* place in complex human–natural systems: Spatial and multi-level models of tropical land use and cover change (LUCC) in Guatemala. *Ecol. Model.* **2012**, *229*, 64–75.
23. Alvarez, R.R., Jr. The Mexican–US border: The making of an anthropology of borderlands. *Annu. Rev. Anthropol.* **1995**, *24*, 447–470.
24. Giersch, C.P. *Asian Borderlands: The Transformation of Qing China's Yunnan Frontier*; Harvard University Press: Cambridge, MA, USA, 2006.

25. Baker, J. *Small Town Africa: Studies in Rural-Urban Interaction*; Seminar Proceedings No. 23; Nordic Africa Institute: Uppsala, Sweden, 1990.
26. Bonnin, C.; Turner, S. At what price rice? Food security, livelihood vulnerability, and state interventions in upland northern Vietnam. *Geoforum* **2012**, *43*, 95–105.
27. Walker, R.; Homma, A.K.O. Land use and land cover dynamics in the Brazilian Amazon: An overview. *Ecol. Econ.* **1996**, *18*, 67–80.
28. Carvalho, G.O.; Nepstad, D.; McGrath, D.; de Carmen Vera Diaz, M.; Santilli, M.; Barros, A.C. Frontier expansion in the Amazon: Balancing development and sustainability. *Environ. Sci. Policy Sustain. Dev.* **2002**, *44*, 34–44.
29. Moller-Jensen, L.; Knudsen, M.H. Patterns of population change in Ghana (1984–2000): Urbanization and frontier development. *GeoJournal* **2008**, *73*, 307–320.
30. Dillman, C.D. Urban growth along Mexico's northern border and the Mexican national border program. *J. Dev. Areas* **1970**, *4*, 487–508.
31. Kusakabe, K. The politics of "opening up": Female traders on borderlands of Cambodia, Laos, Thailand and Burma (Myanmar). In *On the Border of State Power: Frontier in the Greater Mekong Sub-Region*; Gainsborough, M., Ed.; Routledge: Abingdon, UK, 2008; pp. 60–74.
32. Anderson, J.B.; Gerber, J. *Fifty Years of Change on the US-Mexico Border: Growth, Development, and Quality of Life*; University of Texas Press: Austin, TX, USA, 2009.
33. Coxhead, I.; Shively, G.; Shuai, X. Development policies, resource constraints, and agricultural expansion on the Philippine land frontier. *Environ. Dev. Econ.* **2002**, *7*, 341–363.
34. Jepson, W. Producing a modern agricultural frontier: Firms and cooperatives in eastern Mato Grosso, Brazil. *Econ. Geogr.* **2006**, *82*, 289–316.
35. Agergaard, J.; Fold, N.; Gough, K.V. Global–local interactions: Socioeconomic and spatial dynamics in Vietnam's coffee frontier. *Geogr. J.* **2009**, *175*, 133–145.
36. Eilenberg, M. Frontier constellations: Agrarian expansion and sovereignty on the Indonesian-Malaysian border. *J. Peasant Stud.* **2014**, *41*, 157–182.
37. De Koninck, R. The theory and practice of frontier development: Vietnam's contribution. *Asia Pac. Viewp.* **2000**, *41*, 7–21.
38. Lary, D. Introduction. In *The Chinese State at the Borders*; Lary, D., Ed.; UBC Press: Vancouver, BC, Canada, 2007; pp. 1–10.
39. Tan, S.B.-H. The struggle to control land grabbing: State formation on the Central Highlands frontier under the First Republic of Vietnam (1954–63). In *On the Borders of State Power: Frontiers in the Greater Mekong Sub-Region*; Gainsborough, M., Ed.; Routledge: Abingdon, UK, 2008; pp. 50–66.
40. Ellis, F. *Rural Livelihoods and Diversity in Developing Countries*; Oxford University Press: Oxford, UK, 2000.
41. Scoones, I. Livelihoods perspectives and rural development. *J. Peasant Stud.* **2009**, *36*, 171–196.
42. Bebbington, A. Capitals and capabilities: A framework for analysing peasant viability, rural livelihoods, and poverty. *World Dev.* **1999**, *27*, 2020–2044.
43. Long, N. *Development Sociology: Actor Perspectives*; Routledge: London, UK, 2001.
44. Sen, A. *The Standard of Living*; Cambridge University Press: Cambridge, UK, 1987.

45. Turner, S. Making a living the Hmong way: An actor-oriented livelihoods approach to everyday politics and resistance in upland Vietnam. *Ann. Assoc. Am. Geogr.* **2012**, *102*, 403–422.
46. McCusker, B.; Carr, E. The co-production of livelihoods and land use change: Case studies from South Africa and Ghana. *Geoforum* **2006**, *37*, 790–804.
47. Rigg, J. Land, farming, livelihoods, and poverty: Rethinking the links in the rural South. *World Dev.* **2006**, *34*, 180–202.
48. General Statistics Office of Vietnam. *The 2009 Vietnam Population and Housing Census: Completed Results*; Central Population and Housing Census Steering Committee: Hanoi, Vietnam, 2010.
49. Dien, K. *Population and Ethno-demography in Vietnam*; Silkworm: Chiang Mai, Thailand, 2002.
50. MacKerras, C. Ethnic minorities in China. In *Ethnicity in Asia*; MacKerras, C., Ed.; RoutledgeCurzon: London, UK, 2003.
51. Michaud, J. *Historical Dictionary of the Peoples of the Southeast Asian Massif*; Scarecrow Press: Lanham, MD, USA, 2006.
52. Leisz, S.J.; Ha, N.T.T.; Yen, N.T.B.; Lam, N.T.; Vien, T.D. Developing a methodology for identifying, mapping and potentially monitoring the distribution of general farming system types in Vietnam's northern mountain region. *Agric. Syst.* **2005**, *85*, 340–363.
53. Corlin, C. Hmong and the land question in Vietnam: National policy and local concepts of the environment. In *Hmong/Miao in Asia*; Tapp, N., Michaud, J., Culas, C., Lee, G.Y., Eds.; Silkworm: Chiang Mai, Thailand, 2004; pp. 295–320.
54. World Bank. *Vietnam Development Report 2008: Social Protection*; World Bank: Washington, DC, USA, 2008.
55. Sikor, T. Forest policy reform in Vietnam: From state to household forestry. In *Stewards of Vietnam's Upland Forests*; Poffenberger, M., Ed.; Asian Forestry Network: Berkley, CA, USA; Hanoi, Vietnam, 1998; pp. 18–38.
56. McElwee, P. Reforesting "bare hills" in Vietnam: Social and environmental consequences of the 5 Million Hectare Reforestation Program. *Ambio* **2009**, *38*, 325–333.
57. McElwee, P. Becoming socialist or becoming Kinh? Government policies for ethnic minorities in the Socialist Republic of Vietnam. In *Civilizing the Margins: Southeast Asian Government Policies for the Development of Minorities*; Duncan, C., Ed.; Cornell University Press: Ithaca, NY, USA, 2004; pp. 182–213.
58. Hardy, A. *Red Hills: Migrants and the State in the Highlands of Vietnam*; Nordic Institute of Asian Studies Monographs: Copenhagen, Denmark, 2005.
59. Clement, F.; Amezaga, J.M. Linking reforestation policies with land use change in northern Vietnam: Why local factors matter. *Geoforum* **2008**, *39*, 265–277.
60. Castella, J.C.; Boissau, S.; Thanh, N.H.; Novosad, P. Impact of forestland allocation on land use in a mountainous province of Vietnam. *Land Use Policy* **2006**, *23*, 147–160.
61. Clement, F.; Orange, D.; Williams, M.; Mulley, C.; Epprecht, M. Drivers of afforestation in Northern Vietnam: Assessing local variations using geographically weighted regression. *Appl. Geogr.* **2009**, *29*, 561–576.
62. Meyfroidt, P.; Lambin, E.F. The causes of the reforestation in Vietnam. *Land Use Policy* **2008**, *25*, 182–197.

63. Sikor, T. The allocation of forestry land in Vietnam: Did it cause the expansion of forests in the northwest? *For. Policy Econ.* **2001**, *2*, 1–11.
64. Müller, D.; Munroe, D.K. Tradeoffs between rural development policies and forest protection: Spatially explicit modeling in the Central Highlands of Vietnam. *Land Econ.* **2005**, *81*, 412–425.
65. Leisz, S.J. Dynamics of land cover and land use changes in the upper Ca river basin of Nghe An, Vietnam. *Southeast Asian Stud.* **2009**, *47*, 287–308.
66. Dao, M.T.; Kono, Y.; Yanagisawa, M.; Leisz, S.J.; Kobayashi, S. Linkage of forest policies and programs with land cover and land use changes in the Northern mountain region of Vietnam: A village-level case study. *Southeast Asian Stud.* **2009**, *47*, 244–262.
67. Pham, T.T.H.; Turner, S.; Trincsi, K. Applying a systematic review to land use land cover change in northern upland Vietnam: The missing case of the borderlands. *Geogr. Res.* **2015**, doi:10.1111/1745-5871.12133.
68. Cuc, L.T.; Rambo, T. *Bright Peaks, Dark Valleys. Bright Peaks, Dark Valleys: A Comparative Analysis of Environmental and Social Conditions and Development Trends in Five Communities in Vietnam's Northern Mountain Region*; The National Political Publishing House: Hanoi, Vietnam, 2001.
69. Greene, J.C. *Mixed Methods in Social Inquiry*; Jossey-Bass: San Francisco, CA, USA, 2007.
70. Meyfroidt, P.; Lambin, E.F. Forest transition in Vietnam and its environmental impacts. *Glob. Change Biol.* **2008**, *14*, 1319–1336.
71. Anderson, J.R. *A Land Use and Land Cover Classification System for Use with Remote Sensor Data*; US Government Printing Office: Arlington, VA, USA, 1976.
72. Hurni, K.; Hett, C.; Epprecht, M.; Messerli, P.; Heinimann, A.A. Texture-based land cover classification for the delineation of a shifting cultivation landscape in the Lao PDR using landscape metrics. *Remote Sens.* **2013**, *5*, 3377–3396.
73. Cassidy, L.; Binford, M.; Southworth, J.; Barnes, G. Social and ecological factors and land-use land-cover diversity in two provinces in Southeast Asia. *J. Land Use Sci.* **2010**, *5*, 277–306.
74. Baatz, M.; Hoffman, C.; Willhauck, G. Progressing from object-based to object-oriented image analysis. In *Object-Based Image Analysis: Spatial Concepts for Knowledge-Driven Remote Sensing Applications*; Blaschke, T., Lang, S., Hay, G.J., Eds.; Springer: Berlin, Germany, 2008; pp. 29–42.
75. Dorren, L.K.A.; Maier, B.; Seijmonsbergen, A.C. Improved Landsat-based forest mapping in steep mountainous terrain using object-based classification. *For. Ecol. Manag.* **2003**, *183*, 31–46.
76. Vieira, M.A.; Formaggio, A.R.; Rennó, C.D.; Atzberger, C.; Aguiar, D.A.; Mello, M.P. Object based image analysis and data mining applied to a remotely sensed Landsat time-series to map sugarcane over large areas. *Remote Sens. Environ.* **2012**, *123*, 553–562.
77. Puyravaud, J.P. Standardizing the calculation of the annual rate of deforestation. *For. Ecol. Manag.* **2003**, *177*, 593–596.
78. People's Committee of Lào Cai. *Quy hoạch tổng thể phát triển kinh tế - xã hội tỉnh Lào Cai đến năm 2020* [The 2020 Lào Cai economic development plan], 2008. Available online: http://laocai.gov.vn/thongtinktxh/dinhhuongphattrien/quyhoachphattrienktxh/Trang/634046199107564190.aspx (accessed on 27 September 2015).

79. Socialist Republic of Vietnam. *Của Chính Phủ Số 36/2000/Nđ-Cp Ngày 18 Tháng 8 Năm 2000 Về Việc Điều Chỉnh Địa Giới Hành Chính Huyện Bắc Hà Để Tái Lập Huyện Si Ma Cai, Tỉnh Lào Cai* [Decree No. 36/2000 / ND-CP of August 18, 2000 of the Government on the adjustment of administrative boundaries of Bắc Hà district to reestablish Si Ma Cai District, Lào Cai Province], 2000. Available online: http://thuvienphapluat.vn/archive/Nghi-dinh/Nghi-dinh-36-2000-ND-CP-dieu-chinh-dia-gioi-hanh-chinh-huyen-Bac-Ha-de-tai-lap-huyen-Si-Ma-Cai-tinh-Lao-Cai-vb46707t11.aspx (accessed on 27 September 2015).
80. Socialist Republic of Vietnam. *Nghị định số 42/2009/NĐ-CP. Về việc phân loại đô thị* [Decree 42/2009/ND-CP 7th May, 2009 on the grading of urban centers], 2009. Available online: http://www.moj.gov.vn/vbpq/en/_layouts/printeng.aspx?id=10670 (accessed on 27 September 2015).
81. Socialist Republic of Vietnam. *Về Việc Điều Chỉnh Địa Giới Hành Chính Để Thành Lập Đơn Vị Hành Chính Cấp Xã, Cấp Huyện Thuộc Tỉnh Lai Châu* [Adjustments of administrative boundaries to establish Administrative Units at the commune and district Levels, Lai Châu Province], 2012. Available online: http://thuvienphapluat.vn/archive/Nghi-quyet-71-NQ-CP-nam-2012-dieu-chinh-dia-gioi-hanh-chinh-de-thanh-lap-don-vi-vb150870.aspx (accessed on 27 September 2015).
82. General Statistics Office of Vietnam. *Socio-economical Statistical Data of 63 Provinces and Cities*; Statistical Publishing House: Hanoi, Vietnam, 2009.
83. General Statistics Office of Vietnam. *The 1999 Vietnam Population and Housing Census: Completed Results*; Central Population and Housing Census Steering Committee: Hanoi, Vietnam, 2000.
84. Bui, T.M.H.; Schreinemachers, P.; Berger, T. Hydropower development in Vietnam: Involuntary resettlement and factors enabling rehabilitation. *Land Use Policy* **2013**, *31*, 536–544.
85. International Rivers Network. Planned Dams in Vietnam, 2001. Available online: http://www.internationalrivers.org/de/resources/planned-dams-in-vietnam-4079 (accessed on 27 September 2015).
86. Socialist Republic of Vietnam. *Về Việc Thành Lập Thị Xã Lai Châu Và Thành Lập Thị Trấn Thuộc Các Huyện Tam Đường, Phong Thổ, Tỉnh Lai Châu* [On the establishment of the township of Lai Châu and district headtowns of Tam Đường, Phong Thổ , Lai Châu province], 2004. Available online: http://thuvienphapluat.vn/archive/Nghi-dinh/Nghi-dinh-176-2004-ND-CP-thanh-lap-thi-xa-Lai-Chau-va-thanh-lap-thi-tran-thuoc-cac-huyen-Tam-Duong-Phong-Tho-tinh-Lai-Chau-vb5530t11.aspx (accessed on 27 September 2015).
87. VietnamNews. PM Okays Establishment of Lai Châu City. 30 December 2013. Available online: http://vietnamnews.vn/domestic-press-highlights/249576/pm-okays-establishment-of-lai-chau-city.html (accessed on 27 September 2015).
88. Resettlement Problems Put Hydro-Plant behind Schedule. *VietnamNews* 21 August 2014. Available online: http://vietnamnews.vn/society/259092/resettlement-problems-put-hydro-plant-behind-schedule.html (accessed on 27 September 2015).
89. General Statistics Office of Vietnam. *Results of the 2006 Rural, Agricultural and Fishery Census*; Statistical Publishing House: Hanoi, Vietnam, 2007.
90. Turner, S. "Forever Hmong": Ethnic minority livelihoods and agrarian transition in upland northern Vietnam. *Prof. Geogr.* **2012**, *64*, 540–553.

91. Novellino, D. Indigenous highlands in transition: The case of Hà Giang Province in northern Vietnam. *Land Reform: Land Settl. Coop. Bull.* **2000**, *2*, 94–107.
92. Kyeyune, V. Yielding to High Yields: Hybrid Maize and Food Security in Hà Giang Province, Northern Vietnam. Master's Thesis, McGill University, Montreal, Canada, 2014.
93. Duncan, C. *Civilizing the Margins. Southeast Asian Government Policies for the Development of Minorities*; Cornell University Press: Ithaca, NY, USA, 2004.
94. Socialist Republic of Vietnam. *Hiến Pháp Nước Cộng Hòa Xã Hội Chủ Nghĩa Việt Nam (Sửa Đổi)* [The Constitution of the Socialist Republic of Vietnam (Amended)]; Socialist Republic of Vietnam: Hanoi, Vietnam, 2013.
95. Turner, S.; Michaud, J. Imaginative and adaptive economic strategies for Hmong livelihoods in Lào Cai Province, northern Vietnam. *J. Vietnam. Stud.* **2008**, *2*, 154–186.
96. Alther, C.; Castella, J.C.; Novosad, P.; Rousseau, E.; Hieu, T.T. Impact of accessibility on the range of livelihood options available to farm households in mountainous areas of northern Viet Nam. In *Doi Moi in the Mountains: Land Use Changes and Farmers' Livelihood Strategies in Bac Kan Province, Vietnam*; Castella, J.C., Quang, D.D., Eds.; The Agricultural Publishing House: Hanoi, Vietnam, 2001; pp. 121–146.
97. Evans, G.; Hutton, C.; Eng, K.K. *Where China Meets Southeast Asia: Social and Cultural Change in the Border Regions*; St Martin's Press: New York, NY, USA, 2000.
98. Meyfroidt, P.; Vu Tan Phuong; Hoang Viet Anh. Trajectories of deforestation, coffee expansion and displacement of shifting cultivation in the Central Highlands of Vietnam. *Glob. Environ. Change* **2013**, *23*, 1187–1198.
99. Zhang, B.Q. 云南省边境口岸物流发展状况分析：以河口口岸为例 [Analyzing logistics development at border crossing points in Yunnan: A case study on Hekou Border Crossing Point]. *J. Manag.* **2014**, *27*, 48–52.
100. Geoghegan, J.; Pritchard, L., Jr.; Ogneva-Himmelberger, Y.; Roy Chowdhury, R.; Sanderson, S.; Turner, B.L., II. "Socializing the pixel" and "pixelizing the social" in land-use and land-cover change. In *People and Pixels. Linking Remote Sensing and Social Science*; Liverman, D.; Moran, E.F.; Rindfuss, R.R.; Stern, P.C., Eds.; National Academy Press: Washington, DC, USA, 1998; pp. 51–69.
101. Kull, C.A. Politicizing land use change in highland Madagascar: Struggles with air photo analyses and conservation agendas. In *Land Change Science, Political Ecology, and Sustainability*; Brannstrom, C.; Vadjunec, J.M., Eds.; Routledge: London, UK, 2013; pp. 66–83.
102. Turner, S. *Red Stamps and Gold Stars. Fieldwork Dilemmas in Upland Socialist Asia*; UBC Press: Vancouver, BC, Canada, 2013.
103. Oxfam. *Growing a Better Future: Expanding Rights and Choices for Small-scale Farmers in Viet Nam*, Brochure; Oxfam: Hanoi, Vietnam, 2012.
104. Mackenzie, C.A.; Christensen, J.; Turner, S. Advocating beyond the academy: Dilemmas of communicating relevant research results. *Qual. Res.* **2015**, *15*,105–121.

Permissions

All chapters in this book were first published in Land, by MDPI; hereby published with permission under the Creative Commons Attribution License or equivalent. Every chapter published in this book has been scrutinized by our experts. Their significance has been extensively debated. The topics covered herein carry significant findings which will fuel the growth of the discipline. They may even be implemented as practical applications or may be referred to as a beginning point for another development.

The contributors of this book come from diverse backgrounds, making this book a truly international effort. This book will bring forth new frontiers with its revolutionizing research information and detailed analysis of the nascent developments around the world.

We would like to thank all the contributing authors for lending their expertise to make the book truly unique. They have played a crucial role in the development of this book. Without their invaluable contributions this book wouldn't have been possible. They have made vital efforts to compile up to date information on the varied aspects of this subject to make this book a valuable addition to the collection of many professionals and students.

This book was conceptualized with the vision of imparting up-to-date information and advanced data in this field. To ensure the same, a matchless editorial board was set up. Every individual on the board went through rigorous rounds of assessment to prove their worth. After which they invested a large part of their time researching and compiling the most relevant data for our readers.

The editorial board has been involved in producing this book since its inception. They have spent rigorous hours researching and exploring the diverse topics which have resulted in the successful publishing of this book. They have passed on their knowledge of decades through this book. To expedite this challenging task, the publisher supported the team at every step. A small team of assistant editors was also appointed to further simplify the editing procedure and attain best results for the readers.

Apart from the editorial board, the designing team has also invested a significant amount of their time in understanding the subject and creating the most relevant covers. They scrutinized every image to scout for the most suitable representation of the subject and create an appropriate cover for the book.

The publishing team has been an ardent support to the editorial, designing and production team. Their endless efforts to recruit the best for this project, has resulted in the accomplishment of this book. They are a veteran in the field of academics and their pool of knowledge is as vast as their experience in printing. Their expertise and guidance has proved useful at every step. Their uncompromising quality standards have made this book an exceptional effort. Their encouragement from time to time has been an inspiration for everyone.

The publisher and the editorial board hope that this book will prove to be a valuable piece of knowledge for researchers, students, practitioners and scholars across the globe.

List of Contributors

Fraser J. Morgan and Adam J. Daigneault
Landcare Research New Zealand, Private Bag 92170, Auckland Mail Centre, Auckland 1142, New Zealand

Philip Brown
Landcare Research New Zealand, P.O. Box 69040, Lincoln 7640, New Zealand

Hila Sagie and Uri Ramon
Open Landscape Institute (Machon Deshe), Negev 2 St., Tel Aviv 66186, Israel

Bhagawat Rimal and Sushila Rijal
Everest Geoscience Information Service Center, Kathmandu, Nepal

Himlal Baral
Centre for International Forestry Research (CIFOR), P.O. Box 0113 BOCBD, Bogor 16000, Indonesia

Nigel E. Stork
Environmental Futures Research Institute, Griffith School of Environment, Nathan Campus, Griffith University, 170 Kessels Road, Nathan, QLD 4111, Australia

Kiran Paudyal
School of Ecosystem and Forest Sciences, Faculty of Science, the University of Melbourne, 221 Bouverie St., Carlton, VIC 3010, Australia

Sheona Shackleton
Department of Environmental Science, Rhodes University, Grahamstown 6140, South Africa

Marty Luckert
Department of Resource Economics and Environmental Sociology, University of Alberta,
515 General Services Building, Edmonton, AB T6G 2H1, Canada

William J. McConnell
Center for Global Change and Earth Observations, Michigan State University, East Lansing, MI 48823, USA

Andrés Viña
Center for Systems Integration and Sustainability, Michigan State University, East Lansing, MI 48823, USA
Department of Geography, University of North Carolina, Chapel Hill, NC 27599, USA

Christian Kull
Institut de Géographie et Durabilité, Université de Lausanne, 1015 Lausanne, Switzerland

Clayton Batkow
Department of Anthropology, Michigan State University, East Lansing, MI 48823, USA

Ricardo Eaton-Gonzalez and Eric Mellink
Centro de Investigación Cientifica y de Educacion Superior de Ensenada (CICESE),
Carretera Ensenada-Tijuana 3918, Zona Playitas, 22860 Ensenada, B.C., Mexico

Laura Schmitt Olabisi
Department of Community Sustainability and Environmental Science & Policy Program, Michigan State University, 151 Natural Resources Building, 480 Wilson Road, East Lansing, MI 48824, USA

Ryan Qi Wang
Department of Civil and Environmental Engineering, Virginia Tech, 113 Patton Hall (0105), Blacksburg, VA 24061, USA

Arika Ligmann-Zielinska
Department of Geography and Environmental Science & Policy Program,
Michigan State University, Room 121 Geography Building, 673 Auditorium Rd.,
East Lansing, MI 48824, USA

Deng Ding
Department of Geographical and Sustainability Sciences, University of Iowa, 316 Jessup Hall, Iowa City, IA 52242, USA

Esri, 380 New York Street, Redlands, CA 92373, USA

David Bennett
Department of Geographical and Sustainability Sciences, University of Iowa, 316 Jessup Hall, Iowa City, IA 52242, USA

Silvia Secchi
Department of Geography and Environmental Resources, Southern Illinois University,
Carbondale, IL 62901, USA

Christian Brannstrom, Mary Tilton, Andrew Klein and Wendy Jepson
Department of Geography, Texas A&M University, 810 O&M Building, College Station, TX 77843, USA

Farid El Wahidi
Centre de la Recherche Forestière, 10050 Rabat-Agdal, Marocco

Julien Radoux, Quentin Ponette and Pierre Defourny
Earth and Life Institute, Université catholique de Louvain, 1348 Louvain-la-Neuve, Belgium

Daniele Spirandelli
Department of Urban & Regional Planning, University of Hawai`i-Manoa, 2424 Maile Way, Saunders 107, Honolulu, HI 96822, USA

Sarah Turner
Department of Geography, McGill University, 805 rue Sherbrooke Ouest, Montréal, QC H3A 0B9, Canada

Thi-Thanh-Hiên Pham
Département D'études Urbaines et Touristiques, Université du Québec à Montréal, 315 rue Sainte-Catherine Est, Montréal, QC H2X 3X2, Canada

www.ingramcontent.com/pod-product-compliance
Lightning Source LLC
LaVergne TN
LVHW080925210926
848485LV00004B/729
* 9 7 8 1 6 3 2 3 9 7 1 3 3 *